Hazardous Waste Management Engineering

Hazardous Waste Management Engineering

Edited by

Edward J. Martin
Ph.D., P.E.

James H. Johnson, Jr.
Ph.D.

VNR VAN NOSTRAND REINHOLD COMPANY
———————————————————————— *New York*

Copyright © 1987 by Van Nostrand Reinhold Company Inc.
Library of Congress Catalog Card Number: 86-11099
ISBN: 0-442-24439-8

Printed in the United States of America.

Published by Van Nostrand Reinhold Company, Inc.
115 Fifth Avenue
New York, New York 10003

Van Nostrand Reinhold Company Limited
Molly Millars Lane
Wokingham, Berkshire RG11 2PY, England

Van Nostrand Reinhold
480 La Trobe Street
Melbourne, Victoria 3000, Australia

Macmillan of Canada
Division of Canada Publishing Corporation
164 Commander Boulevard
Agincourt, Ontario M1S 3C7, Canada

15 14 13 12 11 10 9 8 7 6 5 4 3 2 1

Library of Congress Cataloging-in-Publication Data
Hazardous waste management engineering.
  Includes bibliographies and index.
  1. Hazardous wastes. I. Martin, E. J. (Edward James)
II. Johnson, James H., 1947–   .
TD811.5.H3994 1986   628.4'4   86-11099
ISBN 0-442-24439-8

Contents

Foreword

The management of hazardous wastes requires an understanding of technical, regulatory, economic, permitting, institutional, and public policy issues. The Resource Conservation and Recovery Act (RCRA) (1976) and the Comprehensive Environmental Response, Compensation, and Liabilities Act (CERCLA) (1980), at the federal government level, along with equivalent state statutes, have been the recent driving forces to deal effectively with the problems of hazardous wastes.

The editors have assembled a group of knowledgeable authors, who have first-hand experience in hazardous-waste management. They have prepared a useful, comprehensive text, particularly for engineers. The text provides a single-source reference of environmental legislation; technical alternatives for storage, treatment, and disposal; risk analysis; and siting of hazardous-waste facilities.

The engineer has a major role in hazardous-waste management, particularly in developing alternative solutions and evaluation criteria; assisting in the selection of the preferred alternative; assisting in the selection and development of hazardous-waste facility sites, if they are necessary; and designing the necessary facilities to implement the selected alternative.

Today's engineer must have an understanding of the statutory and regulatory requirements to carry out his or her responsibilities. These requirements are presented by A. Edelman in Chapter 1.

Chapter 2, by B. P. Smith, covers the extemely important tool of exposure and risk assessment, whose understanding and application by the engineer is becoming more and more critical.

In Chapter 3 Martin, Oppelt, and Smith present the state-of-the-art in chemical, physical, and biological treatment of hazardous wastes. Recent data and treatment effectiveness are reported, which should enable process engineers to develop quantitative solutions to newly identified problems.

Incineration has emerged as one of most effective techniques for destroying some hazardous wastes. Crumpler and Martin, in Chapter 4, provide an in-depth technical presentation of the critical design aspects of incineration. Information is presented on wastes suitable for incineration, the type and capacity of equipment required, and the efficiency of incineration.

A frequently overlooked aspect of hazardous-waste management is storage of the wastes. In Chapter 5 Ball and Johnson present procedures for locating, sizing, operating, and closing hazardous-waste storage facilities.

Loehr, in Chapter 6, presents the scientific and engineering basis for using land as a treatment and disposal method and a remedial action method for hazardous wastes.

One of the major consequences of improper handling of hazardous wastes is the generation of a leachate, which results in contamination of soil, groundwater, and surface water. Shuckrow, Touhill, and Pajak, in Chapter 7, provide guidance in process selection for leachate treatment.

Nothing is more critical in hazardous-waste management than obtaining approval for the siting of the facility. Barclay, in Chapter 8, provides very useful, practical approaches pertaining to the siting process.

Drs. Martin and Johnson have made a valuable contribution in developing this text, which should be of considerable value to engineers.

Leon W. Weinberger

Contributors

Roy Ball
ERM North Central Inc., 835 Sterling Avenue, Palatine, IL 60607

Michael M. Barclay
U.S. Environmental Protection Agency, GRC, Suite 201, 1100 Sixth Street SW, Washington, D.C. 20020

Eugene Crumpler
U.S. Environmental Protection Agency, Office of Water Regulations and Standards, 401 M Street SW, Washington, D.C. 20460

Arnie Edelman
U.S. Environmental Protection Agency, Office of Toxic Substances, Washington, D.C. 20460

James H. Johnson, Jr.
Department of Civil Engineering, Howard University, Washington, D.C. 20059

Raymond C. Loehr
Department of Civil Engineering, University of Texas, Austin, TX 78712

Edward J. Martin
Department of Civil Engineering, Howard University, Washington, D.C. 20059

E. Timothy Oppelt
U.S. Environmental Protection Agency, H-WERL, 26 West St. Clair Street, Cincinnati, OH 45268

Andrew P. Pajak
Baker/TSA Inc., 4301 Dutch Ridge Road, Beaver, PA 15009

Alan J. Shuckrow
Baker/TSA Inc., 4301 Dutch Ridge Road, Beaver, PA 15009

Benjamin P. Smith
U.S. Environmental Protection Agency, Office of Solid Waste, WH 562 B, Washington, D.C. 20460

C. J. Touhill
Baker/TSA Inc., 4301 Dutch Ridge Road, Beaver, PA 15009

INTRODUCTION

Many alternatives are available for the destructive disposal of hazardous wastes. Hence there is no reason to delay applying the technologies immediately for appropriate waste management.

We have identified two stumbling blocks to immediate application of at least some of the technologies: the lack of technical information for design on a widely available basis, and a means for evaluating competing technologies when applied to hazardous-waste management. We hope that this text will partially address these two problems.

The body of the text provides technical data and information that will be useful to the practitioner in developing preliminary design strategies, doing design, and evaluating actual applications—some at the research stage—but most of it comes from pilot- and full-scale facilities. Therefore, the engineer may expect to use the material for guidance in specific applications. Of course, as usual, there is no substitute for site-specific information in the design and evaluation process.

We would like the book to help widen the perspective of engineers in the field of hazardous-waste disposal. The technologies/techniques presented are ready for full-scale application without further research. It is not necessary to wait for additional information to make a decision to apply incineration or treatment for destructive disposal, for example, although it may be desirable to make field evaluations and pilot the technologies against specific waste streams.

In addition to having technical data and information available, it is necessary to have a means of comparing alternatives for solving a given waste problem—a means of comparison that allows matching the capabilities of approaches that appear to be "apples and oranges." In the recent past, for example, practitioners have compared incineration to landfill, usually on unequal bases. Incineration may be expected to decompose organic compounds, landfill to preserve them. A cost-analysis that concludes that landfill is "cheaper" than incineration is therefore not useful. Alternatives or technical options must be compared on bases that are functionally equivalent or the results are meaningless.

CONTROL AND DISPOSAL INDICES FOR COMPARING WASTE-MANAGEMENT ALTERNATIVES

Management and disposal of hazardous wastes are causing a revolution in past sanitary engineering practices. To place chemical, physical, biological, and thermal treatment of hazardous wastes in perspective, one must examine the words that have been used to describe waste management functions.

Ultimate management of hazardous wastes includes *disposal*—that is, elimination of the waste or the hazard potential of the waste. In classical waste management, "disposal" has more often meant transferring the potential adverse impact from one location to another or from one media (air, land, water) to another. In the context of hazardous-waste management it is necessary to eliminate the hazard potential to the environment and human health, a stated goal of the RCRA.

In every case where management of hazardous wastes is being contemplated, a comparison of engineering alternatives should be conducted in detail to assess the potential long- and short-term effectiveness of available options. Often such comparisons are made on the basis of cost estimates, which are purported to be based on comparable options. One of the most frequently used is a comparison of costs for incineration versus landfill.

Costs are not examined here. The intent of this discussion is to present a scheme for comparison of management options based on factors other than cost. Once the comparison scheme has been developed for a given case, costs can then be compared on an equitable basis. There are only two objectives possible for hazardous-waste management:

1. *Control* for some period of time
2. *Disposal* or elimination of hazard potential

Each management option may be treated as a "black box," where:

X = Quantity of hazardous component being managed ($X = 1$ unit)

X_1 = Quantity of hazardous component in effluent from the box

X_2 = Quantity of hazardous component in the air emissions

X_3 = Quantity of hazardous component in residue

The black-box approaches for common management options are summarized qualitatively in Table 1, with a general expectation for X_1, X_2, and X_3 for each approach (based on destruction and/or control efficiencies with $X = 1$).

To develop a useful comparison approach, one must estimate the quantity of expected discharges from the black box and account in some way for the time factor. Also, it is necessary to deal with the relative degree of disposal of hazardous- and toxic-waste components.

In a specific engineering analysis of alternatives for specific wastes at an individual site, estimates could be made for each element necessary in the scheme and applied by using this technique.

Table 1. The Black-Box Approach

Management Option	Effluent X_1	Air Emission X_2	Residue X_3
Incinerator	0	Fugitive stack, PIC[a]	Ash and scrubber water
Secure L.F.	Leaks (leachate)	Air emission	Remainder after lifetime
Land treatment	Leachate and runoff	Air emission	Remainder on land
Storage tanks	Leakage	Air emission	Pump out
Piles	Leaching and runoff	Air emission	Pile residue
S.I.	Leaching	Air emission	Residue in impoundment
Containers	Leakage and spillage	0	Container contents
Treatment removal			
(a) conservative pollutants	Effluent + PIT[b]	Air emission	Residue
(b) nonconservative pollutants	Effluent + PIT[b]	Air emission	Residue
Destruction	Effluent + PIT[b]	Air emission	Residue not destroyed
Fixation/stabilization	Leachate	Air emission	Fixed solids
Encapsulation	Leachate	0	Fixed solids

[a] PIC = Products of incomplete combustion.
[b] PIT = Products of incomplete treatment.

For this general analysis, estimates were made for each fractional destruction, removal and emission rate based on the authors' experience and consultation with others. The actual values of the estimates are not important to the usability of the scheme. Others will have their own estimates, and as we have mentioned, specific waste, site, and management option conditions should be substituted when available; this should be done to derive meaningful *site*-related analyses. As will be seen, even wide variations in discharge estimates will preserve the relative ranking of control and disposal options.

"Control" relates to management of liquid effluents (or discharges) and air emissions without necessarily destroying components of the wastes. Residues are treated separately. Table 2 presents the relatively estimated discharge fractions (assuming $X = 1$ unit) for 26 options.

Table 2. Control—Years and Disposal Indices

Management Option	Effluent X_1	Air Emission X_2	Residues X_3	Time (Yr)	Control — Years Index (yr -1)				Disposal Index $1/X_3$	
					High >10	Med. 1-10	Other 0.1-1	Other 0.01-0.1	High >10	Other 1-10
Incineration	0	0.00009	0.0001	30	370				10,000	1
Secure landfill	0.01	0.01	0.98	50		1				1
Landfill										
volatiles	0.5	0.05	0.45	30				0.06		2.2
NV and metals	0.2	0	0.8	30			0.17			1
Land treatment										
volatiles	0.05	0.9	0.05	10			0.1		20[a]	1.2
NV	0.1	0.05	0.85	10			0.67			1.1
metals	0.1	0	0.9	10		1				
Storage (long term)										
open tank (vol.)	0.01	0.98	0.01	20				0.05	100[a]	1
closed tank (vol.)	0.01	0.05	0.94	20			0.83			1.3
open tank (NV)	0.01	0.20	0.79	20			0.24			1
closed tank (NV)	0.01	0.01	0.98	20		2.5				
Piles										
volatiles	0.15	0.8	0.05	10			0.1		20[a]	
NV	0.15	0.3	0.55	10			0.22			1.8

4

Surface impoundment								
volatiles	0.05	0.94	0.01	10		0.1		100[a]	
NV	0.05	0.25	0.70	10		0.33			1.4
Containers	0.1	0	0.90	5	2				1.1
Storage (short term)									
open tank (vol.)	0	0.05	0.95	20	1				1.1
closed tank (vol.)	0	0.001	0.999	20	50				1
open tank (NV)	0	0.01	0.99	20	5				1
closed tank (NV)	0	0.0001	0.9999	20	500				1
Treatment removal									
a. conservative pollutant 1.7[c]	$(X - FR)$	0.001	$(FR - 0.001)$	30		1.6[b]	0.08[c]		1.7[c], 1[b]
b. nonconservative pollutant	$(X - FR)$	0.01	$(FR - 0.01)$	30		1.1[b]	0.08[c]		1.7[c], 1[b]
Destruction	$(X - FR)$	0.001	$(FR - X(1 - Y))$	30		0.6[d]	0.167[e]	50[d]	4[e]
Fixation/stabilization									
volatiles	0.01	0.98	0.01	50			0.02	100[a]	
NV and metals	0.1	0.03	0.87	50		0.15			1.2
Encapsulation	0.01	0.01	0.98	50	1				1

vol. = Volatiles; NV = nonvolatiles; FR = fraction removed; Y = fractional quantity destroyed.
[a] = "Disposal" is volatilization.
[b] = at FR = 0.60.
[c] = at FR = 0.98.
[d] = at FR = Y = 0.98.
[e] = at FR = Y = 0.60.

5

For hazardous-waste management, control is exercised over the widespread exposure of the hazard component of the given waste for some period of time: for example waste still bottoms are incinerated, or heavy-metal treatment sludges are placed in a landfill for 30 years. Exposure to the environment or exposure in such a way as to impact human health is either through the air or water medium. These exposure routes are expressed in the black-box approach as X_1 (liquid or other discharge) and X_2 (air emissions). One may therefore define an index to describe the degree of control possible through the application of a black box to a waste in terms of these two parameters of the process and time.

The control–years index is defined by

$$CYI = \frac{1}{(X_1 + X_2) \text{ (time in years)}} \tag{1}$$

If the time is different for liquid and air emissions, the index becomes

$$CYI = \frac{1}{X_1 t_1 + X_2 t_2} \tag{2}$$

For risk analysis the index definition may be expanded to include multiple control efficiencies and exposure times for both liquid and air emissions:

$$CYI = \frac{1}{\sum_i X_i t_i + \sum_j X_j t_j} \tag{3}$$

Equation (3) can be used as the basis for a detailed site risk assessment by including all pertinent population or individual exposure routes and compounds.

For the purposes of developing the values for Table 2, general, non-site-specific estimates were used for X_1, X_2, and X_3. Even with these general estimates, the separation among strategies can be seen. For a given application the time used to calculate the index is the period in years during which the option is being used. In the most general case this period corresponds to the lifetime of the facility (Eq. 1). Also presented in the table are the time estimates (yr) used for the comparative analysis.

The results of the CYI calculations ranged over four or more orders of magnitude and are grouped for convenience in the table, into "high," "medium," and "other." High values of CYI reflect a high degree of control of air and other discharges; low values reflect reduced levels of control. It should be emphasized again that the estimates used for this analysis may not be universally applied and are likely to vary significantly for specific cases—wastes and sites.

True disposal is reflected by the amount of toxic or hazardous components remaining after the option of choice is applied to manage the waste. The disposal index $(1/X^3)$ is also tabulated in Table 2. As in the case of the CYI, the disposal index may be generalized to account for multiple destruction efficiencies and thereby becomes the basis for site risk assessment:

$$DI = \frac{1}{\sum_i X_i}$$

A high DI value means that the option exhibits high destruction efficiency or a high degree of true disposal. Some options exhibit a high DI because of high volatilization (i.e., incidental disposal). The same would be true for a leaking landfill. Such options should be evaluated closely to determine the impact on local ambient air concentrations and groundwater concentrations of toxics.

The best options (independent of cost) are those that exhibit high values of *both* CYI and DI. In this analysis the options that exhibit the highest values of both CYI and DI are

	CYI	*DI*
Incineration	370	10,000
Destructive treatment (at 98% effectiveness)	1.6	50

Secure landfill, land treatment for metals removal, long-term storage of nonvolatiles in closed tanks, relatively short-term storage in containers (five years), short-term storage of both volatiles and nonvolatiles in closed tanks, treatment for removal of toxic pollutants, and encapsulation all exhibit medium-range CYIs and low DIs. These values reflect the fact that there are storage options (except for land treatment and treatment by removal, which require further considerations for residue disposal, or in the case of land treatment assurance of permanent retention in the soil medium). These options can and often are combined for total hazardous-waste management systems. For combined options the analysis may proceed in a similar fashion. The last option in the treatment train is likely to be the controlling one for determination of CYI and DI.

Costs may now be superimposed to arrive at what are "normalized" control and disposal comparisons. The cost per unit of control-years purchased is given by

$$\text{Unit control cost} = \frac{\text{Total control cost}}{\text{CYI}}$$

The cost per unit of disposal purchased is

$$\text{Unit disposal cost} = \frac{\text{Total disposal cost}}{\text{DI}}$$

Fractional values of either CYI or DI can be interpreted to indicate negative cost benefit when options are considered separately. Unit cost comparisons involving one or more options that exhibit fractional CYI and DI values will favor those options with high index values. A cost advantage accrues to those options with values greater than unity.

Incineration and destructive treatment were compared favorably in the previous discussion, but there is a very wide disparity in the values of both CYI and DI for the two options. This disparity illustrates a problem that is important to the consideration of data.

Percent removal is the classical treatment parameter and continues as the parameter of choice for monitoring treatment process efficiency. It is the main reporting parameter for processes reported here. Although 98% removal is considered very high for process treatment efficiency, the 99.99% destruction and removal efficiency for incineration (the current RCRA regulation) represents three orders of magnitude better control Even at 99.99%, 1 lb of unburned toxic waste component could be discharged for every 10,000 lb burned. This difference between control effectiveness of the two options is reflected in the index values.

It is important to consider and report in detail process removal and destruction efficiencies in future work. Identification of highly toxic waste components will allow the researcher and designer to limit the range of chemical analyses required to conduct relevant studies and to "track" compounds of most concern. Also, only those treatment processes that are very highly effective on highly toxic compounds—namely, much higher than 99%—will compare favorably to combustion.

1

Hazardous Waste and Chemical Substances—Statutory Review

Arnie Edelman
U.S. Environmental Protection Agency
Washington, D.C.

Management and disposal of hazardous wastes are causing a revolution in sanitary engineering practices. To put the control of hazardous wastes into perspective, we must examine the definition of hazardous waste and the many statutes adopted by Congress that may place constraints on or otherwise affect hazardous waste management.

This chapter gives an overview of the numerous statutes requiring the Environmental Protection Agency (EPA) and other federal agencies [e.g., the Department of Transportation (DOT)] to designate and control toxic and hazardous chemicals. The discussion for each statute focuses on what must be considered by each agency when identifying or listing chemicals to be regulated.

IDENTIFYING HAZARDOUS WASTES

The EPA's Office of Solid Waste (OSW), under authority of the Resource Conservation and Recovery Act (RCRA), has promulgated regulations concerning identification of hazardous waste, has developed standards for generators, transporters, treaters, storers, and disposers of hazardous waste, and has developed criteria for solid waste (nonhazardous) disposal facilities.

A hazardous waste is defined by the RCRA statute to include any solid waste or combination of solid wastes that, because of its quantity, concentration, or physical, chemical, or infectious characteristics, may cause or significantly contribute to an increase in mortality or an increase in serious irreversible or incapacitating reversible illness, or pose a substantial present or potential hazard to human health or the environment when improperly managed.

Clearly, the definition of solid waste is the key to determining if a waste is subject to hazardous-waste regulations. A solid waste is defined by regulation (40 CFR 261.4) as any garbage, refuse, sludge, or other waste material that is not

1. Domestic sewage or any mixture of domestic sewage and other wastes that passes through a sewer system to a publicly owned treatment works for treatment
2. Industrial wastewater discharges that are point-source discharges subject to regulation under the Clean Water Act (CWA)
3. Irrigation return flows
4. Source special or by-product material as defined by the Atomic Energy Act (AEA) of 1954
5. Materials subject to in situ mining techniques that are not removed from the ground as part of the extraction process.

The regulations also specify which solid wastes are not hazardous wastes [40 CFR 261.4(b)]; they exclude, for example, wastes from households, the raising of animals, fly ash, and scrap leather.

A solid waste is also defined [40 CFR 261.2(b)] to include any solid, liquid, semisolid, or contained gaseous material resulting from industrial, commercial, mining, or agricultural operations or community activities that

1. Is discarded, accumulated, stored, or physically, chemically, or biologically treated prior to being discarded; or
2. Has served its original intended use and sometimes is discarded; or
3. Is a manufacturing or mining by-product and sometimes is discarded.

Regulation 40 [CFR 261.2(c)] defines a discarded material as one that is abandoned by being disposed of, burned, or incinerated (unless the material is burned as a fuel for the purpose of recovering usable energy), or physically, chemically, or biologically treated in lieu of or prior to being disposed of.

On April 4, 1983, the EPA proposed to emend the definition of solid waste to include reused or recycled wastes (48 F.R. 14472). The proposed definition of solid waste would include spent materials, sludges, by-products, and commercial chemical products that are

1. Being recycled if use constitutes disposal

2. Burned as a fuel for energy recovery or used to produce a fuel
3. Reclaimed
4. Accumulated for speculation
5. Accumulated without sufficient amounts or stored materials being recycled.

Materials not considered as solid wastes by the EPA in the proposal are

1. Secondary materials used or reused as ingredients or feedstocks in the production process
2. Secondary materials used as substitutes for raw materials in recovery processes that usually use raw materials as feedstocks
3. Secondary materials used or reused as substitutes for commercial products
4. Reclamation conducted at a single plant site when the reclaimed materials are reused within the original process in which it was generated.

To determine if a waste is hazardous, first determine if it is solid. Then ascertain if the solid waste is hazardous. A solid waste is hazardous if it is not excluded from regulation and it meets any of the criteria for identifying hazardous waste.

Section 3001 of RCRA requires the EPA administrator to develop and promulgate criteria and regulations for identifying the characteristics of hazardous waste and for listing hazardous waste, taking into account accumulation in tissue and other hazardous characteristics.

Hazardous wastes are identified by certain characteristics, such as ignitability, corrosivity, reactivity, and extraction-procedure (EP) toxicity. The criteria establishing which wastes are covered and the appropriate test methods to use are found in 40 CFR 261.21–261.24. EP toxicity is determined by a specific extraction procedure to identify the presence of certain toxic materials (e.g., arsenic, barium, cadmium, etc.) at levels greater than those specified in 40 CFR 261.24.

The preamble to the hazardous-waste-listing regulation published in the *Federal Register* of May 19, 1980 (45 *F.R.* 33107), states that "the criteria for listing toxic wastes are intended by EPA to identify all those wastes which are toxic to aquatic species." One of the criteria in 40 CFR 261.11 for listing a hazardous waste is that it contain a constituent listed in Appendix VIII of 40 CFR 261. The constituents listed are those that have been shown in reputable scientific studies to have toxic, carcinogenic, mutagenic or teratogenic effects on humans or other life forms and include such substances as those identified by the agency's Carcinogen Assessment Group. The presence os any of these constituents in the waste is presumed to be sufficient to list the waste. Key sections of the regulations providing a listing of such chemicals are 40 CFR 261.31–261.33. Sections 261.31 and 261.32 list waste from specific and nonspecific sources. These listings are based upon the criteria in 40 CFR 261.11. The inclusion of commercial chemical products, off-specification species, containers, and spill residues as hazardous waste is found in 40 CFR 261.33. This section includes "those materials which are being thrown away in their pure form or as an off-specification species. . . as well as the contaminated residues and debris from those materials."

Regulation 40 CFR 261.5 provides special requirements for hazardous waste produced by small-quantity generators. The law stipulates that persons generating a total of less than 1000 kilograms (kg) of hazardous waste in a calendar month are not subject to the regulations as long as the waste is either treated or disposed of in an on-site or off-site facility permitted (interim or final) by EPA, in a facility permitted, licensed, or registered by a state to manage municipal or industrial solid waste, or in a resource recovery facility.

Although a waste may be identified as hazardous, the regulations allow generators of the waste to obtain an exemption: 40 CFR 260 includes delisting procedures for generators to follow if they can provide data to show that the specific waste does not meet the criteria that caused the agency to list the waste.

FEDERAL STATUTES REGULATING CHEMICAL SUBSTANCES

In addition to the RCRA, twenty-four major pieces of legislation, administered by five federal agencies [Consumer Product Safety Commission (CPSC), Occupational Safety and Health Administration (OSHA), Food and Drug Administration (FDA), Department of Transportation (DOT), and EPA], direct the federal government to regulate chemicals. Table 1-1 lists the statutes regulating chemicals, of which many, when disposed, are also hazardous wastes that may be present as a hazard in the workplace, the environment, households, consumer products, food, drugs, cosmetics, and drinking water.

Because so many statutes and regulations with differing purposes designate hazardous chemicals, confusion may result. One purpose of this section is to lay out the key factors, as mandated in the various statutes, that must be considered by each agency when identifying or listing chemicals to be regulated. These key provisions include statutory purpose, definition of harmful substance, risk criteria, and required considerations. Once all statutes are placed into the proper perspective, a discussion of what is hazardous waste will follow.

Statutory Purpose

Understanding why the statutes were developed is basic to understanding why various statutes cover different hazardous chemicals. The stated purpose of the statute directs the regulatory agency's effort to one or more specific concerns, helps to establish priorities for accomplishing the goals outlined in the statute, and whether explicit or implied, directs the carrying out of the mandate, which results in differences in the ways the chemicals are addressed by each statute. See Table 1-2 for a summary of the purposes of the various acts as they relate to health and safety regulations. Because all statutes do not explicitly state their purpose, statements reflecting legislative history and/or legal decisions were developed. (They are enclosed in brackets in the table.) The specificity of the stated purpose varies from statute to statute. Some list specific methods to achieve their goals (e.g., RCRA, OSHA) and specify how the objectives are to be met; others, such as the Hazardous Materials Transportation Act (HMTA), simply present the general purpose.

Definition of Harmful Substance

A key factor affecting the regulatory coverage of a statute is the definition of the substances subject to regulation. The statutes use several descriptive terms, not necessarily having the same meaning, to identify "harmful substances." These include pollutant, toxic pollutant, hazardous substance, contaminant, hazardous material, and hazardous waste. Evidently not all *statutes* define or use a term to define harmful substances. The Toxic Substances Control Act (TSCA) for example, defines "chemical substances" and "mixtures" subject to regulation if certain criteria are met; the Federal Insecticide, Fungicide, and Rodenticide Act (FIFRA) and the Marine Protection, Research, and Sanctuaries Act (MPRSA) specify categories of substances: FIFRA defining "pesticides," and MPRSA "materials."

Table 1-3 lists the definitions included in the statutes for harmful substances. (For brevity statutes and some words ("section") have been abbreviated.) If a harmful substance is not identified, the definition of the general category of substances is included.

Risk Criteria

Three aspects of risk are addressed in each statute, legislative history, or case law; type of harm, type of risk, and required considerations between the chemical and the harm that may result.

Type of Harm. The type of harm is usually explicitly described by terms that define the chemical or the substances to be addressed (e.g., hazardous waste—hazardous substance that may cause *injury to health or environment*). The harm components of a statute's risk definition generally consist of a description of an undesired outcome (death, injury) and/or a description of the population (public, wildlife) at risk or the objective of the regulation, such as protection of the environment. Table 1-4 presents the description of harm and/or the population (object) to be protected. Description of harm may explain why chemicals are regulated under different statutes.

Type of Risk. Considering risk when developing federal regulations encompasses the probability of harm occurring. The probability of harm presented by a chemical may be considered zero, insignificant, or significant, thus presenting a significant or substantial risk. Table 1-5 presents the types of risk discussed in the statute that must be considered when developing regulations. Such phrases as "significant risk" or "substantial risk" imply that low-probability risk need not be addressed in the rule making of an agency. Note, however, that these terms are ambiguous, hard to define, and subject to legal interpretation. The definition of risk also often employs a certainty requirement (e.g., may present, presents, or will present a risk). Table 1-6 presents a listing of the causal connections as described in the statute.

Required Considerations. The statutory language may guide the designation and setting of technical or control standards for explicitly specifying a basis for making regulatory decisions and also for indicating what factors must be, may be, or may not be considered when developing regulations. As presented in Table 1-7, the statutes discuss the amount of protection or risk reductions to be addressed through the issuance of standards "necessary," "adequate," or "sufficient" to protect health or the environment by providing detailed guidance (e.g., ample margin of safety) and/or by prescribing partial factors (e.g., risk and cost) that must be considered. Table 1-8 presents an overview of the risk, economic, and technical factors that must be addressed as well as the need to conduct cost-benefit analysis.

Table 1-1. Federal Laws Controlling Toxic Substance Exposures

Statute	Agency	Coverage
Bulk Flammable Combustible Liquids Act (1970)	DOT–USCG	Water Shipment of toxic materials
Clean Air Act (1970), amended 1977	EPA	Air pollutants
Comprehensive Environmental Response, Compensation, and Liability Act (1981)	EPA	Hazardous substances, pollutants, contaminants
Consumer Product Safety Act (1972)	CPSC	Dangerous consumer products
Dangerous Cargo Act (1952)	DOT–USCG	Water shipment of toxic materials
Federal Hazardous Substances Amendment Act (1966)	CPSC	Toxic household products
Federal Insecticide, Fungicide, and Rodenticide Act (1948), amended 1972	EPA	Pesticides
Federal Water Pollution Control Act (1972, amended 1977, 1978)	EPA	Water pollutants
Food, Drug, and Cosmetic Act (1932)	FDA	Food, drugs, cosmetics
Color-Additive Amendments (1960)	FDA	Color additives
Food Additives Amendment (1958)	FDA	Food additives
Medical Device Amendments	FDA	Medical devices
New Animal Drug Amendments (1968)	FDA	Animal drugs, feed additives
New Drug Amendments (1962)	FDA	New drugs
Hazardous Materials Transportation Act (1970)	DOT	Transportation of hazardous materials
Lead-Based Paint Poison Prevention Act (1973, amended 1976)	CPSC	Lead paint in federally assisted housing
Marine Protection, Research, and Sanctuaries Act (1972)	EPA	Ocean dumping
Occupational Safety and Health Act (1970)	OSHA	Workplace, exposures
Pesticide Residues Amendment (1954, amended 1972)	EPA	Pesticide residues in food
Poison Prevention Packaging Act (1970)	CPSC	Packaging of dangerous products
Ports and Waterways Safety Act (1972)	DOT–USCG	Water shipment of toxic materials
Resource Conversion and Recovery Act (1976)	EPA	Hazardous wastes
Safe Drinking Water Act (1974, amended 1977)	EPA	Drinking-water contaminants
Toxic Substances Control Act (1976)	EPA	Chemical hazards not covered by other laws

Table 1-2. Statutory Purposes

Statute	Purpose	Source
CAA	To protect and enhance the quality of the Nation's air resources so as to promote the public health and welfare and the productive capacity of its population.	Sec. 101(b)(1)
CERCLA	...to improve the overall quality of the Nation's environment and to protect the health of our citizens....	Cong. Rec. – Sec. 51-5002 – Nov. 24, 1980
CPSA	...to protect the public against unreasonable risks of injury associated with consumer products....	Sec. 2(b)
CWA	...to restore and maintain the chemical, physical and biological integrity of the Nation's waters.... (1) it is the national goal that the discharge of pollutants into the navigable waters be eliminated by 1985; (2) it is the national goal that, wherever attainable, an interim goal of water quality which provides for the protection and propagation of fish, shellfish, and wildlife and provides for recreation in and on the water be achieved by July 1, 1983; (3) it is the national policy that the discharge of toxic pollutants in toxic amounts be prohibited;	33 U.S.C. 1251, Sec. 101(a)(1)–(3)
EPIA	To prevent the movement or sale for human food, of eggs and egg products which are adulterated or misbranded or otherwise in violation of this chapter.	Sec. 1032
FDCA	...to protect the public health.	Meserey v. United States, 447 F. Supp. 548, 553 (D.C. Nev. 1977)

		The purpose of the *new drug provision*...is "very clearly, to keep inadequately tested medical and related products which might cause widespread damages to human life out of interstate commerce.	AMP, Inc. v. Gardner, 389 F. 2d 825, 829–830 (2d Cir. 1968), cert. denied, "393" U.S. 825 (1968), reh. denied, 395 U.S. 917 (1969)
		The *food additive amendments*...are "aimed at preventing the addition to the food our people eat of any substance the ingestion of which reasonable people would expect to produce not just cancer but any disease or disability."	S. Rept. 2422, 85 Cong., 2 sess (1958), 3 U.S. C. Cong. and Admin. News, p. 5310, 1958
FHSA		To provide nationally uniform requirements for adequate cautionary labeling of packages of hazardous substances which are sold in interstate commerce and are intended or suitable for household use.	H. Rept. 1861, 86 Cong., 2 sess., reprinted in U.S. C. Cong. and Admin. News 2833, 2834, 1960
		The purpose of the 1966 Amendment to the Act was: (T)o ban the sale of toys and other children's articles containing hazardous substances; to authorize the Secretary of Health, Education and Welfare to ban the sale of other substances...so hazardous in nature that they cannot be made suitable for use in or around the household by cautionary labeling; to extend coverage of the Hazardous Substances Labeling Act to unpackaged as well as packaged hazardous substances intended for household use; and to make it clear that household products treated with pesticides are not exempt from that act.	H. Rept. 2166, 89 Cong., 2 sess., reprinted in U.S. C. Cong. and Admin. News 4095, 1966
FIFRA		...it was the Congressional intent that potentially dangerous pesticides...be removed from the market without delay and that the EPA be given expanded authority to regulate pesticides.	Dow Chemical Co. v. Blum, 467 F. Supp. 872, 900 (March 1977)
HMTA		...to improve the regulatory and enforcement authority of the Secretary of Transportation to protect the Nation adequately against the risks to life and property...inherent in the transportation of hazardous materials in commerce.	Sec. 102

(continued)

17

Table 1-2. (continued)

Statute	Purpose	Source
MPRSA	To regulate the dumping of all types of materials into ocean waters and to prevent or strictly limit the dumping into ocean waters of any material which would adversely affect human health, welfare, or amenities, or the marine environment, ecological systems, or economic potentialities.	Sec. 1401(b)
OSHA	...to assure so far as possible every working man and woman in the Nation safe and healthful working conditions and to preserve our human resources— ...by authorizing the Secretary of Labor to set mandatory occupational safety and health standards applicable to business affecting interstate commerce; ...by providing medical criteria which will assure insofar as practicable that no employee will suffer diminished health, functional capacity, or life expectancy as a result of his work experience; and ...by providing for the development and promulgation of occupational safety and health standards.	Sec. 2(b)
PPPA	...to provide for special packaging to protect children from serious personal injury or serious illness resulting from handling, using, or ingesting household substances.	H. Rept. 1642, 91 Cong., 2 sess., reprinted in U.S. C. Cong. and Admin. News 5326, 5327, 1970
PWSA	(a) ...navigation and vessel safety and protection of the marine environment are matters of major national importance; (b) ...increased vessel traffic in the Nation's ports and waterways creates substantial hazard to life, property, and the marine environment; (c) ...increased supervision of vessel and port operations is necessary in order to— ...reduce the possibility of vessel or cargo loss, or damage to life, property, or the marine environment;	

18

RCRA	...to promote the protection of health and the environment and to conserve valuable material and energy resources by—	Sec. 1003
	...prohibiting future open dumping on the land and requiring the conversion of existing open dumps to facilities which do not pose a danger to the environment or to health;	
	...regulating the treatment, storage, transportation, and disposal of hazardous wastes which have adverse effects on health and the environment;	
SDWA	...to assure that water supply systems serving the public meet minimum national standards for protection of public health.	H. Rep. 1185, 93 Cong.
TSCA	(1) Adequate data should be developed with respect to the effect of chemical substances and mixtures on health and the environment and that the development of such data should be the responsibility of those who process such chemical substances and mixtures;	Sec. 2(b)
	(2) Adequate authority should exist to regulate chemical substances and mixtures which present an unreasonable risk of injury to health or the environment, and to take action with respect to chemical substances and mixtures which are imminent hazards; and	
	(3) Authority over chemical substances and mixtures should be exercised in such a manner as not to impede unduly or create unnecessary economic barriers to technological innovation while fulfilling the primary purpose of this Act to assure that such innovation and commerce in such chemical substances and mixtures do not present an unreasonable risk of injury to health or the environment.	
WMA	To prevent and eliminate burdens upon such commerce, to effectively regulate such commerce, and to protect the health and welfare of consumers.	Sec. 602, 451, 1031, resp.
WPA	To prevent the movement or sale in interstate or foreign commerce of, or the burdening of such commerce by, poultry products which are adulterated or misbranded.	Sec. 452

Table 1-3. Statutory Definitions of Harmful Substances

Statute [section]	Definition
BFCLA [4417(a)(2)(c)]	**Hazardous material** Any liquid material or substance which is (i) flammable or combustible; or (ii) designated a hazardous substance under sec. 311(b) of the FWPCA, as amended; or (iii) designated a hazardous material under sec. 104 of the HMTA.
CAA [302(g)]	**Air pollutant** Any air pollution agent or combination of such agents, including any physical, chemical, biological, radioactive (including source material, special nuclear material, and byproduct material) substance or matter...emitted into or otherwise enters the ambient air.
[112(a)(1)]	**Hazardous air pollutant** An air pollutant to which no ambient air quality standard is applicable and which judgement of the Administrator causes, or contributes to, air pollution which may reasonably be anticipated to result in an increase in mortality or an increase in serious irreversible, or incapacitable illness.
CERCLA [101(14)]	**Hazardous substance** (A) any substance designated pursuant to sec. 311(b)(2)(A) of the Clean Water Act (FWPCA); (B) any element, compound, mixture, solution, or substance designated pursuant to sec. 102 of this Act (shown below); (C) any hazardous waste having the characteristics identified under or listed pursuant to section 3001 of the Solid Waste Disposal Act (but not including any waste the regulation of which under the Solid Waste Disposal Act has been suspended by Act of Congress); (D) any toxic pollutant listed under sec. 307(a) of the FWPCA; (E) any hazardous air pollutant listed under sec. 112 of the CAA; and (F) any imminently hazardous chemical substance or mixture with respect to (sic) which the Administrator has taken action pursuant to sec. 7 of the TSCA. The term does not include petroleum, including crude oil or any fraction thereof...not otherwise specifically listed or designated as a hazardous substance under...(A) through (F)...[or] natural gas, natural gas

liquids, liquefied natural gas, or synthetic gas usable for fuel (or mixtures of natural gas and such synthetic gas).

[102(a)]

The Administrator shall promulgate and revise, as may be appropriate regulations designating as hazardous substances, in addition to those referred to in sec. 101(14) of this title. . .such this title, such elements, compounds, mixtures, solutions, and substances which, when released into the environment, may present substantial danger to the public health or welfare or the environment.

[104(a)(2)]

Pollutant or contaminant Shall include, but not be limited to, any element, substance, compound, or mixture, including disease-causing agents, which after release into the environment and upon exposure, ingestion, inhalation, or assimilation into any organism, either directly from the environment or indirectly by ingestion through food chains, will, or may reasonably be anticipated to, cause death, disease, behavioral abnormalities, cancer, genetic mutation, physiological malfunctions (including malfunctions in reproduction) or physical deformations in such organisms or their offspring.

CPSA
[3(a)(1)]

Consumer product Any article, or component part thereof, produced or distributed

   (i) for sale to a consumer for use in or around a permanent or temporary household or residence, a school, in recreation, or otherwise, or

   (ii) for the personal use, consumption or enjoyment of a consumer in or around a permanent or temporary household or residence, a school, in recreation, or otherwise; but such term does not include

   (A) any article. . .not customarily produced or distributed for sale to, or use or consumption by, or enjoyment of, a consumer,

   (B) economic poisons (as defined by the FIFRA),

   (C) drugs, devices, or cosmetics (as such terms are defined in sections. 201 (h), and (i) of the FDCA,) or

   (D) food. . .the term "food," as used in this subparagraph means all "food," as defined in sec. 201(f) of the Federal Food, Drug, and Cosmetic Acts, including poultry and poultry products (as defined in sec. 4(e) and (f) of the Poultry Products Inspection Act) meat, meat food products (as defined in sec. 1(j) of the Federal Meat Inspection Act), and eggs and egg products (as defined in sec. 4 of the Egg Product Inspection Act).

DCA

No hazardous substance definition. However, the DCA does define "combustible liquid."

(continued)

Table 1-3. *(continued)*

Statute [section]	Definition
FDCA	A food is considered **adulterated:** or deleterious substance which may render it injurious to health; but in case the substance is not added substance such food shall not be considered adulterated under this clause if the quality of such substance in such food does not ordinarily render it injurious to health; or
[(a)(2)(A)]	if it wears or contains any added poisonous or deleterious substance which is unsafe within the meaning of sec. 406(a); or
[(a)(2)(B)]	if it is a raw agricultural commodity in conformity with an exemption granted or a tolerance prescribed under section 408 and such residue has been removed to the extent possible in good manufacturing practice and the concentration does not exceed the prescribed tolerance; or
[(a)(2)(D)]	if it is, or it bears or contains, a new animal drug (or conversion product thereof)...unsafe within the meaning of sec. 512; or
[c]	if it is, or it bears or contains, a color additive...unsafe within the meaning of sec. 706(a); or
[(a)(3)(7)]	if it has been intentionally subjected to radiation, unless the use of the radiation was in conformity with a regulation or exemption in effect pursuant to sec. 409.
FHSA [2(f)(1)]	**Hazardous substance** Any substance or mixture of substances which: (1) is toxic, corrosive, and irritant, a strong sensitizer, flammable, combustible, or generates pressure through decomposition, heat, or other means; or (2) causes substantial personal injury or illness during or as a proximate result of any customary or reasonably forseeable handling or use, including reasonably forseeable ingestion by children; or (3) is a radioactive substance that requires labeling to protect the public health; or (4) is a toy or other article which presents an electrical, mechanical, or thermal hazard. However, the term shall not apply to (1) economic poisons subject to the FIFRA, nor to foods, drugs, and cosmetics subject to the FDCA, nor to substances intended for use as fuels when stored in containers and used in the heating, cooking, or refrigerations system of a house, but...shall apply to any article...not itself an economic poison

22

within the meaning of the FIFRA but which is a hazardous substance substance by reason of bearing or containing such an economic poison; or

(2) any source material, special nuclear material, or byproduct material as defined in the AEA.

FWPCA
[502(6)]

Pollutants Dredged spoil, solid waste, incinerator residue, sewage, garbage, sewage sludge, munitions, chemical wastes, biological materials, radioactive materials, heat, wrecked or discarded equipment, rock, sand, cellar dirt, and industrial, municipal and agricultural waste discarded into the water.

[502(13)]

Toxic pollutants Those pollutants, or combinations of pollutants, including disease-causing agents, which after discharge and upon exposure, ingestion, inhalation or assimilation into any organism, either directly from the environment or indirectly by ingesting through food chains, will, on the basis of information available to the Administrator, cause death, disease, behavioral abnormalities, cancer, genetic mutations, physiological malfunctions (including malfunctions in reproduction) or physical deformations in such organisms or their offspring.

[311(b)(2)(A)]

Hazardous substance When discharged in any quantity presents an imminent and substantial danger to the public health or welfare, including, but not limited to, fish, shellfish, wildlife, shorelines, and beaches.

HMTA
[103(20)]

Hazardous materials A substance or material in a quantity and form which may pose an unreasonable risk to health and safety or property when transported in commerce.

OSHA

The term "toxic material" is used in ['6(b) (5)] but not defined in the Act.

PPPA

CPSC may require "special packaging" standards for any "household substance" which is also

(A) a hazardous substance under the Federal Hazardous Substances Act (toxic, corrosive, irritant, strong sensitizer, or substance which generates pressure);

(B) a food, drug, or cosmetic according to the FDCA; or

(C) a fuel stored in a portable container used for heating, cooking, or refrigerating in the home;

(D) an "economic poison" as defined in the Federal Insecticide, Fungicide, and Rodenticide Act

PWSA

No hazardous substance definition. 33 U.S.C., sec. 1225(a)(2)(A) does reference definition of hazardous material contained in the BFCLA.

(continued)

23

Table 1-3. *(continued)*

Statute [section]	*Definition*
SDWA [1401(b)]	**Contaminant** Any physical, chemical, biological, or radiological substance or matter in water.
TSCA [3(2)]	**Chemical substance** Any organic or inorganic substance of a particular molecular identity, including (i) any combination of such substances occurring in whole or in part as a result of a chemical reaction or occurring in nature, and (ii) any element or uncombined radical. (B) Such term does not include (i) any mixture, (ii) any pesticide (as defined in FIFRA) when manufactured, processed, or distributed in commerce for use as a pesticide, (iii) any source material, special nuclear material, or byproduct in the AEA of 1954 (and regulations issued under such Act), (iv) any food, food additive, drug, cosmetic, or device (as such terms are defined in sec. 201 of the FDCA) when manufactured, processed, or distributed in commerce for use as food, food additive, drug, cosmetic, or device.

Mixture Any combination of two or more chemical substances if the combination does not occur in nature and is not, in whole or in part, the result of a chemical reaction; except that such term does not include any combination which occurs, in whole or in part, as a result of a chemical reaction if none of the chemical substances comprising the combination is a new chemical substance and if the combination could have been manufactured for commercial purposes without a chemical reaction at the time the chemical substances comprising the combination were combined.

A meat, poultry, or egg product is **adulterated** if

(1) it bears or contains any poisonous or deleterious substance which may render it injurious to health; but in case the substance is not an added substance, such article shall not be considered adulterated under this clause if the quantity of such substance in or on such article does not ordinarily render it injurious to health;

(2) it bears or contains any added poisonous or added deleterious substance;

(3) it is, in whole or in part, a raw agricultural commodity and such commodity bears or contains a pesticide chemical which is unsafe within the meaning of the FDCA;

(4) it bears or contains any food additive which is unsafe within the meaning of the FDCA;

(5) it bears or contains any color additive which is unsafe within the meaning of the FDCA....

(6) it has been intentionally subjected to radiation, unless the use of the radiation was in conformity with a regulation or exemption in effect pursuant to the FDCA.

WMA
[1(j)]
WPA
[4(e)]
EPIA
[4]

25

Table 1-4. Statutory Description of Harm

Statute [section]	Description of Harm and/or Objects of Protection
BFCLA [33 U.S.C. sec. 391a]	Hazards to life and property, for navigation and vessel safety, and for enhanced protection of the marine environment.
CAA [108,111,157,202, 211,231]	Endanger public health or welfare
[109]	[protect] the public health
[112]	Increase in mortality or serious irreversible, or incapacitating reversible, illness.
[157]	Adverse effects on the stratosphere, especially ozone
CERCLA [102]	[substantial danger to] the public health or welfare or the environment
[104]	Death, disease, behavioral abnormalities, cancer, genetic mutation, physiological malfunctions (including malfunctions in reproduction) in...organisms or their offspring
CPSA [7,8]	[unreasonable risk of] injury defined as "death, personal injury, or serious or frequent illness."
CWA [311(b)(2)(A)]	[imminent and substantial danger to] the public health or welfare, including, but not limited to, fish, shellfish, wildlife, shorelines, and beaches.
[311(b)(4)]	[harmful to] the public health or welfare...including, but not limited to, fish, shellfish, wildlife and public and private property, shorelines, and beaches.
[307(a)]	Death, disease, behavioral abnormalities, cancer, mututations, physiological malfunctions, including malfunctions in reproduction, or physical deformations in any organisms or their offspring.
FDCA [406, 408]	[protect] the public health
[402, 601]	Injurious to health
[409(c), 512, 706(b)]	Cancer in man or animal
FHSA [2(f), 3(a)]	Substantial personal injury or serious illness
[3(b), 2(q)]	[protect] the public health and safety
FIFRA [3(c),3(d),6(b)]	Unreasonable adverse effects on the environment
[25(c)]	Serious injury or illness to children and adults resulting from accidental ingestion or contact

T able 1-4. *(continued)*

Statute [section]	Description of Harm and/or Objects of Protection
HMTA [104]	[unreasonable risk to] health and safety or property
MPRSA [102(a)]	Unreasonably degrade or endanger human health, welfare, or amenities, or the marine environment, ecological systems, or economic potentialities
OSHA [6(b)(5)]	No...material impairment of health or functional capacity
PPA [3]	Serious personal injury or serious illness affecting children
PWSA [33 U.S.C. sec.1223]	Vessel or cargo loss, or damage to life, property, the marine environment...to structures in, on, or immediately adjacent to the navigable waters or the resources within such waters
[33 U.S.C. sec.1225]	Damage to, or the destruction of, any bridge...[harm to] the navigable waters and the resources therein
RCRA [1004(5)]	An increase in mortality or in serious irreversible, or incapacitating reversible, illness or [substantial hazard] to human health or the environment
[3002,3,4]	[protect] human health and the environment
SDWA [1401(1), 1412(b)(1)(B)]	Any adverse effect on the health of persons
[1421]	The presence of any contaminant Affect adversely the health of persons
TSCA [4(a)]	Injury to health or the environment or Significant or substantial human exposure
[4(f)]	Serious, widespread harm to human beings from cancer, gene mutations, or birth defects
[5(b)(4)(A), 5(f), 6(a)]	Injury to health or the environment

Table 1-5. Special Types of Risk Incorporated in Statutory Definitions

Statute [section]	Type of Risk
CAA [108(a), 111(b), 157(b). 202(a), 211, 231]	Endanger
CERCLA [102]	Substantial danger
CPSA [15 U.S.C., sec.2056, 2057]	Unreasonable
CWA [311(b)(2)(A)	Imminent and substantial danger
FDCA (med. dev.)	Unreasonable and substantial
FIFRA (3,6)	Unreasonable
HMTA [104]	Unreasonable
MPRSA [102(a)]	Unreasonable
RCRA [1004(5)]	Substantial present or potential hazard
TSCA [4(a), 5(b)(4)(A), (f), 6(a)]	Unreasonable
[4(f)]	Significant

Table 1-6. Types of Causal Connections

Statute [section]	Causal Connection
CAA [112]	May reasonably be anticipated to result
[157]	May reasonably be anticipated to affect
[108, 111, 112, 202, 211, 231]	Causes or contributes to
[108, 111, 157, 202(a), 211, 231]	May reasonably be anticipated to endanger
CERCLA [102]	May present [danger or risk]
[104]	Will or may reasonably be anticipated to cause
CPSA [8]	Presents [danger or risk]; [is] associated with
CWA [307(a)]	Will cause
[311(b)(4)]	May be harmful to
[311(b)(2)(A)]	Presents [danger or risk]
FDCA [402(a)]	Ordinarily render(s)
[402(a), 601]	May render [food injurious]
FHSA [2(f), 3(a)]	May cause
FIFRA [3(c)(5)]	Will not [unreasonably degrade or endanger]
[3(c)(5), (d)]	Will not generally cause
[3(c)(7)(C)]	Will not cause
[3(c)(7)(A), (B)]	Would not [significantly increase the risk of]
[3(d)(1)]	May generally cause
[3(d)(3)]	Would not cause
[6(b)]	Generally causes
SDWA [1421]	May result in; may effect [adversely]
[1401, 1412]	May have [any adverse effect]
TSCA [4(a)]	May present [danger or risk]
[4(f), 5(b)(4)(A), (f), 6(a)]	Presents or will present
USDA (PPIA, FMIA, and EPIA)	May make [product injurious]
(PPIA, FMIA, and EPIA)	Ordinarily render(s)

Table 1-7. Protection Afforded by Chemical Control Laws

Statutory Provisions [section]	Level of Protection
BFCLA	Increased protection against hazards through application of BAT[a] unless undue economic impacts would result which are not outweighed by the benefits
CAA [109]	Adequate margin of safety
[112(b)(1)(B)]	Ample margin of safety to protect the public health
[211]	Ample margin of safety
CPSA	Reasonably necessary to prevent or reduce an unreasonable risk of injury
CWA [307(a)]	An ample margin of safety [307(a)(4)] through applying...BAT[a] economically achievable [307(a)(2)]
FIFRA	No unreasonable adverse effects
HMTA	Necessary or appropriate
MPRSA	No unreasonable degradation
OSHA [6(b)(5)]	Adequately assures, to the extent feasible, that no employee will suffer material impairment of health or functional capacity
[3(8)]	Reasonably necessary or appropriate to provide safe or healthful employment...
PWSA	Necessary to prevent/protect...
RCRA [3002–3004]	Necessary to protect human health and the environment
SDWA [1412(a)(2)]	As close to the maximum contaminant level to the extent feasible... (taking costs into consideration)
TSCA [6(a)]	To protect adequately against (unreasonable risks) using the least burdensome requirement

[a]BAT = best available technology.

Table 1-8. Factors to Consider for Developing Regulations

Statute [section]	Risk Factors	Economic Factors[a]	Technical Factors[a]	Cost-Benefit Analysis[a]
CAA [104(a)]	Risk to public health and welfare is statutory basis for listing decision. Listing is a discretionary decision: *Thompson v. Chicago*, 7 ERC 1682 (N.D. Ill. 1975)	NA	NA	NA
[109(b)(1)]	Primary ambient air quality standards must be based on air quality criteria, must allow an "adequate margin and safety," and must be "requisite to protect the public health" from adverse health effects Primary standard must protect against uncertain as well as certain harms: *LIA v. EPA.* 14 ERC 1906 (D.C. Cir. 1980)	D.C. Circuit Court of Appeals held that the costs and technical feasibility of attaining the standards are not to be considered in establishing them, *LIA v. EPA.* 14 ERC 1906 (1980), referencing the legislative history Sec. 108–110 are "technology-forcing" provisions, the attainment of the primary, health-based standards takes precedence over the cost and present technological feasibility of achieving the requisite control: *Ethyl Corp. v. EPA.* 541 F.2d 1, 14 (D.C. Cir. 1976), citing legislative history and other decisions	NA	Not required by statute, *LIA v. EPA.*
[111(b)]	New source categories listed on basis of health risk New source performance standards ['111(a)] must also take into account "any non-air quality health and environmental impacts . . . ,"	Not mentioned for listing Economic and technological feasibility must be considered for standards For standards, the cost must be taken into account when setting emission limits, as well as energy requirements: *Portland Cement Ass'n. v. Ruckelshaus*, 513 F.2d 506 (D.C. Cir. 1976)	Not mentioned for listing. Economic and technological feasibility must be considered for standards.	Formal cost-benefit study not required: *Portland Cement Ass'n. v. Ruckelshaus*, 486 F.2d 375 (D.C. Cir. 1973)
[111(a)]			New source performance standard must be based on the "best technology system of continuous emission reduction . . . which has been adequately demonstrated." This is a technology-forcing provision	

(continued)

Table 1-8. (continued)

Statute [section]	Risk Factors	Economic Factors[a]	Technical Factors[a]	Cost-Benefit Analysis[a]
[112(a)]	Risk to public health is statutory basis for listing	NA	NA	NA
[112(b)]	Emission standard must be set at the level which provides an "ample margin of safety" to protect the public health	NA. EPA has incorporated these factors administratively for carcinogen standards	NA. EPA has incorporated these factors administratively for carcinogen standards	NA. EPA has used a form of cost-benefit analysis in setting standards for carcinogens
[157(b)]	Risk rationale is statutory basis for regulation	Regulations must take costs and feasibility of controls into account		NA
[202(a)(1),(2)]	Risk rationale is statutory basis for regulating motor vehicle emissions	Regulations must take effect after the time necessary to permit the development and application of the requisite technology, giving appropriate consideration to the cost of compliance		NA
[202(a)(4)]	Emission control devices shall not be used to comply with standards if the device will cause or contribute to an "unreasonable risk" to public health, welfare, or safety			
[211(c)(1), (2)(A)(B)]	Risk rationale is statutory basis for regulation of fuels and fuel additives	EPA must consider "other technologically or economically feasible means" of achieving 202 emission limits before regulating fuels or fuel additives		"Available economic data" must be considered, "including a cost-benefit analysis" of controlling controls
[211(c)(2)(A)(B)]	"All relevant medical and scientific evidence available" must be considered	"Available economic data" must be considered		

[211(c)(2)(C)]		To prohibit fuels of additives, EPA must find that the prohibition "will not cause the use of any other fuel or . . . additive which will produce emissions which will endanger the public health or welfare in the same or greater degree"		NA
[23(a)(2),(1)(b)]	Risk rationale is statutory basis for regulating aircraft emissions	Regulations must take effect after the period of time "necessary" (after consultation with the Secretary of Transportation) to permit the development and application of the requisite technology, giving appropriate consideration to the cost of compliance"		
CERCLA [102]	Risk rationale is statutory basis for designation of hazardous substances	NA	NA	NA
CPSA	"Unreasonable risk" is statutory basis for regulation. A standard must be "reasonably necessary" to prevent or reduce the unreasonable risk [7(a)], or no "feasible" standard "would adequately protect" against the risk in case of a regulatory ban [8]	NA	NA	Unreasonable risk determination requires balancing risk, technical, and economic factors but not a cost-benefit analysis: Legislative History, *Aqua Slide 'N' Dive v. CPSC*, 569 F.2d 831 (5 Cir. 1978)
[9(c)(1)(A),(C),(D)]	The CSPC must consider the degree and nature of the risk the rule is to reduce or eliminate	The CSPC must consider the public's need for the product and the probable effect of a rule on the utility, cost, or availability of products to meet the need	The CSPC must consider any means of minimizing adverse efforts on competition or disruption or dislocating of manufacturing and other commercial practices consistent with the public health and safety	
DOT/USCG	Risk rationale is statutory basis for regulation (PWSA, DCA, BFCLA)	NA	NA	NA

(continued)

Table 1-8. (continued)

Statute [section]	Risk Factors	Economic Factors[a]	Technical Factors[a]	Cost-Benefit Analysis[a]
FDCA [406,400, 402(a),601]	Risk rationale is statutory basis for regulation	NA	NA	NA
[409,505,512, 706]	Safety is statutory basis for regulation	Economic and technological factors must be considered for [408]		NA
		NA	NA	For medical devices, new drugs and new animal drugs, therapeutic risks and benefits are weighed
				For food additives [409], color additives [706], and new animal drugs [512], no carcinogens may be approved
FIFRA [3(c)(5),(7),3(d).6(b)]	Unreasonable adverse effects is the statutory basis for regulations, which is defined to mean "any unreasonable risk to man or the environment, taking into account the economic, social and environmental costs and benefits"			NA
				Implicit in balancing approach
[25(c)]	Risk rationale is basis for packaging standards	NA	Packaging requirements must be consistent with those established by the PPPA	
FISHA	Risk rationale is statutory basis for regulation	NA	If no feasible cautionary labeling can adequately protect the public health, a ban may be imposed	NA
HMIA [104,105]	Unreasonable risk is statutory basis for regulation	NA	Regulations must be "necessary or appropriate"	NA

MPRSA [102(a)]	No "reasonable" degradation or endangerment of "human health, welfare, or amenities, or the marine environment, ecological systems, or economic potentialities" is statutory basis for issuing ocean dumping permits EPA must consider enumerated risk factors in establishing ocean dumping permit criteria	EPA must consider the need for dumping and its economic effects in establishing permit criteria		Use of term "unreasonable" implies some sort of balancing of interests, although not necessarily a formal cost-benefit analysis
OSHA [6(b)(5)]	Risk rationale is statutory basis for regulation of toxic materials	Standards for toxic materials must protect health "to the extent feasible" which includes both economic and technological feasibility. [6(b)(5)] interpreted by *IUD v. Hodgson,* 499 F.2d 467, 478 (D.C. Cir. 1974)		Cost-benefit balancing is not allowed or required for setting standards: *ATMI v. Donovan*
[3(b)]	Risk must be "significant" to justify any regulation: *Industrial Union v. API,* 100 S. Ct. 2844, 1900	A standard must be "necessary or appropriate" to provide healthful places of employment		[a][3(8)] does not incorporate cost-benefit analysis requirement: *AIMI v. Donovan,* 9 OSHC 1913 (1981)
PPPA	Risk rationale is statutory basis for requiring special packaging	NA	Technical feasibility is a statutory basis for regulation	NA
RCRA [3001(b)]	Risk rationale is statutory basis for identifying hazardous wastes characteristics and listing such wastes	NA		NA
[3002–3004]	Risk rationale is statutory basis for standards	NA		NA

(continued)

Table 1-8. (continued)

Statute [section]	Risk Factors	Economic Factors[a]	Technical Factors[a]	Cost-Benefit Analysis[a]
SDWA [1412(b) (1)(B)]	EPA must establish, based on an NAS study, the recommended *b* . . . MCL–for each contaminant which may have an adverse effect on health. The MCL shall be set at a level at which "no known or anticipated adverse (health) effects" occur and which allows an "adequate margin of safety"	NA	NA	NA
[1412(b)(3)]	Revised primary drinking water regulations shall be set as close to the MCLs as is "feasible"			
[1412(b)(3)]		"Feasible" means "with the use of the best technology treatment techniques, and other means, which . . . are generally available (taking costs into consideration)"		NA. EPA must strike a balance between promotion of public welfare and avoidance of unnecessary expense: *dicta* in *EDF v. Costle*, 11 ERC 1209 (D.C. Cir. 1978)
[1412(b)(4)]			Drinking water regulations must be emended "whenever changes in technology, treatment techniques, and other means permit greater protection of the health of persons"	
[1421]	Risk rationale is statutory basis for regulating underground injection which "endangers" drinking water	NA	NA	NA

[a] NA = Not addressed in statute.
[b] MCL = Maximum contaminant level.

2

Exposure and Risk Assessment

Benjamin P. Smith
U.S. Environmental Protection Agency
Washington, D.C.

> Nothing would be done at all if a man waited till he could
> do it so well that no one could find fault with it.
> —Cardinal Newman

Cardinal Newman's statement is most pertinent to the subject of this chapter. The assessment of health risks posed to humans by anthropogenic activities is an incredibly complex undertaking. Because it is complex, it is impossible to do much more than identify sources and effects of potential concern and, in a rough fashion, to quantify transport along major pathways. Nevertheless, although we may be years away from developing general models for characterizing the total exposure milieu, we cannot afford to bury our heads in the sand. The health and social costs of inaction and the economic costs of overzealous regulation demand that we try to impose reason upon current undertakings. Consequently, whereas the "pure scientists" who review this chapter may protest loudly the simplifications invoked and the imprecision of some terminology, it is hoped that the engineers will find it quite useful as a tool to assist them in performing preliminary evaluations of the risks posed by hazardous-waste treatment and disposal activities.

Before proceeding with the discussion of risk and exposure assessment, we give a further resolution of the scope of this chapter. Risks posed by hazardous-waste treatment, storage, and disposal activities are due to emissions of a wide variety of pollutants. However, most models are not designed to deal with pollutant interrelationships; in many cases the relationships are not even known. On the other hand, specific compound studies can lead to a serious underestimation of risk unless efforts are made to complete the exposure picture. Furthermore, single compound studies will often be further in error if the impacts of natural background levels, extraneous sources and antagonistic/synergistic compounds are not properly evaluated. Nevertheless, one must crawl before one can walk, and we believe that the typical reader would be better served by a discussion of the basic elements of the exposure and risk assessment rather than an esoteric consideration of everything that could go wrong. Therefore, this chapter, although not a complete rendering of the engineer's needs vis-a-vis risk assessment, should serve as an effective introduction to the area. Furthermore, the methods and equations should be useful to those charged with performing preliminary identification of critical sources and setting up sampling requirements.

In summary, our intent is to identify tools for bounding health risks to humans near hazardous-waste treatment, storage, or disposal facilities (TSDs). For each tool the underlying principles and necessary inputs are defined. Although these tools are mostly conservative estimators, remember that our understanding of complex interactions and our ability to catalogue all sources are still limited. The importance of these assessment tools is that they provide a semiquantitative basis for the evaluation of health risks posed by TSDs.

Risk and exposure assessment can be viewed as an eight-step process:

1. Defining source terms
2. Transport and transformation analyses
3. Locating receptors 2nd defining dose
4. Health effects projection
5. Evaluating risk acceptability
6. Identifying control options
7. Recomputing exposure levels
8. Monitoring results

Each of these areas is investigated.

DEFINING SOURCE TERMS

The first step in any assessment of environmental risks posed by TSDs is a thorough inventory of all present and planned sources at the site. Although emissions data are highly desirable, they frequently depend on the operational and environmental conditions prevailing at the time of data acquisition. This is especially true for air emissions from a TSD. This consideration and the high cost of characterizing an existing site (e.g., the site sampling cost for Love Canal is estimated to approach $6 million) are strong incentives for first performing a theoretical analysis of the site to direct sampling activities to the main operations of concern. Nevertheless, not all sampling activities can be postponed. Thorough characterization of the wastes fed to all processes in terms of chemical composition is a necessary precondition to any exposure assessment. If this source characterization has been conducted, principal waste constituents and acutely toxic compounds of concern can be identified for further analysis.

Table 2-1 presents the sources of off-site releases that may be present at a TSD. A review of the table illustrates the complexity and difficulty of air-emissions assessment. The potential for air emissions exists at any point on a site. Therefore, it is prudent to immediately examine the vapor pressures of the constituents of concern. The presence of volatile, toxic compounds in a treatment facility is strongly suggestive of pollutant transport, rather than chemical destruction, from a liquid phase to the atmosphere. But before a detailed discussion of air emissions estimators is provided, some discussion of groundwater and surface-water contamination is necessary.

Groundwater and Surface-Water Contamination

Emissions of pollutants to the groundwater are difficult to quantify and are poorly understood. Compounds with a high octanol–water partition coefficient are easily absorbed by the soil matrix and tend to be immobile even in a leaking landfill. On the other hand, some of these compounds are almost completely solubilized in the presence of selected solvents. This can result in leachate concentrations that are considerably higher than those found in the original waste. It is this latter tendency and our incomplete understanding of the chemodynamics that mitigate against any land disposal of highly toxic compounds like the tetrachlorodibenzo-*p*-dioxins.

Table 2-1. Sources of Off-Site Releases at a TSD

| | Media Directly Contaminated | | |
Source	Groundwater	Surface Water	Air
Open treatment tanks[a]			
stirred			X
aerated			X
clarifiers		Y	X
Closed treatment tanks		Y	X
Storage tanks			X
Surface impoundments	X	Y	X
Landfills	X		X
Land treatment	X		X
Incinerators		X	

Note: Y indicates source only if the indicated process is at the end of the treatment train.

[a] Note that tank leakage can be the major groundwater contamination vector at some facilities. It is not discussed in this section because source characterization for transport analysis is fairly straightforward.

The best current approach to leachate composition estimation is to conduct laboratory leachate tests. Unfortunately, field validation of such test results is not available. It is hoped that this will change when the Environmental Protection Agency's (EPA) Municipal Environmental Research Laboratory releases the results of several ongoing studies. Even then, definition of the appropriate leachate will still be an issue. For the interim, the only "reasonable" approach is to use available leachate tests [1] coupled with field data for estimating leachate composition or to assume complete solubilization of the target compounds within the source.

With respect to surface-water discharges, source-term estimation is relatively straightforward. Discharges from treatment facilities are generally few in number and readily sampled. Furthermore, the literature contains numerous models for predicting the effectiveness of chemical, physical, and biological treatment facilities (see chap. 3).

Estimating Air Emissions: Fick's Model

Air-emissions quantification from a facility is a more complex undertaking. Nevertheless, models exist for predicting emissions in the absence of data. These models are based on mass transfer principles and use Fick's first law of diffusion. Because Fick's model is the basis for the subsequent equations, we give some details about it.

In practice, the boundary layer mass transfer coefficients are difficult or impossible to determine. Assumptions are generally made about which overall mass transfer coefficients are controlling.

Fick's law states that the rate of pollutant transfer across an interface is equal to the product of the concentration difference across the boundary and an overall mass transfer coefficient (k). The coefficient represents the resistance to mass transfer in each phase. Figure 2-1 depicts the steady-state situation for an air–water interface. Here

$$F = K_l(C_{1i} - C_1) = k_g (C_g - C_{gi}) \tag{2-1}$$

where

$$
\begin{aligned}
F &= \text{Flux of the pollutant through the boundary} \\
&\quad \text{layers in the } z \text{ direction} \\
k_i &= \text{the individual-layer mass transfer coefficients} \\
k_l &= \text{liquid} \\
k_g &= \text{gas}
\end{aligned}
$$

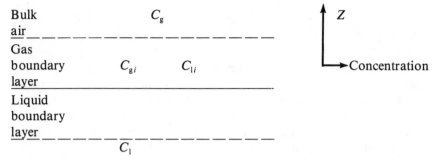

Figure 2-1. Schematic of Fick's Law for steady-state of an air-water interface.

If we assume further that Henry's law is obeyed at the air–water interface, then

$$C_{gi} = HC_{li} \qquad (2\text{-}2)$$

where H is Henry's constant, and Eq. (2-1) can be reduced to

$$F = \frac{C_g/H - C_l}{1/k_l + 1/Hk_g} = \frac{C_g - HC_l}{1/k_g + H/k_l} \qquad (2\text{-}3)$$

or

$$F = K_l(C_g/H - C_l) = K_g(C_g - HC_l) \qquad (2\text{-}4)$$

where K_i are the overall mass transfer coefficients. Note that

$$\frac{1}{K} = \sum_i \frac{1}{k_i} \qquad (2\text{-}5)$$

Therefore, if one phase presents a particularly strong resistance to gas movement, its mass transfer coefficient (k_j) will be much smaller than the other k's. In that case ki would be approximately equal to K and would control the overall rate of pollutant diffusion. This simplification, in addition to being easily computed, can be validated by straightforward experimentation, which has shown that the Fickian model has broad applicability.

One must be careful when applying Fickian-based analyses to complex mixtures containing several volatile compounds. For those cases the diminished activity of the individual compounds results in large over-estimation of the emission rate for each compound. Although this might be suitable if the objective is below the level of air emissions, we must be careful to not underestimate water and sludge discharges.

Henry's constant may be approximated by Pgi/S, where Pgi is the pure component vapor pressure at the temperature of interest and S is the solubility many compounds use; published Henry's constants, which should be used as a first resort. Lyman, Reehl, and Roscubla [2] discuss extensively the computation of Pgi and S from fundamental properties, such as heat of vaporization boiling point and actual water partition coefficients. Neely [3] also provides an approach for estimating Henry's constant.

Although the Fickian model is the basis for many of the air-emissions estimations discussed herein, each TSD activity still requires separate model development to account for varying boundary conditions. Fortunately, many unit operations have general similarities that can be grouped for air-release modeling purposes. Table 2-2 is a functional regrouping of potential activities to account for these similarities. Following the table are equations for estimating emission rates of individual pollutants from the grouped TSD activities. Note that many of these equations were developed either for relatively uncomplicated, low-concentration wastes in wastewater treatment plants or for pure products. Hence, they overestimate actual, atmospheric release rates when applied to complex, concentrated wastes where there is diminished chemical activity in the liquid phase.

Table 2-2. Grouping of TSD Activities

TSD Activity	*Applicable Air Emissions Modeling Equation*
Sedimentation/clarification	Nonaerated surface impoundment
Equalization	Nonaerated surface impoundment
Neutralization	Nonaerated surface Impoundment
Flocculation/coagulation	Nonaerated surface impoundment
Activated sludge and other processes with high-rate mixing	Aerated tank models
Aerated lagoons	Aerated surface impoundment
Closed treatment tanks	Can usually ignore
Storage tanks	Storage Tank models
Anaerobic digestor	Storage tank models
Surface impoundment	Applicable surface-impoundment model
Landfill	Landfill
Land treatment	Land treatment

Surface Impoundments: Thibodeaux, Parker, and Heck Model. Only the model developed by Thibodeaux, Parker, and Heck [4] is recommended for use in aerated and nonaerated surface impoundments. The model assumes two-film resistance and handles different surface characteristics by providing separate mass transfer coefficients for turbulent and quiescent regions. The equations follow:

$$Q_I = \mathrm{MW}_i(A)K_{1i}(X_i - X^*_i) \qquad (2\text{-}6)$$

The mass transfer coefficients (K_{1i}) in Eq. (2-6) are evaluated separately for the turbulent and quiescent regions. For each pollutant i,

$$\frac{1}{K_1^t} = \frac{1}{k_1^t} + \frac{1}{Hk_g^t} \qquad (2\text{-}7)$$

and

$$\frac{1}{k_1^n} = \frac{1}{k_1^n} + \frac{1}{Hk_g^n} \qquad (2\text{-}8)$$

The terms are defined below along with the source of the information:

Q_i = computed rate of air emission of compound i in g/s

MW_i = molecular weight of component i g/g-mol (from periodic table)

A = open-surface area of the operation in cm² (measured)

K_{1i} = overall liquid-phase mass transfer coefficient for component i in mol/cm²-s [computed by Eq. (2-7) and 2-8)]

X_i = mole fraction of component i in the aqueous phase (measured)

X^*_i = liquid-phase, equilibrium mole fraction based on bulk air-pollutant concentration (usually assumed to be zero)

K_1^t, K_1^n = overall liquid-phase mass transfer coefficients for turbulent (aerated) and natural (non-aerated) zones of the impoundment, respectively, in mol/cm²-s [computed from Eq. (2-7) and (2-8)]

A_t, A_n = surface areas of the turbulent and natural zones in cm² (measured)

k_1^t, k_1^n = individual liquid-phase mass transfer coefficients for the turbulent and natural zones in mol/cm²-s [computed by Eq. (2-9), (2-11), or (2-12)]

k_g^t, k_g^n = individual gas-phase mass transfer coefficients for the turbulent and natural zones in mol/cm^2-s [computed by Eq. (2-10), (2-12), (2-13)]

H' = Henry's law constant in mole fraction form = (P_gHC_1) (chemical property literature)

To apply the model to surface impoundments, solve Eq. (2-7) and (2-8) and then Eq. (2-6) for the quiescent and turbulent regions, separately. The major problem with this model is that four mass transfer coefficients must be determined. Fortunately, for the nonaerated case, the "t" terms drop out, and the computation is considerably simplified. Laboratory determination of the coefficients can be avoided if certain approximations are invoked. One such approach is credited to Hwang [5]. Hwang's approximations for the nonturbulent mass transfer coefficients are

$$k_{1i}^n = (MW_{O_2}/MW_i^{1/2} \, \theta k_{1,O_2}/298 \qquad (2-9)$$

and

$$k_{gi}^n = (MW_{H_2O}/MW_l)^{0.335}(\theta 298)^{1.005} k_{g,H_2O} \qquad (2-10)$$

where

θ = temperature of concern in kelvins
k_{1,O_2} = liquid-phase mass transfer coefficient for oxygen at 25°C in g-mol/cm^2-s
k_{g,H_2O} = gas-phase mass transfer coefficient for water vapor at 25°C in g-mol/cm^2-s

and other terms are as before. However, k_{1,O_2}^t and k_{g,H_2O}^t still need to be computed. For the aerated surface impoundment the mass transfer coefficients can be computed as follows:

$$k_{1i}^t = \frac{J\alpha(POWR)(1.024)^{\theta-20}}{1.65 \times 10^{-4} \, (\theta_V)V} \sqrt{\frac{D_{i,H_2O}}{D_{O_2,H_2O}}} \qquad (2-11)$$

$$k_{gi}^t = 1.35 \times 10^{-2} \frac{\varrho_a D_{ia}}{d} \, Re^{1.42} P^{0.4} Sc^{0.5} Fr^{-0.21} \qquad (2-12)$$

An alternative formulation for the mass transfer coefficients in the nonturbulent region is

$$k_{gi}^n = 1.3 \times 10^{-5}U_A^{0.78}Sc^{-0.67}d_e^{-0.11}\frac{\varrho_a}{MW^a} \tag{2-13}$$

$$k_{li}^n = 4.24 \times 10^{-4} (1.024)^{\theta - 20}U_0^{0.67}h_0^{-0.85}\frac{D_{i,H_2O}}{D_{O_2,H_2O}} \tag{2-14}$$

where

J	=	oxygen-transfer rating of the mechanical aerator (typically 3 lb O_2/hp-h)
α	=	correction factor for wastewater oxygen transfer (0.80–0.85)
POWR	=	working horsepower input to aerators
θ	=	water temperature in °C
D_{i,H_2O}	=	diffusion coefficient for i in water in cm²/s
D_{O_2,H_2O}	=	diffusion coefficient for oxygen in water in cm²/s
A_V	=	surface area/unit volume in impoundment in ft⁻¹
V	=	volume of impoundment in region of aerators in ft³
ϱ_a	=	density of air in g/cm³
D_{ia}	=	diffusion coefficient of i in air in cm²/s
d	=	diameter of aerator impeller in cm
Re	=	Reynolds number = $d^2 w\, \varrho_a/\mu_a$
μ_a	=	viscosity of air in g/cm-s
ω	=	impeller rotational speed in rad/s
P	=	power to impeller in ft-lb/s
Sc	=	Schmidt number = $\mu_a/D_{ia}\varrho_a$
Fr	=	Froude number = $d\omega^2/32.17$
U_A	=	wind speed (m/h, typically taken at a height of 10 m)
d_e	=	effective diameter of impoundment quiescent area in m
U_0	=	surface wind velocity = $0.035U_A$
h_0	=	effective mixing depth of impoundment in ft

Although it is tedious, Eqs. (2-11)–(2-14) can be evaluated for many compounds. Furthermore, the previously mentioned work by Lyman, Reehl, and Roscubla can be used for extending the equations to other compounds. As stated previously, Eqs. (2-11)–(2-14) provide all case-specific estimates of the gas- and liquid-phase mass transfer coefficients for the compound under study. Armed with those values, we can estimate impoundment air emissions by Eqs. (2-6)–(2-8). Although this appears tedious for multiple-compound evaluations, only the D_i change from compound to compound. Work by Thibodeaux, Parker and Heck [4] suggests that these equations are accurate within a factor of two to five for wastes with a limited number of volatile species and no additional surface films.

Aerated Treatment Tank: Freeman's Model.
Activated sludge units, dissolved air flotation units, and other treatment devices where there is a high degree of subsurface mixing and/or biological activity cannot, in most cases, be evaluated by using the aerated surface-impoundment equations of the previous section. In these cases, the introduction of air bubbles near the bottom of the tank creates a large number of tiny air–liquid interfaces for gas transfer. The net result is a greatly increased emission rate for volatile compounds. Biological activity, on the other hand, sometimes reduces air emissions levels when pollutants are absorbed into the waste sludge. Unfortunately, only very complex models exist for evaluating this absorption pathway and these models have little supportive evidence. Since the net effect of ignoring microbial absorption is a conservative estimate of health risks due to emissions, absorption will not be discussed here. See Hwang [5] and Freeman [6] for a fuller treatment. What will be presented is Freeman's model for estimating the air stripping of pollutants in an aerated wastewater basin in the absence of biological activity. This model has been validated for selected compounds on a laboratory scale [7,8].

Freeman's model enables one to compute the pollutant gas concentration as the bubble breaks the surface. Then, simply multiply by the overall gas flow rate to derive the anticipated emissions. The equation is

$$y_2 = HC_i \left[1\text{-}exp \left(\frac{-6K_x t_b MW_a}{d_b \varrho_a H'} \right) \right] \qquad (2\text{-}15)$$

where

y_2	=	concentration of compound i in the gas bubble as it breaks the surface
H	=	Henry's law constant
C_i	=	concentration of compound i in liquid in ppm
K_x	=	overall mass transfer coefficient in g-mol/h-cm^2
t_b	=	rise time of bubbles in h
MW_a	=	molecular weight of air
D_b	=	bubble diameter cm

$$= 4.15 \ \frac{\sigma^{0.6}}{(P/V)^{0.4}\varrho_1^{0.2}} \ \chi^{1/2} + 0.09$$

ϱ^a	=	density of air in g/cm^3
σ	=	liquid surface tension in g/s^2
ϱ_1	=	liquid density in g/cm^3
P/V	=	power input to liquid per unit volume, where $P/V =$ $v_g\varrho_1^g$, in g-cm^2/s^3-cm^3
χ	=	gas holdup fraction in basin = $v_g/(0.9 + 2v_g)$
v_g	=	superficial gas velocity (cm/s) = Q_g/(area of tank)
Q_g	=	volumetric flow rate of aeration gas in cm^3/s
g	=	acceleration due to gravity (980.6 cm/s^2)
μ_1	=	liquid viscosity in g/cm-s

When applying the diffused-air mass transfer model of Eq. (2-15), assumed that the overall mass transfer was controlled by the liquid film [i.e, $K_x = k_l$, the liquid-phase mass transfer coefficient of Eq. (2-8)]. This enabled him to use Calderbank's relationship for the liquid-film mass transfer coefficient in the following form:

$$k_1(Sc)^{1/2} = 0.42 \left[\frac{(\varrho_1 - \varrho_a)\mu_1 \cdot g}{\varrho_1^2} \right]^{1/3} \times 3600 \ \frac{\varrho_1}{MW_{H_2O}} \qquad (2\text{-}16)$$

where all terms are previously defined.

Storage Tanks.   There are four main types of waste storage tanks: open-roof tanks estimating emissions from open-tank storage of hazardous waste—hopefully, an infrequent practice—should be performed by using Eqs. (2-13) and (2-14) for quiescent regions in a surface impoundment), open tanks with a floating roof, closed tanks with a fixed roof and closed tanks with an internal floating roof. A fifth type of tank—one pressurized with an inert gas blanket—produces so few emissions and has such limited application for hazardous wastes that it is not included.

Covered storage tanks produce air emissions due to breathing and working losses. Breathing losses occur when the ambient temperature rises, causing an increase in the vapor pressure of volatile constituents and an expansion of the gases present. Working losses occur when a tank is refilled, resulting in displacement of the saturated gases present. The bulk of the work conducted on these sources has been conducted to evaluate total, volatile, organic, compound emissions from stored petrochemical wastes. Specific constituent emission information is almost nonexistent except for work on benzene storage emissions conducted by EPA's Office of Air Quality Planning and Standards (OAQPS). What follows is an effort to extend the validated approaches developed by OAQPS and the American Petroleum Institute to the analysis of specific organic compounds.

The following equations on storage tank-emissions will provide emission estimates based on unpolluted vapor space above each tank. Obviously, the situation is one where several compounds are simultaneously equilibrating in a single space, and compound reaches Henry's law equilibrium vapor pressure. However, since most storage operations allow for a relatively long period of contact between a given volume of gas and waste, it is reasonable to assume near steady state concentrations of constituents in the gas and liquid phases. The concentration of a particular constituent in the gas above the liquid could then be approximated by Henry's law ($P_i = H'C_i$). Calculating Henry's law pressures for all constituents with significant vapor pressures and summing the partial pressures give a first approximation for the gas composition. This above approach works fine unless there are several volatile constituents in the waste, in which case the combined pressures could exceed the total pressure (P_t). One approach to adjusting the pressures is to multiply P_i by the factor $P_t / \Sigma P_i$ to obtain a rough gas composition. In each of the following storage-emissions equations, this adjustment should be made when emissions estimates are being produced for several compounds from a single tank.

Fixed Roof Tanks.   Emissions from fixed-roof tanks may be computed from the following equations [9,10]:

$$L_B = 2.26 \times 10^{-2}(MW_i)\left(\frac{P_i}{14.7 + P_i}\right)(0.68 D^{1.13} H^{0.51} T^{0.50} F_p^{CK_c})$$

$$(2\text{-}17)$$

and

$$L_W = 2.40 \times 10^{-2}(MW_i)P_i K_n K_c \qquad (2\text{-}18)$$

where

L_B, L_W = fixed-roof breathing and working losses, respectively in lb/yr
MW_i = molecular weight of compound i in lb/lb-mole
P_i = adjusted compound vapor pressure in psia (Note: $P_i = H_i C_i$ for $\Sigma P_i \leq P_t$ and $H_i C_i P_t / \Sigma P_i$ for $\Sigma P_i > P_t$)
d = tank diameter in ft
H = average vapor-space height in tank, including roof volume in ft

T = average ambient day to night temperature change in °F
F_p = paint factor
K_c = crude-oil factor developed in original equation applications
[use K_c = 0.65 in Eq. (2-17) and K_c = 0.84 in Eq. (2-18)]
K_n = turnover factor
C = adjustment factor for small-diameter tanks

These equations are modifications of two developed specifically for estimating total volatile organic carbon (VOC) emissions. The quantities F_p, C, and K_n are defined further in Table 2-3, Figure 2-2, and Figure 2-3.

When Eqs. (2-17) and (2-18) were used to evaluate an organic in the petrochemical industry, field data indicated an accuracy of approximately 10%. Although equivalent accuracy is unlikely in the modified equations, 50% accuracy for each compound is probably achievable.

External Floating Roof.   Open tanks with floating roofs are designed to eliminate the standing storage loss associated with an ordinary open tank. Standing storage loss emissions from external floating roof tanks are controlled by either a primary seal or a primary seal in conjunction with a secondary seal. The three main types of primary seals used on floating roofs are mechanical, resilient, and flexible wipers. Resilient seals are mounted to eliminate the vapor space between the seal and liquid surface (liquid mounted), to allow a vapor space between the seal and the liquid surface (vapor mounted), or to allow a vapor space between the edge of the floating roof and the tank wall.

Table 2-3. Paint Factors for Use in Equation (2-17)

Tank Color		Paint Condition	
Roof	*Shell*	*Good*	*Poor*
White	White	1.00	1.15
Aluminum (specular)	White	1.04	1.18
White	Aluminum (specular)	1.16	1.24
Aluminum (specular)	Aluminum (specular)	1.20	1.29
White	Aluminum (diffuse)	1.30	1.38
Aluminum (diffuse)	Aluminum (diffuse)	1.39	1.46
White	Gray	1.30	1.38
Light gray	Light gray	1.33	1.44[a]
Medium gray	Medium gray	1.40	

[a] Estimated from the ratios of the seven preceding paint factors.

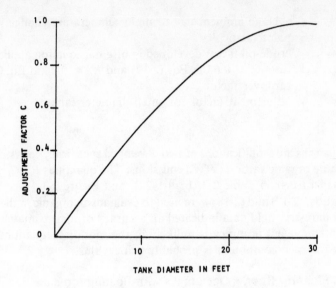

Figure 2-2.   Adjustment factor for small diameter tanks.

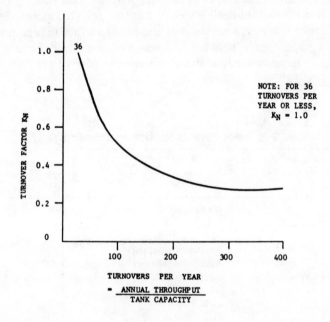

Figure 2-3.   Adjustment factor for frequency of tank turnover.

The equation for standing losses from external floating-roof tanks is taken from Supplement 12 to EPA's AP-42 [9].

$$L_s = K_S V^N P^* d M_V K_c E_f \qquad (2\text{-}19)$$

where

L_s = standing storage loss in lb/yr
K_S = seal factor [use 1.00 for tight-fitting seals and 1.33 for older loose-fitting seals in lb-mole/ft$(mi/hr)^N$yr]
v = average wind speed at tank site in mi/hr
N = seal-related wind-speed exponent (see AP-42, Table 4.3-3)
P^* = vapor pressure function (dimensions) [from AP-42, Tables 4.3-1, 4.3-5, and 4.3-6]
D = tank diameter in ft
M_V = average vapor molecular weight in lb/lb-mole
K_c = product factor (dimensionless) [use 1.0 for volatile organic liquids and 0.4 for crude-oil-like wastes]
E_f = secondary seal factor (use 1.0 for primary seal only and 0.75 for dual seal system)

The withdrawal loss in a storage tank must also be evaluated for an external floating roof tank. These losses result from liquids clinging to the side of the tank when the roof is lowered. The equation for estimating these emissions is the same as the equation recommended for internal floating roofs. It is also taken from AP-42 [9].

$$L_w = \frac{(0.0225) \, QCP_D}{d} \qquad (2\text{-}20)$$

where

L_w = withdrawal loss in lb/yr
Q = average tank throughput in gal/yr
C = shell clingage factor in bbl/1000 ft^2 (from AP-42, Table 4.3-5)
P_D = average organic liquid density in lb/gal
d = tank diameter in ft

Closed Tanks with Floating Roof.   The internal floating roof is a step above the external floating roof for emission control. Although it is subject to the same types of emissions as the external floating roof, the wind does not play as big of a role since it does not directly impinge upon the floating roof. In fact, unless the number of these tanks is large, their emissions can usually be ignored when conducting an overall facility risk assessment. In any event, the appropriate loss equation for storage losses is modified in the following manner.

$$L_s = K_S P^* dM_V K_c E_f \tag{2-21}$$

where Ks = 0.7 for all types of seals, and other terms are as previously given. Note that all of the floating roof equations are under review by EPA to expand their applicability for volatile organic liquid storage.

Landfills: Farmer's Model.   Pollutant emissions to the atmosphere from a landfill are difficult to quantify. This problem arises from the variable and typically unknown conditions with respect to moisture found in a hazardous waste landfill. Although Thibodeaux has developed an emissions model for the case of codisposal with gas generating municipal wastes, Farmer's model [11], as modified by Shen [12] and Hwang [5] is being recommended. Farmer's model assumes that emissions are controlled by gas diffusion through the air-filled pores. The basic equation is

$$Q = \frac{D_0 C_s A}{L} \left(\frac{W_i}{W}\right) \left(\frac{\epsilon_a^{10/3}}{\epsilon_t^2}\right) \tag{2-22}$$

where

Q	=	pollutant emission rate in g/s
D_0	=	diffusion coefficient of the chemical in the air in CM^2/s
C_s	=	saturation vapor concentration in g/cm^2
A	=	surface area of the landfill in cm^2
W/W	=	weight fraction of pollutant i in the waste
ϵ_t	=	soil porosity = $1 -$ (soil bulk density) / (soil particle density)
ϵ_a	=	air-filled soil porosity = $\epsilon_t - F_w$
F_w	=	volume fraction of water in the soil in cm^2/cm^2
L	=	depth of soil cover in cm

All values are easily measured or computable except C_s. Depending on the point in transport being modeled, C_s may be computed by the following equations:

1. If present in a chemical mixture,
   $$C_s = \gamma \, XP^0 \, (MW)/760RT$$

2. If present in groundwater,
   $$C_s = C_w H$$

3. If absorbed on soil,
   $$C_s = C_d H / K_p \varrho RT$$

where

$$
\begin{array}{lll}
P^0 & = & \text{vapor pressure of pure chemical in mm Hg} \\
MW & = & \text{molecular weight in g/mole} \\
R & = & \text{gas constant (82.06 atm } = \text{cm}^2/\text{mole } = \text{deg)} \\
T & = & \text{temperature in kelvins} \\
\gamma & = & \text{activity coefficient of chemical (unitless)} \\
X & = & \text{mole fraction of chemical in mixture (unitless)} \\
C_w & = & \text{concentration of chemical in groundwater in ppb} \\
C_d & = & \text{concentration of chemical in soil in g/cm}^2 \\
K_p & = & \text{soil adsorption coefficient in cm}^2/\text{H}_2\text{O/g adsorbent} \\
\mu & = & \text{soil density in g/cm}^3 \\
H & = & \text{Henry's law constant (nondimensional)}
\end{array}
$$

For a new landfill site this information is readily available. The problem lies in assessing existing landfills without using a very expensive sampling program.

Land Treatment: Thibodeaux and Hwang's Model. Although based on an idealization of a pesticide application model, the recommended model has not been validated for hazardous wastes. This model, developed by Thibodeaux and Hwang [13], is based on the assumption that vapor diffusion through the air-filled pores of the treated soil is the rate-limiting step.

$$q_i = \cfrac{D_{ei} C_{ig}}{h_s^2 + \left[\cfrac{2D_{ei} t A \, (h_p - h_s) C_{ig}}{M_{i0}} \right]^{1/2}} \tag{2-23}$$

and

$$C_{ig} = \left[\cfrac{H_c}{1 + \cfrac{6D_{ei}Z_oH_c}{D_{wi}A_s\,(h^2 + h_ph^2 - 2h^2)}} \right] C_{i0} \quad (2\text{-}24)$$

where

q_i = flux of component i from the soil surface in g/cm²-s
D_{ei} = effective diffusivity of component i in the air-filled soil pore spaces in cm²/s = $D_{ai}\epsilon/\tau$
ϵ = soil porosity
τ = tortuosity (assumed = 3)
C_{ig} = effective wet zone–pore space concentration of component i in g/cm³
h_s = depth of subsurface injection in cm
h_p = depth of soil contaminated or wetted with land-treated waste in cm
t = time after application in s
A = surface area over which waste is applied in cm²
M_{i0} = initial mass of component i incorporated into the zone ($h_p - h_s$) in g
H_c = Henry's law constant in concentration form in cm³ oil/cm³ air
Z_o = oil-layer diffusion length in cm
D_{wi} = effective diffusivity of species i in the oil in cm²/s
A_s = interfacial area per unit volume of soil for oily waste in cm²/cm³
D_{ai} = molecular diffusivity of component i in air
C_{i0} = concentration of i in waste at time of application

Remember that these equations were derived for land-treatment applications involving oily sludges. Thus, aqueous wastes might perform quite differently. The sole use of this equation should be as an estimating tool.

TRANSPORT AND TRANSFORMATION ANALYSES

Having presented approaches in lieu of detailed sampling for estimating emissions from hazardous waste treatment, storage and disposal operations, we now proceed to the next step—evaluating the movement of pollutants away from the point of generation. This section discusses the basic equations for computing this dispersion, the types of transformations occurring along the way, and the applicable computer models currently available for assessing transport. But first a brief discussion of accuracy and model use is necessary.

Accurate modeling of transport and environmental transformation can be complex and expensive. Thus we should think about the eventual use of the results. If validating transport theories and physiochemical processes is the aim, then extensive input data and complex computer analyses will be required. Furthermore, the results obtained frequently will be of limited use for evaluating other situations. This problem arises in part from the complexity of the data manipulated and the resultant loss of meaning of individual elements of the analysis.

To avoid the problems associated with complex models, a number of less precise screening models have been developed over the years. These models make simplifying assumptions about some processes and ignore nonquantifiable degradation mechanisms. The resultant environmental concentrations may be an order of magnitude high for a particular receptor location, but the models have noteworthy advantages. First, observed concentrations will be conservatively high. The benefit of this overestimation is that cases that clearly pose no environmental risk can be removed from further consideration. In addition, other TSD operations can be prioritized for more detailed analyses. The second advantage of screening models is that they are inexpensive and easy to apply. This facilitates immediate direction of resources to the areas of concern. The third advantage is that these models facilitate the manipulation of output to identify optimal corrective approaches. For all of these reasons, engineers will frequently find it advantageous to initiate environmental analyses with screening models. They should turn to analytical models only when they are in the position of trying to resolve just how much additional environmental cleanup might be necessary.

The Basic Transport Equations

The basic transport equations are typically perceived as not very useful to the engineer concerned with obtaining results. This is a big mistake because failure to understand the forces at work leaves the engineer without an appreciation of the sensitivity and precision of the modeling tools. However, space limitations prevent us from fully exposing the elements of the underlying mathematical analyses. Nevertheless, one will always be merely pulling numbers out of a hat unless the basic principles underlying the movement within a given media are understood. Modeling the behavior of a pollutant in the environment is really a two stage process. It is the modeling of the movement of a given element of fluid coupled with a compartmental analysis to examine the compositional change of the pollutant in that compartment over time. Basic transport equations for groundwater movement are thoroughly described in Mercer and Faust [14] and Jevandel, Doughty, and Tsang [15]. Thomann [16] provides an excellent discussion of surface-water transport equations. Stern [17] and Pasquill [18] describe the transport equations for airborne material. The reader is strongly encouraged to seek out and become familiar with these sources prior to attempting any serious pollutant-dispersion analysis in a given media.

Transformation Analysis

Pollutant behavior in the environment is governed by two mechanisms: transport and transformation. Pollutant transformation analysis, like transport analysis, requires considerable background in fluid mechanics, chemistry, physics, and mathematics. It is not possible to define within the context of this chapter sufficient background so that an engineer would be able to thoroughly evaluate the transformation characteristics of a given chemical. Consequently, the emphasis in this section is on identifying the mechanisms and providing some references that should be particularly useful to those needing to perform an analysis.

Table 2-4 identifies the major transformation processes at work in the various media; however, it is a generalization, and certain chemical groups might perform quite differently. The table may be a useful starting point for identifying sinks for given pollutants. It is important to remember that transformation is not always a conservative influence on the risk posed by environmental emissions.

Hydrolysis.   Hydrolysis is a nucleophilic displacement-type reaction between a compound and water. The nucleophiles are provided by the water hydrolysis products, H_2O^+ and OH^-. In a typical reaction

$$RX + H_2O \rightarrow ROH + HX$$

Hydrolysis is highly dependent on pH and can be significantly influenced by the presence of other ions. Table 2-5 provides a breakout of the hydrolysis susceptibility of various chemical groups. A compound half-life of one day in water has been used to separate the groups, but there are exceptions within many of the groups. Readers requiring specific values are referred to the CRC Handbook [19]. Lyman, Reehl, and Roscubla [2] also discuss techniques for deriving these values when they are not available.

Volatization.   Although volatization from surfacewater and groundwater can be a major pollutant attenuation mechanism for many compounds in

Table 2-4. Major Transformation Processes

Process	Significance in			Factors Influencing Rates
	Air	Water	Soil	
Hydrolysis	S	S	S	pH, temperature, compound solubility, solution ionic strength, presence of catalysts
Volatization	M	S	S	Pollutant vapor pressure, concentration vis-a-vis solubility, background concentration in air, temperature, soil properties
Absorption/ adsorption	M	S	S	Octanol–water partition coefficient, available organic carbon and/or biomass, compound polarity and solubility, carrier solvents present
Photolysis/ photooxidation	S		M	Solar intensity, transmissivity of water, availability of free radicals
Chemical oxidation	S	S	S	Concentration of oxidants, oxygen concentration, temperature
Biodegradation	M	S	M	Microbial population, pH, temperature, extent of acclimation
Sedimentation	S	S	M	Size, density, carrier fluid velocities, and turbulence

S = significant mechanism; M = minimal or not applicable.

Table 2-5. Hydrolysis Susceptibility of Some Chemical Groups

Resistant	Reactive
Alcohols	Acid esters
Aldehydes	Amides
Alkanes, alkenes, and alkynes	Amines
Benzene, toluene, and most halo- and nitrosubstituted benzenes	Carbamates
Glycols	Epoxides
Heterocyclic compounds	
Ketones	
Nitrosamines	
Organ halides	
Organs phosphorous compounds	
Phenol, cresols, and halo- or nitrosubstituted phenols	
Polynuclear aromatic hydrocarbons	

transit, it simultaneously is a major contributor to the air emissions from a TSD. Volatization has been thoroughly discussed under the subject of source characterization. The only additional point to recognize is that volatization from groundwater occurs at a much slower rate than that from surface water. The principal factors are laid out in Table 2-6. However, the overall mass of pollutants transferred from groundwaters in a locality could easily exceed the transfer from surface waters due to the slower pollutant dispersion from a given point.

Absorption/Adsorption.   Although absorption and adsorption are two distinct processes, the environmental modeling of sorption processes can easily be handled as one topic. It is wise to do this because most studies to date have concentrated on the combined effects of those mechanisms.

A variety of sorption mechanisms occur in the environment. Ion and liquid exchanges are major attention mechanisms for metals moving through clays and humic materials, respectively. Van der Waals forces and charge transfer can lead to adsorption that retards the movement of a chemical through the soil. Chemisorption can also occur as the pollutants are chemically bonded to the soil organic material. In general, sorption is more important for groundwater transport, but sorption of hydrophobic compounds like polychlorinated biphenyls (PCBs) onto sediments and water-surface organic microlayers can be very important mechanisms for producing equilibrium concentrations in excess of solubility.

Hamaker [20] has analyzed sorption in a variety of soils and has found that variation was primarily due to the organic content of the soil. His empirical relationship for the pollutant soil absorption coefficient is

$$K_{0c} = \frac{(\mu g \text{ absorbed}) \, / \, (g \text{ of soil organic carbon})}{(\mu g \text{ dissolved}) \, / \, (g \text{ of solution})}$$

Table 2-6. Factors in Reduced Volatization from Groundwater

Factor	Effect
Decreased liquid mixing	For high-volatility compounds converts bulk liquid diffusivity to the rate-limiting step
Lack of transporting wind	Results in $C_g > 0$, which decreases gas-phase flux
Decreased air interface area	Air-filled porosity is the evaporation pathway, which is very small compared to surface waters
Soil organic carbon content	Sorption of pollutant gases as they move through the unsaturated zone

Numerous regression equations exist for estimating K_{OC}. They have been documented by Lyman, Reehl, and Roscold.

The soil sorptive ability is defined by

$$q_s = f_C K_{0c} C^n \qquad (2\text{-}25)$$

where

q_s = pollutant sorbed per unit weight of soil in $\mu g/g$
f_C = g soil organic carbon per g of soil
K_{0c} = soil sorption coefficient in mL/g
C = pollutant concentration
n = constant

Photolysis/Photooxidation.   Photolytic degradation occurs when a compound absorbs light energy of a wavelength corresponding to an increased energy level in the molecule. Absorption of the light energy can break apart a molecular bond if competing nondestructive processes do not deenergize the molecule. The degree to which conversion results from light absorption is known as the quantum yield of the reaction. Photooxidation, on the other hand, refers to chemical oxidation, resulting from molecule interaction with a free radical created by solar irradiation. The basic equations of photochemistry boil down to the relation that at least one quantum of radiation must be absorbed to transform one molecule. It is important to note that the maximum energy level available from sunlight is approximately 95 kcal/mole. Consequently, compounds like benzene (C_6H_5—H bond energy = 112 kcal/mole) will be very stable, whereas compounds such as trichloroethylene (C—Cl bond energy = 80 kcal/mole) will be more susceptible to photolytic reaction. Trichloroethylene is in fact photochemically degraded to phosgene and acetyl chlorides. Zepp and Cline [2] have written a computer program to estimate photolysis half-lives based on the quantum yield and ultraviolet absorption spectrum of a compound. The literature is replete with information on photochemical reactions and should be examined in given compound modeling because by-products that are quite toxic and stable can be generated through this mechanism. One example is HCl formations in degradation of phosgene.

Chemical Oxidation.   The primary oxidative reaction occurring in most waters is reaction of compounds with oxygen. The rate of chemical oxidation and susceptible compounds are well understood. The following list, adapted from Weber [22] is provided for completeness; see [22] for further discussion of this topic. Chemical reaction in the atmosphere was previously discussed.

High reactivity: phenols, aldehydes, aromatic amines, certain organic sulfur compounds.

Moderate reactivity: alcohols, alkyl-substituted aromatics, nitro-substituted aromatics, unsaturated alkyl groups, aliphatic ketones, acids, esters, and amines

Low reactivity: halogenated hydrocarbons, saturated aliphatic compounds, benzene and chlorinated insecticides

Biodegradation.   Biodegradation is a major removal mechanism for many toxic compounds. However, as was the case in volatization, microbial degradation in soil differs substantially from that in surface waters. Soil microorganisms are more diverse and generally limited to the top few feet of the earth. Hamaker [23] and Rittman and Kobayashi [24] have provided an extensive compendium of the microbial degradation potential for a variety of compounds.

Sedimentation.   Sedimentation coupled with solid sorption can attenuate the dispersion of pollutants from a source and result in serious localized accumulation and long term exposure problems. Sedimentation in the atmosphere—whether it be due to the size of the pollutants (metals), high sorption onto particulates, or rainout—is a factor that must be addressed in dispersion analysis. Similarly, for surface water modeling, concentration in the sediment phase can result in large deposits that pose a risk for years to come. The reader is referred to the American Society of Civil Engineers (ASCE) [25] and Stern [17] for detailed discussions of settling phenomena. It is important to keep in mind, however, that the effect of sorption on increasing pollutant concentration in the solid phase is an integral part of the solid phase transport analysis.

State of the Art Modeling

Although there may be more powerful models than the ones identified herein, these models are being recommended because of their superior ratings in the following four basic criteria:

1. Broad range of applicability to the heterogeneous sources present at a hazardous waste TSD facility
2. Comprehensive users' manuals exist to instruct others in their application.
3. Availability to interested users.
4. Utility for both screening and theoretical applications.

Surface Water Models. The most advanced of these models in terms of its completeness in describing chemical fate and transport processes is the *Exposure Analysis Modeling System* (EXAMS) model [26]. Its major shortcomings are the oversimplified hydraulics employed and that it only deals with one chemical at a time. Although the hydraulics problem is not critical for analytical rankings, one should consider using a more rigorous hydraulic model if field validation is planned or needed. With respect to the second disadvantage, putting aside the considerable cost that can be introduced by the single-chemical approach, there is a potentially more serious problem. Assessing one chemical at a time can result in a failure to account for chemical interactions that can modify observed concentrations. Although EXAMS allows one to incorporate these effects, no guidance is provided for the user.

EXAMS is designed to analyze the behavior of a compound in aquatic environments (lakes, rivers, estuaries, etc.). When supplied with the physical-chemical properties of the chemical and the ecosystem under study, the pollutant loadings, and the system configuration, EXAMS will compute (steady-state) expected environmental concentrations as pollutants disperse from the site for constant, long-term loadings. Information can also be obtained on persistence after source termination and on the consumption of pollutants by each of the transformation processes discussed in the previous section. This latter factor, coupled with EXAMS interactive mode, facilitates sensitivity analyses and the direction of resources to the processes of concern.

EXAMS is a compartment model; that is, it divides the aquatic environment around the site into a connected series of control volumes and then examines the fluxes in and out of each element of the overall system. The system elements are tied together by conservation of mass considerations, and decay processes are mostly treated as first order reactions. The modeler can obtain many of the rate constants from the chemical compounds published literature. A general listing of the processes considered by EXAMS is displayed in Table 2-7. Equilibrium processes are separated because they tend to go to completion well within the time scale of hazard assessment.

Air Models. Single-pollutant, air-pollution fate and transport models exist that are as impressive as EXAMS. Unfortunately, their availability and/or applicability is quite limited. For modeling atmospheric dispersion and fate within 50 km of a TSD facility, the Industrial Source Complex (ISC) models [27] are the best available for short- and long-term analyses. The ISC model is a Gaussian PLUME model incorporating the best features of EPA's preceding models of choice. A major advantage of ISC is that it is possible to run the program in anything from a screening to a rigorous mode. The main features of the ISC program are

Table 2-7. Physical–Chemical Transformations Modeled by EXAMS

Process	Capabilities
Equilibrium	
Chemical ionization	Five chemical species for all subsequent analyses
Sediment sorption	Linear isotherms for aquatic sediments and soils
Biosorption	Uptake factor for combined organisms
Kinetic	
Transport	Water, sediment, and planktonic movements tracked
Hydrology	Rainfall, stream, nonpoint, and groundwater flows of water and pollutants may be input at any point
Volatization	Uses two-film resistance model for interface flux
Reaeration	Allows reintroduction of oxygen within any compartment
Photolysis	Specific absorption spectra or a rate constant may be entered and attenuation is considered. Separate rates for biota and sediment are also possible
Hydrolysis	Rates can be entered to evaluate neutral, specific acid, and/or specific base hydrolysis at varying temperatures
Oxidation	Designed for indirect photolysis but usable for transformation resulting from chemical oxidation by other pollutants.
Microbial transformation	Second-order equations can be evaluated in both water and benthic compartments

Multiple source terms located on a coordinate grid

Elevated or nonelevated; point, area, and line sources

Gravitational settling and by deposition of particles

Output concentrations for durations from one hour to one year

Terrain adjustment for elevated terrain

A time-dependent, pollutant decay factor

Beyond these features ISC is similar to most air dispersion models, differing only in the greater flexibility possible with respect to meteorological inputs. The big disadvantage of ISC is its failure to provide the option to specifically address atmospheric sinks like photolysis, hydrolysis, particle sorption, and rainout. However, the option does exist to lump these mechanisms into a single pollutant-decay coefficient if the processes are significant.

Groundwater Models. Identifying a single state-of-the-art model for groundwater pollution assessment is not possible for many reasons, not the least of which is the considerable expense associated with the collection of even superficial information on the flow fields. In addition to field definition, a limited understanding of constituent behavior in the soil, the effects of flow field variability, a lack of model validation, poor model documentation, and large differences in transport characteristics between the saturated and unsaturated zones all contribute to make modeling a difficult process. These problems, coupled with the great expense of underestimates and the nonconservative nature of many current models, leave the engineer little choice but to seek experienced help for even basic studies. Excellent initial contact points for the latest information on modeling are EPA's Robert S. Kerr Environmental Research Laboratory in Ada, Oklahoma, and the Holcomb Research Institute of Butler University. Both groups also offer training courses related to groundwater modeling.

Even if the engineer hands over the analysis of groundwater problems to another firm, certain key points need to be well defined. Bachmet and others [28] provide an excellent discussion of the points requiring definition if one is to obtain a cost-effective study with appropriate levels of precision and accuracy.

LOCATING RECEPTORS AND DEFINING DOSE

Unfortunately, for the exposure modeler, humans tend to not confine themselves to one place for any great length of time. Consequently, whereas plant- and animal-exposure definitions are relatively straightforward, defining the exposure of a human for other than acute exposure during an 8-h workday or other short-term basis is much more difficult. Place of residence is not a perfect indicator of exposure. Some people spend at least eight hours per day closer to the source (for working purposes), whereas others spend a portion of the day further from the source than from their residence. People also change jobs and/or move from the area. In addition, there will be variations in intake even if two people spend all of their time in the same area. These problems are of concern because the critical health effects modeled for most facilities will be related to chronic exposure over a number of years.

The difficult problem of finding someone for exposure assessment of chronic effects that take years to manifest themselves is not easily resolved. Although some researchers are investigating human movement patterns, most modeling has been conducted under the assumption that individuals are continually exposed at their current residences. The merits of this approach are that it is easy to obtain residence information from census data and that, if cancer risk is considered to be linearly related to dose, risk assessments will account for the fact that someone lives in that house at all times. The importance of the latter fact for cancer incidence studies derives from the numerical equivalence in most studies of an estimated maximum incidence of one out of three cancers for each of three people and one out of six cancers for six people. The other reason for considering the residence is that it often corresponds to the location of the maximally involuntarily exposed individual.

For defining residence, the Census Bureau is typically the best resource. However, the modeler faced with a low-population-density area will probably find a direct survey the best approach in terms of speed and cost. For those faced with urban or multiside analyses, the resource has been developed by EPA. EPA's Office of Solid Waste and Office of Research and Development now have on EPA's national computer the 1980 census data based on groups of approximately 800 people. This information is in a program that evaluates proximate cell density and then proportions that 800 people over the area around a source. It is thus possible, by specifying coordinates, to assign a population to an exposure area at any point in the country.

The other exposure group to be located in a thorough exposure assessment is working populations. This can be done by locating them at the point of maximum concentration within the plant boundaries and multiplying exposure levels by 40/168 to convert from a continuous exposure basis.

The foregoing discusses the questions of what people and what level of exposure, but they do not tell us the actual dose received. Exposure level does not equal body dose. To define dose, one must examine uptake and retention. Although this is not much of an issue for gaseous air emissions, it can become very complex as one turns to drinking water and food doses. Looking first at air exposure, the International Council on Radiation Protection (ICRP) [29] has defined breathing rates of 1.2 M^3/h for an active adult over 16 h and 0.4 M^3/h during sleep. EPA's Carcinogen Assessment Group, on the other hand, uses 20 M^3/day. The real issue with respect to inhaled pollutants (as with other routes of administration) is the percent retention of pollutant in the lungs. In acute exposure assessments it is common to use 50%, but many compounds are retained at much higher levels. Consequently, 100% retention is the best approach for screening purposes unless data to the contrary are found. Stern [17] gives information on particulate retention in the lungs as a function of particle size.

With respect to water dosages, 2 l/day is considered the typical adult level of water consumption. Table 2-8 depicts some of the dose assumptions currently under review by EPA [30].

Table 2-8. Adult Dosage Assessment Factors

Respiratory rate (M³/h)[a]	Male (70 kg)	Female (60 kg)
Resting	0.4	0.27
Light work	1.2	1.0
Medium or heavy work	1.8	1.5
Size of respirable particles[b]	Reach Alveoli	Retained in Nasal
(aerodynamic diam. in μn)	(%)	Passage (%)
1	100	0
2	80	20
5	50	50
10	~0	~100

Note: Based on the following parameters: skin surface area (m₂) = 1.85 (totally exposed 180-cm tall male), = 0.294 (wearing short-sleeved shirt, pants, shoes), = (wearing long-sleeved shirt, gloves, pants, shoes); effective pore size of skin and other external membranes = 4 A° (0.4 nm); food consumed = 1500 g/day (excluding beverages).

[a]For an 8-h workday (6-h light work, 8-h heavy work) or a 24-h day (12-h rest, 10-h light work, 2-h heavy work).

[b]Mouth breathers can inhale particles up to 15 μm aerodynamic diameter.

The real problem area in terms of dosage is in estimating food intake of pollutants that have entered the food chain. Because many foods receive a high level of processing, accurate exposure estimates necessitate quite a complex study. ICRP [29] has done some work in this area, especially with respect to diet definition. Perhaps the key element with respect to food intake is the examination of plant and animal dosages near the facility. Compounds being emitted that have high octanol–water partition coefficients will tend to bioaccumulate in the food chain and biomagnify as they go up the chain. For example, Neely has shown [4] that the concentration in some fish (C_f) above that observed in water (C_w) can be described by

$$\log \frac{C_f}{C_w} = 0.542 \mid \log(PC) + 0.124 \qquad (2\text{-}26)$$

where PC is the octanol–water partition coefficient. Therefore, if a compound has a high partition coefficient and it is persistent, modeling should consider biomagnification and food uptake as well as secondary and tertiary sources of pollutants.

IDENTIFYING TOXIC EFFECT OF INTEREST

The emission of toxic chemicals to the environment can produce a wide range of health effects. They can be acute, such as responses to hydrogen sulfide, or they can be chronic, such as cancer or pulmonary fibrosis from asbestos exposure. Table 2-9 displays some major health effects categories of concern. However, the table's title should also make another point—that any chemical, if taken in excess, will result in the manifestation of a toxic effect. This is even true for everyday items like sugar, salt, and alcohol. The purpose of a risk assessment is to insure, with a reasonable margin of safety, that human health effects do not result from exposures caused by production or waste management practices.

Much of the work to date in the human health effects area for toxic pollutant emissions has been conducted under the auspices of the Occupational Safety and Health Administration (OSHA), the National Institute of Health (NIH), the National Academy of Sciences (NAS) or EPA. In general, the focus has been on the development of adverse-effect levels and carcinogen potency values. Adverse-effect levels are described in a variety of forms, ranging from the OSHA 8-h threshold limit values (TLVs) for workplace exposure to EPA's acceptable daily intakes (ADIs).

ADIs are based on no observed adverse effect levels in animal studies or on adjustments to TLVs that account for the longer duration of exposure for the nonworkplace population. The ADI is a projected dosage from animal or epidemiological studies that insures no observable adverse effects, whether they be weight loss, lachrymation, coughing, emphysema, or organ damage. This full range of effects is included because frequently it has been found that minor symptoms, even in an animal study, are highly correlated with serious systemic harm at higher doses. In any event the user of such health data should give serious scrutiny to the basis of the numbers in their application. Appendix A contains some preliminary discussion values for ADIs and carcinogens that have been developed by various groups in EPA. Many of these values are still under experimental investigation and/or agency review. Consequently, some values might change considerably as new data are developed, and some chemicals may cease to be considered carcinogenic. The numbers are included here solely for illustrative purposes. They in no way reflect EPA policy.

Numerous sources exist for identifying exposure risks posed by various compounds. After waste analysis, the first step in any exposure assessment should be a review of these sources to establish critical effects. Some of the more conventional data sources are the Registry of Toxic Effects of Chemical Substances [31], the International Agency for Research on Cancer Monographs [32] and Sax [33].

One must remember that any chemical can produce a whole spectrum of results, depending on a number of factors (see Table 2-10). As an example of this variation, consider the case of chloroform. Chloroform is a suspected carcinogen when administered at low doses over time, but it also has produced death in high-level exposures. For years it was also used as a form

Table 2-9. Health Effects Associated with Pollutant Overdoses

Observed Health Effect	Sample Producer or Effect	Types of Responses
Cancer	Vinyl chloride	Tumors, death
Fibrosis	Cotton dust	Irritations, scarring of lungs, emphysema, pneumonia
Asphyxiation	Hydrogen sulfide	Dizziness, nausea, disorientation, death
Mutagenesis	DDT	Reproductive effects due to chromosomal changes
Teratogenesis	Dimethylacetimide	Birth defects, fetal death
Systemic poisoning	Lead	Organ or brain damage, anemia, death

Table 2-10. Factors Influencing Human Response to Toxic Compounds

Factor	Effect
Dose	Larger doses correspond to more immediate effects
Method of administration	Some compounds nontoxic by one route and lethal by another (e.g., phosgene)
Rate of administration	Metabolism and excretion keep pollutant concentrations below toxic levels
Age	Elderly and children more susceptible
Sex	Each sex has hormonally controlled hypersensitivites
Body weight	Inversely proportional to effect
Body fat	Fat bioconcentrates some compounds (large doses can occur in dieters due to stored pollutants)
Psychological status	Stress increases vulnerability
Immunological status	Influences metabolism
Genetic	Different metabolic rates
Presence of other diseases	Similar to immunological status; could be a factor in cancer recurrence
Pollutant pH and ionic states	Interferes or facilitates absorption into the body
Pollutant physical state	Compounds absorbed on particulates may be retained at higher rate
Chemical milieau	Synergisms, antagonisms, cancer "promoters," enhanced absorption
Weather conditions	Temperature, humidity, barometric pressure, and season enhance absorption

of anesthesia prior to surgery. Clearly, one number does not tell the whole story. Consider also the case of asbestos, which is not carcinogenic when swallowed but very much so if inhaled. One must perform a comprehensive review of the literature if all effects are to be documented.

Without such a review, there is one possibility for economies in the study of many compounds, namely to investigate the doses that would pose a low-level cancer risk. It follows from the methods of derivation that the dose of a carcinogen that would produce a low-level (say 10^{-5} or 10^{-6}) cancer risk is smaller than the dose producing direct, clinical manifestations. Consequently, if one designs a source of carcinogens to minimize the cancer risk, minimal additional pollutant control will be necessary. The remaining gap, emissions of noncarcinogens, can be treated by insuring that the lower observed adverse effect levels (LOAEL) are not exceeded. Although this approach might seem overly conservative, it serves to minimize the possibility that a costly equipment retrofit might ultimately be necessary.

Wide variation in human response to a dose must also be considered when using a single number to represent toxicity. This point is discussed further under "Evaluating Risk Acceptability."

HEALTH EFFECTS PROJECTION

Having examined the range of health effects and some of the factors in varied human response, we now back up and discuss how toxicity numbers are derived. We will then be in a better position to discuss application of estimated risk values.

Development of Health Effects Numbers

One exciting area of study in the realm of risk assessment is the interest in performing massive lifestyle inventories in an attempt to develop information on cancer and circulatory disease factors. Unfortunately, it will be years before most results are available. It is this long period and the questionable ethics of waiting idly that lead researchers to continue to use a more timely species for toxic chemical assessment. Although pigs, dogs, monkeys, and other animals are used, the mouse, the rat, and the hamster are the species of choice because

Their natural life spans are short (approximately two years)

They are easier to breed and handle in large numbers

They are inexpensive

Genetically homogeneous groups are easily bred

In spite of all the controversy, the use of rat data to assess toxic chemicals really involves only two tenets of faith. First, effects observed in an animal due to a chemical will, at the proper dose, produce the same effects in a human. Although there are notable exceptions both ways, this assumption has stood the test of time. (Humans typically are about a factor of 10 more sensitive to given chemicals on a body-weight basis [34]. Second cancers still occur at a reduced rate as the dose decreases. This assumption has been repeatedly questioned. There really is no viable alternative, though, because testing at the lowest possible levels of concern would involve hundreds of thousands of rats. Given the large number of suspected carcinogens, the National Cancer Institute (NCI) [35] estimated in 1979 that as many as 600 to 800 compounds may be carcinogenic out of the 7000 chemicals investigated to that time), the gross national product would need to be funneled into the health effects research area. The logistics problem of caring for and disposing of the rats is likewise imponderable. Lacking a more sensitive tool, we should accept what has, for the most part, been demonstrated to be an accurate, albeit conservative, indicator.

The other principal type of health effect member developed by EPA is frequently not based on independent studies of compounds. Rather the ADIs were developed based on noncancer effects observed in cancer studies. Since cancer is the primary effect under study and is generally of greater interest, the rest of this section discusses items in terms of cancer research. Readers interested in other effects should consult the literature; however, keep in mind that many other health effects are studied as a subset of cancer research.

Demonstration of carcinogenicity requires strict observance of analytical protocols. In [35] NCI has presented some criteria for evaluating experimental designs. These criteria are presented in Table 2-11. Laboratory data not developed in compliance with these protocols are questionable.

If we have data on animals from a well-conducted experiment, we would be ready to develop numbers for assessing carcinogenicity. Such computations are performed by developing a log-log plot of dosages and cancer incidence. The following discussion focuses on rigid risks at different exposures, but remember that degrees of cancer risk in any species vary among the population. Fortunately, it has been observed that these variations are normally distributed, and a statistical treatment is valid.

The problem with human risk estimation is that the area of interest frequently is beneath the doses shown on the rat dosage curve. Consequently, potency is being estimated in a region that is really a tail of the rat susceptibility distribution. The sensitivity of calculations in that region has inhibited the development of a single risk-projection model. Three approaches to extrapolating data downward are currently in vogue: the one-hit model with linear extrapolation, the log-probit model, and the multistage model. The

Table 2-11. Criteria for Evaluating Carcinogen Experiments on Animals

Criteria	Recommendations
Experimental design	Two species of rats, and both sexes of each; adequate controls; sufficient animals to resolve any carcinogenic effect; treatment and observation throughout animal lifetimes at range of doses likely to yield maximum cancer rates; detailed pathological examination; statistical analyses of results for significance
Choice of animal model	Genetic homogenity in test animals, especially between exposed and controls; selection of species with low natural-tumor incidence when testing that type of tumor
Number of animals	Sufficient to allow for normal irrelevant attrition along the way and to demonstrate an effect beyond the level of cancer in the control group
Route of administration	Were tumors found remote from the site of administration? No tumors observed should demonstrate that absorption occurred
Identity of the substance tested	Exposure to chemicals frequently involves mistures of impurities. What effect did this have on results? What is the significance of pure compound results? Also, consider the carrier used in administration
Dose levels	Sufficient to evoke maximum tumor incidence
Age of treatment	Should be started early
Conduct and duration of bioassays	Refer to NCI's "Guidelines for Carcinogen in Small Rodents"

one-hit model is discussed in detail because it was used to develop the numbers in Appendix 2-1. (See [36] for further descriptions.)

The one-hit model assumes that a tumor will be induced after a single receptor is exposed to a single effective dose [35]. The probability of contracting the cancer due to exposure is then

$$P(x) = 1 - e^{-h(x)} \qquad (2\text{-}27)$$

where h(x) is by definition a hazard rate function. Researchers typically use a simple polynomial to describe h(x). The one-hit model with linear extrapolation assumes a linear hazard rate function of the form

$$h(x) = ax + b \qquad (2\text{-}28)$$

This expression is evaluated by conducting a linear regression on a log-log plot of P(x) versus x.X

If we analyze Eqs. (2-27) and (2-28) at zero, we see that $1 - e^{-b}$ is the background level of cancer in the control rats. The actual probability of cancer contraction is then $1 - e^{ax}$. In the low-dose region where we are examining the risk, P(x) is approximately linear. Therefore, the risk can be approximated by

$$P(x) = ax \qquad (2-29)$$

This linearity is assumed for assessing the risk. The term a in Eqs. (2-28), (2-29) when normalized by division by the body weight, is the carcinogenic potency, in $(mg/kg\text{-}day)^{-1}$. However, it is not uncommon in some applications to use a slightly different value. Some risk assessors develop a potency based on a downward projection of the 95% confidence interval of the slope of the dose-response curve to the origin. The rationale for this approach is that it provides a number corresponding to the maximum number of cancers (with 95% confidence) that might be expected in a given exposure situation. Such a number can be quite useful to demonstrate the lack of risk associated with a particular facility.

Risk Calculation

We now show how the numbers could be applied in a specific exposure scenario. Suppose Table 2-12 represents a groundwater sample from a well and the corresponding health effects data. Exposure risks may be calculated by assuming a water consumption of 2 L/day, a 50% uptake of the chemicals in that water, and a 70-kg person.

As can be seen, tetrachlorodibenzo-*p*-dioxin (TCDD) poses a substantial risk. With this level we would conclude that as many as six cancer cases might occur for every 1000 people drinking this water over their lifetime but that level does not correspond to the total risk. We must also consider that there are several pollutants in this water and, more importantly, that all of the carcinogens cause hepatocellular carcinomas. Because these compounds attack the same part of the body, it is reasonable to add the risks [34] to develop a composite individual risk of 6.1×10^{-3}. In a population analysis the number of people becoming ill is frequently computed as the sum of those subject to each effect, unless information is available on synergistic effects.

We make two other comments before closing. First, notice that the level of TCDD creating a large cancer risk is one part per trillion (ppt). This level would not even be detected in most pollutant scans. Even though few compounds are anywhere near this potency, the resultant risks illustrate the im-

Table 2-12. Health Effects Associated With a Contaminated Groundwater

Constituent Present	Concentration (µg/L)	Health Effect of Interest	Potency (mg/kg-day)$^{-1}$	Dose (mg/kg-day)	Cancer Risk
Chloroform	10	Hepatocellular	0.18	1.4×10^{-4}	2.6×10^{-5}
2,3,7,8-tetra-chloro-dibenzo-p-dioxin	0.001	Hepatocellular carcinoma	4.25×10^5	1.4×10^{-8}	6×10^{-3}
Toluene[a]	1000	Central nervous system depression cardiac arythmias	29.057	0.14	
1,1,2-tri-chloro-ethane	100	Hepatocellular carcinoma	0.057	1.4×10^{-3}	8×10^{-5}

[a]Toluene has an ADI; therefore, the potency is in mg/kg-day.

portance of obtaining a detailed characterization of the waste-generating source in a facility risk assessment.

The second point concerns the indicated toluene level. Because the toluene level in the water is considerably less than the ADI, no observable health effects from this pollutant would be expected.

EVALUATING RISK ACCEPTABILITY

Until now, risk assessment has been at least a semiquantitative process. Methods for defining source-emission rates, modeling dispersion and degradation, and quantifying resultant risk levels have been provided. But we still must determine observed risk acceptability. This step is the least quantitative and most politicized element of risk assessment. Nevertheless, it must be undertaken if we are to avoid the generally unattainable, zero discharge demands that result from a failure to define a safe level. Perhaps the best place to start this discussion is by defining risk.

Risk is the probability that some negative outcome will occur. Although risk could involve several indirect human problems (such as fish or animal death or contamination, visibility impairment, decreased stream usability, etc.), this discussion concentrates on direct human health effects. Table 2-13 displays some common risks. The key point is seen by comparing an involuntary activity (pedestrian killed in an auto accident) with a voluntary one (drinking two beers per day). Clearly, the individual's decision to accept a risk and then to pursue a course that by his or her own hand could lead to

Table 2-13. Everyday Activities with a Risk of Death

Activity	Annual Level of Risk (voluntary activity)	Annual Level of Risk (involuntary activity)
Smoking 10 cigarettes/day	1.25×10^{-3}	—
Coal-mining–railroad work accidents	1.3×10^{-3}	—
Motor vehicle accident	2×10^{-4}	—
Drinking groundwater of Table 2-12		1.2×10^{-4}
Manufacturing work accidents	8×10^{-5}	
Pedestrians hit by automobiles		4×10^{-5}
Drinking two beers/day (cirrhosis only)	4×10^{-5}	
Person in a room with a smoker	—	1×10^{-5}
Living near the storage tanks of a large benzene producer	—	6×10^{-6}

Sources: W. D. Rowe, 1977, *An Anatomy of Risk,* Wiley Interscience, New York, pp. 290, 354, 356, except the last item, which is from U.S. Environmental Protection Agency, 1980, *Benzene Emissions from Benzene Storage Tanks—Background Information for Proposed Standards,* EPA-450/3-80-034a, Durham, N.C.

death has a strong bearing on our perception of acceptability. An acceptable level is determined by several factors. These factors and their effects are presented in Table 2-14. The point of this discussion is that what is an acceptable level of risk depends upon the individual concerned (as anyone who has attended a public hearing could tell you.)

Not much work has been done to determine specific numbers for acceptable risks due to TSD activities. The one example that comes to mind is an incinerator burning PCBs that was allowed to operate at a 10^{-6} risk level. Nevertheless, EPA and OSHA have long been performing risk and exposure assessments. Although costs are a key mitigating factor, the general level of regulatory acceptance of risk for cancer contraction has been in the range of one case for every 100,000 to 1,000,000 (10^{-5} to 10^{-6}) people.

With respect to noncarcinogenic effects, safety factors of 10 to 1000 have demonstrated that lowest observable adverse-effects levels are customary. In the past these numbers have generally been applied to the maximally exposed individual (MEI) around the largest source. However, there are signs that this may be changing. Given the conservatism of health effects data and the manner in which it was developed, this would be a welcome change in some quarters. The approach that is gaining in popularity in environmental assessment is the evaluation of aggregate population risk. The basic assumption in these studies—linearity of dose with response—is the same assumption made in generating the health effects data. Simply put, the assumption

Table 2-14. Factors in Risk Acceptability

Factor	Effect
Voluntary vs. nonvoluntary	Chosen risks more acceptable to individual unless mitigated by necessity (coal mining)
Degree of control	Greater personal control enhances willingness to accept risks (airplanes feared over autos)
Extent of assignability	Long latency period that hinders ability to identify exposure reduces concern about exposure
Magnitude of the outcome	People tolerate occasional obnoxious odors, but want wide margin between them and cancer. This factor differs from catastrophic potential because it is a personal assessment
Awareness	Failure to understand nature of risks involved or assessment methods used to evaluate risks black/white type of decisions that have larger factors of safety
Catastrophic potential	Another factor in contradictory perception of airplane vs. auto risks. Big disaster requires greater margin of safety
Needs met by source of risk	The rationale for continued pesticide use and people's rationale for saccharin consumption
Group involvement	The "risky shift" phenomenon. Group definition of acceptable level can result in considerably higher level for represented parties
Cost of alternatives	The individual-at-risk personal cost decreases required margin of safety
Trust and ability to monitor compliance	Lack of trust is key element in large factor of safety required in hazardous-waste handling

is that the same maximum number of cancers would be expected to occur in 1000 people exposed to 1000 mg/day of a compound as in 1 million people exposed to 1 mg/day.

Both approaches have their merits, but each can underestimate the problem if improperly applied. A more reasonable approach is to assess both risk levels. Although ultimate acceptability hinges on the regulatory groups' assessment and public opinion, ample conservatism for human health should typically be provided by an MEI exposure risk of less than 10^{-6} and an aggregate excess cancer prediction of less than one person.

IDENTIFYING CONTROL OPTIONS

If all of the foregoing procedures have been followed, a complete risk and exposure assessment for the TSD will have been conducted. In some cases this analysis leads to the conclusion that there is a need for additional control of emissions. There are three possibilities for corrective action:

Change the generating industrial process to eliminate or to reduce the level of the constituent

Change waste-management equipment to alternative devices mentioned in this book

Change practices to reduce factors of safety (e.g., reroute toxic wastes to storage or cease dissolved air flotation operations on valley inversion days)

However, before doing anything, one must resolve two issues. The first is the extent to which conservatism in the various stages of the assessment contributed to the unacceptable level observed. It may be immensely more cost effective to reduce uncertainty in the analytical tools than to replace equipment, especially for large, multisite corporations faced with marginally unacceptable levels. The second issue is the extent to which improvements are economically and technologically feasible. Obviously, this also should have been the first step in the risk assessment. In other words, some compounds have very strong driving forces pushing them to one media. This most commonly occurs with volatile organic compounds that will partition to the atmosphere. It may not be within the capabilities of the facility to meet any criteria for selected compounds. In such a case removal to a different site for disposal may be the only answer. This treatability problem frequently arises with TCDD.

MONITORING RESULTS

Although it was shunted to a side rail early in the discussion, the assessing of bioaccumulation, biocentration, and background levels is a serious unresolved problem for the modeler. Furthermore, TSD facilities are often not designed well and do have frequent operation problems resulting in high emissions levels. In addition, the occurrence of environmental concentration zones makes continuous performance monitoring a critical element of risk assessment. Monitoring not only alerts us to perturbations that cause concern, but it can provide the basis for sizable reductions in the uncertainty of the modeling tools.

CLOSURE

It is hoped that this chapter clarifies the elements of a risk and exposure assessment. The process is a tedious one, but considerable economies are possible. Waste analyses are not the place to scrimp, however. Thorough waste analyses, coupled with semiquantitative analyses of fugacity,* pollutant mobility, and health effects, can reduce the many compounds to be modeled at a facility to a handful. In addition, many elements of these calculations involve linear assumptions and collectable constants. For example, single source runs of the air dispersion model can be used to determine ambient concentrations for all compounds by simple proportionality among emission rates. The key rules in risk assessment are consider the use of the data and think before you leap into computations.

APPENDIX 2-1

As stated in the chapter, the following numbers for carcinogens (Table A-1) are still undergoing technical review. These values are based on EPA's preliminary findings and is of the literature and secondary source. Changes in estimated potency might occur as these values undergo further agency review or as additional information becomes available. Also, some compounds could be removed from this list. The values are from a variety of sources. Values based on animal studies are the 95% upper confidence intervals of the linear multistage model. Human exposure based values are point estimates from application of the linear, nonthreshold model at 10^{-5} risk level.

The ADIs (Table A-2), on the other hand, are merely based on the best available literature information to date. Large changes in acceptable levels may occur as the compounds receive further study.

REFERENCES

1. U.S. Environmental Protection Agency, 1978 (May), *Compilation and Evaluation of Leaching Test Methods,* EPA-600/2-78-095, Durham, N. C.
2. W. J. Lyman, W. F. Reehl, and H. D. H. Roscubla, 1982 *Handbook of Chemical Property Estimation Methods,* McGraw-Hill, New York.
3. W. Brock Neely, 1980, *Chemicals in the Environment—Distribution Fate Transport Analysis,* Marcel Dekker, New York.
4. L. J. Thibodeaux, D. G. Parker, and H. H. Heck, *Measurement of Volatile Chemicals Emissions from Wastewater Basins,* draft report to U.S. Environmental Protection Agency, Industrial Environmental Research Laboratory. Cincinnati, Ohio.
5. S. T. Hwang, 1980, Treatability and pathways of priority pollutants in the biological wastewater treatment, *A.I.Ch.E. Symposium Series, Water-1980,* vol. 77, no. 209, American Institute of Chemical Engineers, New York, N.Y.

---

* Fugacity, the escaping tendency of a compound from a phase, is a powerful tool for predicting environmental fate. An excellent discussion of this principle and its application to screening analyses is provided in MacKay [37].

Table A-1. Potency Slope Values for Selected Carcinogens

Carcinogen	Potency (mg/kg-day^{-1} [a]	
Acrylonitrile	0.24	(W)
Aflatoxin B	2924	
Aldrin	11.4	
Arsenic	14	(D)
Benzene	0.052	(W)
Benzo(a)pyrene	11.5	
Bis(chloromethyl)ether	9400	(I)
Cadmium	6.7	(W)
Chloroform	0.11	
DDT	8.4	
Dimethylnitrosamine	26	
Dinitrotoluene	0.31	
Ethylene dibromide	1.17	(I)
1,1-ethylene dichloride	1.04	(I)
Formaldehyde	0.021	(I)
Hexachlorodioxins	1.5×10^5	
Nickel	1.15	(W)
PCBs	4.4	
2,3,7,8-tetrachlorodibenzo-*p*-dioxin	1×10^3	
Tetrachloroethylene	0.053	
Trichloroethylene	0.013	
Vinyl chloride	0.018	(I)

Source: Interagency Regulatory Liaison Group (IRLG), 1979, Scientific bases for identification of potential carcinogens and estimation of risk, *Journal of The Cancer Institute* 63(1).

[a] W = human workplace exposure based; D = human drinking-water exposure based; I = inhalation value.

Table A-2. Acceptable Daily Intake (ADI) for Some Chemicals

Compound	Acceptable Daily Intake (mg/kg-day)
Acetonitrile	0.013
Bromomethane	0.411
Chlorobenzene	0.144
Chloromethane	0.543
Endrin	0.001
Hydrogen sulfide	0.077
Naphthalene	0.257
Phenol	0.098
Toluene	0.771

6. R. A. Freeman, 1980, Laboratory Study of Biological System Kinetics, unpublished report to U.S. Environmental Protection Agency's Effluent Guidelines Division, Catalytic Inc., Washington, D.C.

7. R. A. Freeman, et. al., 1980 (June), Experimental Studies on the Rate of Air Stripping of Hazardous Chemicals from Wastewater Treatment Systems, Paper No. 80-16.7, presented at the 73rd Air Pollution Control Association Convention, Montreal, Canada.

8. R. A. Freeman, 1982, Comparison of Secondary Emissions from Aerated Treatment Systems, Paper No. 5C, presented at the March 3, 1982 American Institute of Chemical Engineers Meeting, Orlando, Fla.

9. U.S. Environmental Protection Agency, 1981 (April) *Compilation of Air Pollution Emission Factors, Supplement No. 12,* Research Triangle Park, N. C.

10. American Petroleum Institute, 1962, *Evaporation Loss from Fixed Roof Tanks,* Bulletin 2518, Washington, D.C.

11. W. J. Farmer, M. S. Yang, and J. Letey, 1980 (Aug.), *Land Disposal of Hexachlorobenzene Wastes—Controlling Vapor Movement in Soil.* EPA-600/2-80-119, Durham, N. C.

12. T. T. Shen, 1981, Estimating hazardous air emissions from disposal sites, *Pollution Engineering,* August.

13. L. J. Thibodeaux, and S. T. Hwang, 1982 Landfarming of petroleum wastes— Modeling the air emissions problem, in *Environmental Progress.* February.

14. J. W. Mercer, and C. R. Faust, *Ground Water Modeling,* National Water Well Association, Worthington, Ohio.

15. J. Javandel, C. Doughty, and C. F. Tsang, 1984, *Groundwater Transport: Handbook of Mathematical Models,* Water Resources Monograph 10, American Geophysical Union, Washington, D.C.

16. R. V. Thomann, 1972, *Systems Analysis and Water Quality Management,* McGraw-Hill, New York.

17. A. C. Stern, 1968, *Air Pollution,* Academic Press, New York.

18. F. Pasquill, 1974, *Atmospheric Diffusion,* John Wiley & Sons, New York.

19. R. C. Weast, and M. J. Astle, *CRC Handbook of Chemistry and Physics,* 63rd Ed. CRC Press, Boca Raton, Fla.

20. J. W. Hamaker, 1975, *The Interpretation of Soil Leaching Experiments in Environmental Dynamics of Pesticides,* R.Hague and V. H. Freed, Plenum, New York.

21. R. B. Zepp, and C. M. Cline, 1977, Rates of direct photolysis in aquatic environments, *Environmental Science & Technology 11:*359.

22. W. J. Weber, Jr., 1972, *Physicochemical Processes for Water Quality Control,* John Wiley & Sons, New York.

23. J. W. Hamaker, 1972, Decomposition: Quantitative aspects in *Organic Chemicals in the Soil Environment,* by C. A. T. Goring and J. W. Hamaker, Dekker, New York.

24. H. Kobayashi, and B. E. Rittman, 1982, Microbial removal of hazardous organic compounds *Environmental Science & Technology.* March.

25. American Society of Chemical Engineers, 1975, *Sedimentation Engineering,* Manuals and Reports on Engineering Practice No. 54, New York,

26. L. A. Burns, D. M. Cline, and R. R. Lassiter, 1982, *Exposure Assessment*

Modeling System (EXAMS). User Manual and System Documentation, EPA-600/3-82-023, U.S. Environmental Protection Agency, Athens Research Laboratory, Athens, Ga.
27. J. F. Bowers, J. R. Bjorklund, and C. S. Cheney, 1979, *Industrial Source Complex (ISC) Dispersion Model Users Guide,* EPA-450/4-79-030, U.S. Environmental Protection Agency, Office of Air Quality Planning and Standards, Durham, N. C.
28. Bachmet, et. al., 1980, *Groundwater Management: The Use of Numerical Models,* Water Resources Monograph 5, American Geophysical Union, Washington, D.C.
29. International Commission on Radiological Protection, 1975, *Report of the Task Force Group on Reference Man,* ICRP Publ. Ser. No. 23, New York, N. Y.
30. U.S. Environmental Protection Agency, 1981, *Handbook for Performing Exposure Assessments* (draft), Washington, D.C.
31. U.S. Department of Health, Education and Welfare, 1977 *Registry of Toxic Effects of Chemical Substances,* vol 2, Rockville, Md.
32. International Agency for Research on Cancer, *Monographs:* Lyon, France.
33. I. N. Sax, 1979, *Dangerous Properties of Industrial Materials,* 5th Ed., Van Nostrand Reinhold, New York.
34. L. J. Casarett, and J. Doull, 1980, *Toxicology, The Basic Science of Poison,* Macmillan, New York.
35. Interagency Regulatory Liaison Group (IRLG), 1979, Scientific bases for identification of potential carcinogens and estimation of risk, *Journal of The Cancer Institute 63*(1).
36. Anonymous, 1979, Guidelines and methodology used in the preparation of health effects assessment chapter of the Consent Decree Water Criteria Documents, *Federal Register* **44**(52).
37. D. MacKay, 1979, Finding fugacity feasible *Environmental Science & Technology* **13**(10).

3

Chemical, Physical, and Biological Treatment of Hazardous Waste

Edward J. Martin
Howard University, Washington, D.C.

E. Timothy Oppelt
U.S. Environmental Protection Agency
Cincinnati, Ohio

Benjamin P. Smith
U.S. Environmental Protection Agency
Washington, D.C.

Treatment technology performance and operating data for industrial process wastes containing a variety of pollutants often are reported by using a surrogate parameter such as total organic carbon (TOC) or chemical oxygen demand (COD). Specific compound removal data are relatively limited compared to data on classical pollutants. This has the effect of obscuring process capability analysis for unit process application to the removal of specific toxic compounds.

Several treatment processes are promising for hazardous wastes. However, it is unlikely that any single unit process will be adequate for the high degrees of control necessary on specific toxic-waste components. In most instances process trains in treatment plant configurations will be required for effective hazardous-waste treatment.

For both aqueous and concentrated hazardous wastes, design and operation of treatment plants are waste specific. Each unit process train is engineered to meet the site specific requirements.

Chemical, physical, and biological unit processes in common use, with the exception of incineration, include wet air oxidation and, to a limited extent, biological and land treatment, which are *removal* rather than destructive in nature. A more appropriate term for nondestructive processes might be "concentration technologies." Residues therefore require further treatment and ultimate disposal. Under Resource Conservation and Recovery Act (RCRA) regulations, residues from treatment processes that are used for hazardous wastes are considered hazardous as well.

Active carbon adsorption, sedimentation, and biological treatment have been used extensively for large-scale treatment of aqueous hazardous-waste streams. Activated carbon has been used almost exclusively for concentration of organics in the limited number of larger-scale hazardous-waste treatment operations. However, because of pretreatment requirements, carbon adsorption is only applicable to the polishing of hazardous-waste streams.

Reverse osmosis, stripping, and ultrafiltration are believed to have specialized applicability to treatment of hazardous wastes.

Much of the available data on specific compound removal have been generated in laboratory and pilot-scale experimental studies. On the other hand, much of the experimental data on chemical treatability has been generated for pure compound systems. Removal from multicomponent systems may differ substantially.

Research on treatment technology is oriented toward solution of waste-specific problems, especially for highly toxic wastes. Current research orientation tends to be along three major avenues:

1. Application of production process technology to waste treatment (e.g., solvent extraction and steam stripping)
2. Application of natural degradation phenomena (e.g., photolysis, long-term microbial degradation)
3. Application of principles of analytical techniques (e.g., ultraviolet destruction, electrowinning, electromagnetic separation)

REMOVAL OF TOXIC POLLUTANTS
FROM AQUEOUS WASTES

Table 3-1 summarizes the data for removal efficiencies of hazardous waste by various processes presently available [1]. A more detailed listing is provided in Appendix 3-1. The pollutants covered by Table 3-1 and Appendix 3-1 may be organized into the following categories:

Metals and inorganics
Ethers
Phthalates
Nitrogen compounds
Phenols
Aromatics
Polynuclear aromatic hydrocarbons
Polychlorinated biphenyls (PCBs) and related compounds
Halogenated hydrocarbons
Pesticides
Oxygenated compounds
Miscellaneous

Table 3-1. Median Removal Efficiencies (Percent)

	Sedimentation with Chemical Addition							Gas Flotation with Chemicals												
	Sedimentation	Polymer	Lime	Lime Polymer	Alum, Coag., Aid	Alum	Alum, Lime	Gas Flotation	Alum, Polymer	Ca Cl2, Polymer	Polymer	Filtration	Activated Sludge	Aerated Lagoons	Steam Stripping	Solvent Extraction	Granular	Powdered Carbon with Sludge	Reverse Osmosis	Ozonation
Classical pollutants																				
BOD 5	25	50	52		37	61	41	4	60	64	47	24	93	86			52		87	
COD	93	71	32	99	59	10	86	48	51	66	20	24	67	62	62	50	50		92	50
TKN														77						
TOC	32	82	18	22	47	63	80		25	50	36	13	69	45	72	35	55		92	9
TSS	97	99	71	99	66	84	93	77	68	68	33	67	44	45			38		87	15
Toxic pollutants																				
Acenapthene												73	99				93		73	
Acenaphthylene	17																			
Acrolein												86						30		
Alpha-BHC								45				38					47			
Anthracene/phenanthrene	64									83		44	68				67		77	48
Aroclor 1232												16								
Aroclor 1248												16								
Aroclor 1254												20								
Aroclor 1260												16								
Asbestos	99		95	99								99								

(continued)

83

Table 3-1. (continued)

	Sedimentation	Polymer	Lime	Lime Polymer	Alum, Coag., Aid	Alum	Alum, Lime	Gas Flotation	Alum, Polymer	Ca Cl2, Polymer	Polymer	Filtration	Activated Sludge	Aerated Lagoons	Steam Stripping	Solvent Extraction	Granular	Powdered Carbon with Sludge	Reverse Osmosis	Ozonation
			Sedimentation with Chemical Addition					Gas Flotation with Chemicals												
Benzene	9				49		50				33	14	81	65	80	96	64	95	50	
Benz(a)anthracene		48	92	81															50	
Benzidene																				
Benzo(a)pyrene	80													41			93			90
Benzo(b)fluoroanthene	83													33			80			80
Benzo(ghi)perylene	17						48							97						
Bis(2-chloroethyl)ether													47					53		
Bis(2-chloroethyoxy)methane														60						
Bis(chloromethyl)ether												36	83							
Bis(2-ethylhexyl)phthalate	16	48		49	78				25	72	51	64	24	61				97	67	
4-bromophenyl phenyl ether	48			99	54		99	99		99	99	89	95	78						
Butyl benzyl phthalate									76	50		37	98				97			97
Carbon tetrachloride																				
Chlordane																				
Chlorine, total residual															94					
Chlorobenzene													84				96			
p-chloro-m-cresol				44									80	99			83			
Chlorodibromomethane	77					50														
Chloroethane													78	50			99			
Chloroform								0		20	41				99		74			

(continued)

84

Compound											
2-chloronaphthalene	44							50			
2-chlorophenol								46	81	75	77
Chrysene			92								
4,4'-DDT			99	52		3					
Di-n-butyl phthalate		50	99		39		84	96	76		
1,2-dichlorobenzene			78		61	55	86		99		
1,4-dichlorobenzene							93	81			
Dichlorobromomethane				85	52			99			
1,2-dichloroethalene			98					98			
1,1-dichloroethane								99	89		
1,2-dichloroethane	70						9	99	97	81	
1,2,trans-dichloroethylene											
2,4-dichlorophenol			30	59	30	34	25		64		
1,2-dichloropropane			28	17		38	67			93	
Diethyl phthalate		98	99	99	99	98	85		89		
2,4-dimenthylphenol	0						99			30	
Dimethyl phthalate			76			98	23	25			
4,6-dinitro-o-cresol	49						99				
2,4-dinitrophenol	48										
2,4-dinitrotoluene	0										
2,6-dinitrotoulene	80						83				
Di-n-octyl phthalate	80		92		61	64			91		
Ethylbenzene	49	81	75		59	82	89		84		
Fluoranthene	55		97			29	99		82		
Fluorene	46		50	29			99				
Heptachlor	79		79				76				
Hexachlorobenzene							45				
Indeno(1,2,3-cd)pyrene							99				
Isophorone	49				95		33		97		
Methyl chloride	77						91			97	50

(continued)

Table 3-1. (continued)

	Sedimentation with Chemical Addition							Gas Flotation with Chemicals												
	Sedimentation	Polymer	Lime	Lime Polymer	Alum, Coag., Aid	Alum	Alum, Lime	Gas Flotation	Alum, Polymer	Ca Cl2, Polymer	Polymer	Filtration	Activated Sludge	Aerated Lagoons	Steam Stripping	Solvent Extraction	Granular	Powdered Carbon with Sludge	Reverse Osmosis	Ozonation
Methylene chloride	31				90	78	13	36	84		61	70		97	81		22		10	
Naphthalene	41			49			70		52	80	65		95	28				96		
Nitrobenzene	52					68														
2-nitrophenol	47												99							
4-nitrophenol	0			9									99							
N-nitrosodiphenylamine	77									66	9		84	67			82			
Pentachlorophenol	55				96								89				76			
Phenol	20	14	22	18		86		51		57	36	17	98	71		80	50	83	25	24
Phenols	69	29	70		26	19	11	4	13	1	42	8	65	65		99	69		2	67
Pyrene														67		98	93			
1,1,2,2-tetrachloroethane							30						22							
Tetrachloroethylene	28		0				95		10				93	60	99		68		60	
Toluene	0	20				55	76		10	6	59	21	62	90	95		24	79		15

(continued)

Pollutant	1	2	3	4	5	6	7	8	9	10	11	12	13	14	15	16
1,2,4-trichlorobenzene	72			90						37	95			99		
1,1,1-trichloroethane	33	98		46	91	74	22	4	88	85	96	9	99			
1,1,2-trichloroethane										9		99	99			
Trichloroethylene	35			10			43		40	96		92				
2,4,6-trichlorophenol					70			14	80	18	47		29			
Other pollutants																
Cyanide			65			74	61				45		63	67		
m-p-cresol															41	93
o-cresol														90		
Fluoride		72	92											51		
Methyl ethyl ketone		74				74								51		
Oil and grease	99	84	80	99	98	49	79	59	20	86	98		24	54	43	
Phosphorus total	3		77	14	75		48		30	31						
Styrene												93				97
TDS															95	
Xylenes												97				

87

Although data on individual process removal capabilities is useful, the treatment processes listed are, in general, combined in a unit process train to match requirements for complete treatment of specific waste streams. Unit process trains may be designed by drawing from the array shown on Figure 3-1 and assembling one or more processes from each of the main steps: conditioning or pretreatment; primary, secondary, or tertiary treatment; sludge treatment; and disposal. The data presented should not be used for detailed design but to aid in judgment and assessment of proposed schemes or to assist in preliminary design considerations. Pilot-plant studies should always be performed on specific waste streams in order to develop design data.

Unfortunately, data were not gathered for each pollutant present in the waste during a given study. Therefore wastes containing different arrays of pollutants in differing amounts are likely to impact removal efficiencies observed in other studies. Care should be exercised also in generalizing removal efficiency values for a compound to other compounds in the same class (e.g., phthalates). Table 3-1 illustrates wide variability.

Process removal effectiveness depends significantly on the concentration of given pollutants in the plant influent waste stream. Waste characteristics are not presented here, but the removal efficiencies can be assumed to be applicable to waste streams typically found in relevant industrial wastes. The effluent concentrations presented in Appendix 3-1 can be used to make general estimates of concentration effects.

It is desirable to generalize conclusions about removal effectiveness related to compounds or classes (see Table 3-1). The data reflect the current application of processes to wastes on the full scale as well as research efforts. The data were drawn from about 400 full-scale plants and about 300 pilot- and bench-scale studies. The largest quantity of information is available on sedimentation (plain), activated sludge, and granular activated carbon adsorption. This undoubtedly reflects the state-of-the-art of these three processes as applied to industrial waste treatment.

Plain sedimentation appears to work surprisingly well for removal of a wide variety of organics from waste streams. Data on activated sludge removals reflect both clarification and activated sludge treatment.

Granular activated carbon is effective in a general sense also, but where comparison data exists it is generally less effective than an activated sludge treatment plant. In all cases the process residues—clarification solids, waste-activated sludge, and spent activated carbon—must be disposed of in an environmentally sound manner.

To date, much information in the technical literature is not usable as a guideline for selecting unit processes to remove specific toxic materials. Although waste characterization analyses included toxic organics, in many cases treatment efficiency data for the compounds were apparently not

Figure 3-1. Treatment technology summary.

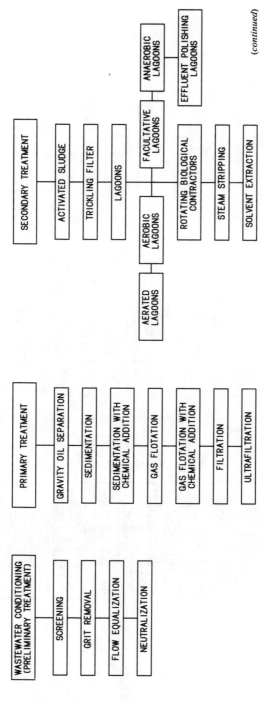

(continued)

89

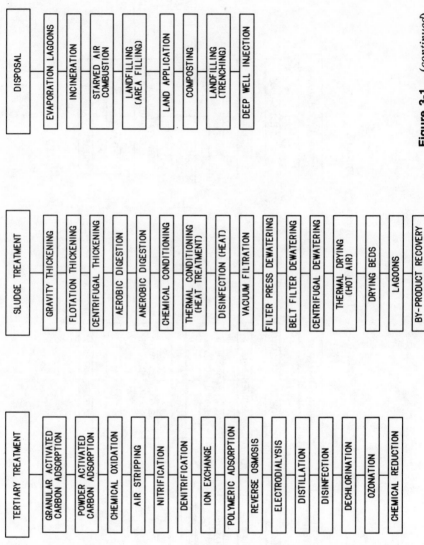

Figure 3-1. (continued)

developed. Treatment efficiency data instead were reported in terms of BOD, (biochemical oxygen demand) COD, TOC, and other classical parameters. This is a serious deficiency in studies and engineering analysis conducted in the recent past. It is necessary to design unit process operations and treatment systems effectiveness around removal of toxic components. Use of generalized removal efficiency values for classical parameters ignores the individual pollutants and could be misleading. Even very small concentrations of highly toxic pollutants can be deleterious in certain environmental settings.

Removal of Toxic Pollutants in Publicly Owned Treatment Works

The U.S. Environmental Protection Agency (EPA) through contractors studied about 50 publicly owned treatment works (POTWs) in 1980 and 1981. Data on toxic organic pollutants and metals were gathered largely for influent, effluent, and raw sludges [2,3].

Priority pollutants were detected in more than 50% of all influent samples analyzed at the 50 POTWs. In calculations of percent removal, pollutant concentrations reported below the detection limit were averaged as zero in influents and as the detection limit in effluents. This convention leads to conservative estimates of removal efficiencies.

Table 3-2 summarizes the removal data for the 23 priority toxic pollutants analyzed. Table 3-2 shows that 50% of the POTWs sampled achieved minimum toxic-metal removals, when present at any measurable level, ranging between 33 and 90%. For common organic priority toxic pollutants, 50% of the POTWs achieved minimum removals ranging roughly between 55% and 92%.

As previously discussed, removal efficiency values can be misleading. For example, for 1,1,1-trichloroethane and methylene chloride, both of which are suspected carcinogens, most of the plants (90%) achieved removals of only 67% of the former and less than 20% of the latter. Depending on starting concentrations, resultant quantities of these compounds in the effluent could still be environmentally significant.

Similarly for metals, where the common view is that incidental removals in secondary treatment are quite high, for most of the plants (90%) removals are significantly less than half, and removals of nickel, lead, cyanide, and cadmium are quite low.

In the same study, pollutant removals through separate treatment processes were examined; these are tabulated in Table 3-3.

Twenty-three priority toxic pollutants included in Table 3-3 are the most prominent ones, and they occurred in at least 50% of all influent samples. Percentage removals shown for a treatment process are those achieved

Table 3-2. Percentage of Removals by Secondary Treatment

Parameter	Number of Plants Included	Percentage of Plants			
		50	75	80	90
BOD	48	91	83	82	79
TSS	48	90	84	79	72
COD	47	82	71	70	62
TOC	43	70	61	57	49
Benzene	22	78	50	49	41
Bis(2-ethylhexl)phthalate	43	55	32	30	22
Butyl benzl phthalate	19	64	41	37	29
Cadmium	22	54	37	36	28
Chloroform	38	56	37	26	19
Chromium	44	77	65	63	47
Copper	47	83	65	56	50
Cyanide	48	90	33	25	10
Di-N-butyl phthalate	49	53	33	27	19
1,2-trans-dichloroethylene	25	71	41	30	17
Diethylphthalate	14	38	25	7	3
Ethylbenzene	35	83	60	56	46
Lead	28	58	46	41	10
Mercury	30	58	41	38	25
Methylene chloride	39	55	35	30	18
Nickel	36	33	19	17	13
Phenol	31	87	62	60	51
Silver	34	69	56	51	40
Tetrachloroethylene	41	81	67	60	38
1,1,1-trichloroethane	39	87	80	78	67
Trichloroethylene	43	85	72	61	51
Toluene	46	92	78	70	53
Zinc	48	78	64	63	43

Note: Plant averages used as basis for percent removal calculation. Values reported below their detection limit were averaged as zero in influents and as the detection limit in effluents.

between the raw influent to the POTW and the effluent from the tabulated process. Primary effluents, tertiary effluents, activated sludge effluents, and trickling filter effluents were evaluated. In addition, the database includes two lagoons, one rotating biological contactor, and one infiltration-percolation process.

Table 3-3 shows that among the treatment processes examined, trickling filter removals for 15 of the priority toxic pollutants were better than the corresponding activated sludge removals. On the other hand, activated sludge removals were higher than trickling filter removals for only four of the pollutants. Adsorption potential on trickling filter media is probably greater than for activated sludge floc.

Table 3-3. Removing Pollutants by POTW Treatment Processes (Percent)

Parameter	Primary Treatment[a]	Trickling Filter[a]	Activated Sludge[b]	Lagoon[c]	Rotating Biological Contactor[d]	Infiltration Percolation[d]	Tertiary Treatment[e]
BOD	—	82	91	90	96	95	94
COD	8	74	83	75	93	93	75
Oil and grease	38	74	78	71	86	90	81
Total susp. solids	48	88	90	92	93	98	95
Ammonia nitrogen	—	4	25	53	37	87	62
Benzene	—	92	81	75	96	96	91
Bis(2-ethylhexyl)phthalate	44	60	50	—	94	91	27
Butyl benzyl phthalate	44	55	—	—	72	72	—
Cadmium	35	93	80	56	96	96	56
Chloroform	44	78	74	72	97	94	76
Chromium	35	85	87	86	98	98	91
Copper	9	72	86	79	99	93	91
Cyanide	—	62	72	—	82	94	—
Di-n-butyl benzyl phthalate	—	40	35	—	62	62	—
1,2-trans-dichloroethylene	25	65	52	25	86	86	75
Diethyl phthalate	—	—	—	—	—	—	—

(continued)

Table 3-3. (continued)

Parameter	Primary Treatment[a]	Trickling Filter[a]	Activated Sludge[b]	Lagoon[c]	Rotating Biological Contactor[d]	Infiltration Percolation[d]	Tertiary Treatment[e]
Ethylbenzene	10	87	78	72	95	94	88
Lead	—	58	57	—	56	56	29
Mercury	11	40	26	76	71	71	74
Methylene chloride	—	96	—	99	99	99	98
Nickel	23	66	36	72	92	92	61
Phenol	—	91	78	47	99	99	82
Phenols, total	—	75	76	60	86	97	74
Silver	—	44	63	63	47	47	58
Solids, diss., total	—	8	6	39	—	2	13
Solids, diss., volatile	—	37	11	38	9	—	32
Solids, settleable	91	96	93	96	96	96	96
Solids, sus., total vol.	51	87	90	92	95	98	96
Solids, total	—	25	26	62	8	30	33
Solids, volatile, total	4	55	53	76	—	55	68
Tetrachloroethylene	11	92	80	95	98	96	96
TOC	5	70	73	65	84	94	81
Toluene	—	98	94	92	100	99	99
1,1,1-trichloroethane	—	97	83	97	99	99	99
Trichloroethylene	16	96	81	92	99	98	93
Zinc	—	90	78	87	98	98	78

[a] 14 = plant database.
[b] 36 = plant database.
[c] 2 = plant database.
[d] 1 = plant database.
[e] 10 = plant database.
Removal is calculated on the basis of average influent to the POTW and effluent of each particular process.

The only rotating biological contactor included in the database showed higher percentage removals when compared to activated sludge or trickling filter processes for 21 of the 23 priority toxic pollutants. Primary treatment was consistently less effective than secondary treatment processes in removing the priority toxic pollutants. The 10 plants that utilized tertiary treatment showed higher average removals of most pollutants when compared to trickling filters or the activated sludge process.

The two lagoons in the database were nearly equivalent to trickling filters or the activated sludge process in removing the conventional pollutants. However, lagoons showed considerable variations in the removal of non-conventional and priority toxic pollutants when compared to other secondary treatment processes.

For toxic-metal pollutants, removals by trickling filter and activated sludge systems vary, depending on the individual metal pollutant. Trickling filters achieved higher removals of cadmium, mercury, nickel, and zinc, and activated sludge systems achieved higher reductions of copper, silver, and cyanide. Both systems were almost equally effective in removing chromium and lead.

In general, the data collected were for relatively short intervals of time (six days). For the Moccasin Bend Wastewater Treatment Plant in Chattanooga, Tenn., the study was conducted for 36 days [3].

Table 3-4 presents tabulated data for removal efficiencies of organics during the 36-day study. Metal removals are not included. Erroneous waste-activated sludge pumping rates caused the data to show metal increases rather than removals.

The total treatment mass calculations are based on the influent load without the load from the recycle lines; the effluent load is equal to the sum of the primary sludge, waste-activated sludge, and secondary effluent waste-stream loads minus the vacuum filtrate and digester decant loads. Similar to the secondary treatment mass balance, the overall treatment mass balance analysis indicates that volatiles, acids, and base–neutral fractions are reduced significantly.

The effects of selected effluent priority pollutant concentrations as a result of varying corresponding influent concentrations were examined by correlating the influent priority pollutant concentrations to the secondary effluent concentrations. Correlations for all priority pollutants that occurred in over 50% of the combined 36-day influent samples were determined. When calculating the linear correlations, the influent and effluent concentrations were assumed to equal zero if the pollutant was not detected in the waste stream; if the pollutant was reported as less than the analytical detection limit, the concentration was also assumed to equal zero.

The correlations were generally fair, with wide variations observed within

Table 3-4. Mass Balance Analysis

Parameter	Primary Treatment	Secondary Treatment	Total Treatment
Volatiles			
Benzene	7	77	79
Chloroform	1	50	51
1,2-*trans*-dichloroethylene	22	100	100
Ethylbenzene	20	86	82
Methylene chloride	16	45	55
Tetrachloroethylene	23	85	88
1,1,1-trichloroethane	11	79	89
Trichloroethylene	38	75	82
Toluene	9	84	85
Acids			
2,4-dichlorophenol	2	46	47
Phenol	5	92	91
Base–Neutrals			
Bis(2-ethylhexyl)phthalate	166	42	85
Di-*n*-butyl phthalate	39	7	22
1,3-dichlorobenzene	14	30	40
1,4-dichlorobenzene	7	88	80
Diethyl phthalate	53	35	7
Naphthalene	29	75	58
Phenanthrene	62	70	25
1,2,4-trichlorobenzene	1	54	45

the metal, volatile, acid, and base-neutral fractions. Correlation coefficients for volatile priority toxic pollutants ranged from 0.804 for 1,1,1-trichloroethane to 0.081 for trichloroethylene. The chlorination of wastes can be a concern when potentially toxic compounds are formed during the process.

A summary of occurrences where the secondary or tertiary chlorinated effluent had a higher concentration of chlorinated hydrocarbon than the corresponding prechlorinated secondary effluent is presented in Table 3-5. Inconsequential data points were excluded from consideration by specifying that the post-chlorinated sample had to exceed the detection limit of the prechlorinated sample. Most of the pollants in Table 3-5 with less than 10 reported occurrences of higher chlorinated hydrocarbon levels can be explained by some normal degree of variation between concentrations within the same order of magnitude. Control tests run on some frequently occurring nonchlorinated compounds also indicated several instances of concentration increase through chlorination. For chloroform there were 55 instances where it increased in concentration through the chlorination process. There were 34 occurrences of increase in Gamma-BHC. For dichlorobromomethane there were 27 instances where it increased through chlorination. Table 3-5 lists the

Table 3-5. Chlorinated Hydrocarbons Formed during Chlorine Disinfection

Parameter	No. of Occurrences	Prechlorinated Effluent Avg[a]	Chlorinated Effluent Avg[a]
Aldrin (μg/L)	20	0	145
Alpha-BHC (μg/L)	18	0	72
Beta—BHC (μg/L)	20	0	74
Delta-BHC (μg/L)	23	2	187
Gamma-BHC (μg/L)	34	49	138
Bis(2-chloroethoxy)methane (μg/L)	4	0	5,521
Bis(2-chloroethyl)ether (μg/L)	4	0	5,521
Bis(2-chloroisopropyl)ether (μg/L)	4	0	5,521
Chlordane (μg/L)	19	0	1,516
Chlorodibromomethane (μg/L)	13	0.3	8
Chloroethane (μg/L)	4	68	230
Chloroform (μg/L)	55	3	9
2-chloronaphthalene (μg/L)	4	0	5,521
2-chlorophenol (μg/L)	3	0	97
4,4'-DDD (μg/L)	18	0	409
4,4'-DDE (μg/L)	19	0	334
4,4'-DDT (μg/L)	20	0	730
1,2-dichlorobenzene (μg/L)	5	1	4,420
1,3-dichlorobenzene (μg/L)	4	0	5,521
1,4-dichlorobenzene (μg/L)	5	1	4,419
3,3$_9$-dichlorobenzene (μg/L)	4	0	5,521
Dichlorobromomethane (μg/L)	27	0.4	7
Dichlorodifluoromethane	3	0	31
1,1-dichloroethylene (μg/L)	3	1	3
2,4-dichlorophenol (μg/L)	3	0	97
Dieldrin (μg/L)	20	0	218
Alpha-endosulfan (μg/L)	19	0	227
Beta-endosulfan (μg/L)	19	0	455
Endosulfan sulfate (μg/L)	20	0	1,460
Endrin (μg/L)	20	0	365
Endrin aldehyde (μg/L)	20	0	1,440
Heptachlor (μg/L)	20	0	148
Heptachlor epoxide (μg/L)	21	0	161
Hexachlorobenzene (μg/L)	4	0	5,521
Hexachlorobutadiene (μg/L)	4	0	5,521
Hexachlorocyclopentadiene (μg/L)	4	0	5,521
Hexachloroethane (μg/L)	4	0	5,521
Methyl chloride (μg/L)	2	195	390
Methylene chloride (μg/L)	26	19	32
Parachlorometacresol (μg/L)	3	0	97
PCB-1016 (μg/L)	19	0	1,453
PCB-1221 (μg/L)	19	0	2,232
PCB-1232 (μg/L)	19	0	531

[a] All units (μg/L) unless otherwise noted.

(continued)

Table 3-5. (continued)

Parameter	No. of Occurrences	Prechlorinated Effluent Avg[a]	Chlorinated Effluent Avg[a]
PCB-1242 (μg/L)	19	0	1,537
PCB-1248 (μg/L)	19	0	2,463
PCB-1254 (μg/L)	19	0	2,968
PCB-1260 (μg/L)	19	0	3,874
Pentachlorophenol (μg/L)	8	0.8	40
TCDD (μg/L)	19	0	779
Tetrachloroethylene (μg/L)	18	6	9
1,2,4-trichlorobenzene (μg/L)	4	0	5,521
1,1,1-trichloroethane (μg/L)	8	10	12
1,1,2-trichloroethane (μg/L)	1	0	14
Trichloroethylene (μg/L)	11	3	5
2,4,6-trichlorophenol (μg/L)	3	0	97

[a] All units (μg/L) unless otherwise noted.

various other instances of higher chlorinated hydrocarbons increased in concentration during chlorine disinfection.

The data are in micrograms per liter or nanograms per liter, so effluent quantities are, in general, small. Nevertheless, where contributions of industrial or other wastes to the raw waste streams are high, removals low, and chlorination effectiveness high, this effect could be significant.

Chemical Treatability Summary

It is useful to consider the impact of these unit processes on wastes of higher concentration than are typically associated with dilute industrial waste streams [4]. Appendix 3-2 summarizes data from about 140 references on wastes generally higher in influent concentration than those represented in Appendix 3-1 and Table 3-1. The entries present treatment characteristics and qualitative inhibitory information for

Process	Appendix 3-2 Process Code No.
Biological	I
Coagulation/precipitation	II
Reverse osmosis	III
Ultrafiltration	IV
Stripping	V
Solvent extraction	VII
Carbon adsorption	IX
Resin adsorption	X
Miscellaneous adsorbents	XII

About 100 of the entries are for wastes containing the subject compound in concentrations of 100 mg/L or greater. The summary statements for most compounds provide information for more than one concentration technology. On the basis of this review, the process application potential for various wastes is mixed and summarized below.

Biological Treatment. Biological processes are, in general, the most cost-effective techniques for treating aqueous waste streams containing organic constituents. Moreover, biological processes have been applied successfully at full scale to a wide variety of industrial wastes. Environmental impacts associated with biological processes are limited. Probability of greatest concern in this regard is the potential release of volatile organic compounds to the atmosphere as a result of aeration.

For biological decomposition of organic materials of a hazardous nature (many of which are toxic to microbial flora at high concentrations, the system must be allowed to acclimate tot he waste to be treated prior to routine operation of the process.

The activated sludge process, in one of its modifications, appears to have the greatest potential for the application of interest because it can be controlled to the greatest extent and best lends itself to the development of an acclimated culture. However, anaerobic filtration, because of ease of operation, minimal sludge production, and energy efficiencies, merits consideration in many situations. Thus, biological treatment is judged a viable technology to be considered for treating hazardous aqueous wastes containing organic constituents.

Carbon Adsorption. Activated carbon adsorption is a well-developed technology that has a wide range of potential waste-treatment applications. It is especially well suited for removing mixed organic contaminants from aqueous wastes. Numerous examples of full-scale waste-treatment applications exist.

No serious environmental impacts are associated with carbon systems employing regeneration. However, if regeneration is not carried out, impacts could result from the disposal of carbon contaminated with hazardous materials.

Energy requirements for systems employing thermal reactivation could be significant—approximately 14,000–18,600 kJ/kg of carbon (6,000–8,000 Btu/lb).

Unit costs for carbon adsorption can vary widely, depending upon the waste to be treated, the adsorption system, and the regeneration technique. This treatment has been shown to be an economical approach in numerous instances.

In the current context, carbon adsorption must be considered a viable candidate for treating hazardous aqueous wastes containing organic contaminants. Granular activated carbon is the most well-developed approach. However, combined biological–carbon systems appear promising for this application.

Catalysis. Several potential applications of catalysis to waste treatment have been identified, but commercial practicality has not been demonstrated.

Catalyst application is generally very selective, but although potentially applicable to destruction or detoxification of a given component of a complex waste stream, it does not have broad spectrum applicability. Centrifugation is a viable ancillary process for sludge dewatering in an overall wastewater processing train. It may also have limited application for separating liquids of different densities. Its chief application would be as an ancillary process to support primary concentration technique.

Chemical Precipitation. Precipitation processes have been in full-scale operation for many years. The techniques can be applied to almost any liquid waste stream containing a precipitable hazardous constituent. Required equipment is commercially available. Associated costs are relatively low, and thus precipitation can be applied to relatively large volumes of liquid wastes. Energy consumption also is relatively low.

Precipitation processes result in the production of a wet sludge that must be further processed prior to ultimate disposal. In some instances the potential for material recovery from this sludge exists. However, very often nontarget materials are precipitated together with the material of interest, thus complicating or foreclosing the potential for materials recovery.

Usually treatability studies must be carried out prior to applying the process to a waste stream to determine the chemical of choice, the degree of removal, and the required chemical dose.

Precipitation is the technique usually chosen to remove heavy metals from aqueous hazardous wastes.

Freeze Crystallization. AVCO Corporation has stated that the inability of the process to respond to changing wastewater characteristics and its operations complexity were primary reasons for abandoning its research efforts conducted largely during the mid-1970s [3].

Because this process has not been reduced to practice, there are no known research plans. Past efforts have not been successful, and this process is judged to have little potential application. The Japanese are successfully using freezing–dehydration for treatment of metal-containing sludges prior to solidification [5].

Density Separation.   Sedimentation processes have been used for many years: they are easy to operate, inexpensive, and energy efficient. Required equipment is relatively simple and commercially available. The process can be applied to almost any liquid waste stream containing settleable material. It is considered to have high potential for application. However, it is an ancillary process which will be utilized primarily in conjunction with some other concentration technique, such as chemical precipitation. Alternatively, it may be used as a pretreatment technique prior to another process, such as carbon or resin adsorption.

Flotation is a proven solid–liquid separation technique for certain industrial applications. It has higher operating costs and more skilled maintenance requirements than does gravity sedimentation. Power requirements also are higher. This technique is judged to be potentially applicable but probably only in situations where the wastewater contains high concentrations of oil and grease.

Dialysis and Electrodialysis.   Neither process has been judged to have much applicability to aqueous hazardous-waste treatment in the current context. They are not well suited to mixed-constituent waste streams, and both rely heavily on recovery and reuse of at least one product stream to offset costs. Dialysis should not be considered to be concentration technology.

Distillation.   Distillation is expensive and energy intensive and can probably be justified only in cases where valuable product recovery is feasible (e.g., solvent recovery). Distillation is judged to have limited applicability to treatment of dilute aqueous hazardous wastes.

Evaporation.   Evaporation is not expected to have broad application to the treatment of aqueous hazardous wastes containing moderately volatile organic constituents (b.p. 100–300 °C). These organics cannot be easily separated in a pretreatment stripper and will appear to some extent in the condensate from the evaporator, depending on their volatility. Therefore, good clean separation of these organics is not possible without post-treatment of the condensate.

The major disadvantages of evaporation are high capital and operating costs and high energy requirements. This process is more adaptable to wastewaters with high concentrations of pollutants.

Filtration.   Filtration is a well-developed economical process used in many applications. A wide spectrum of filtration systems is commercially available. Energy requirements are relatively low, and operational parameters are well defined. Therefore, filtration is judged to be a good candidate for application. However, it is not a primary treatment process but is used to support other processes either as a polishing step subsequent to precipitation and sedimentation or as a dewatering process for sludges generated in other processes.

Flocculation.   Flocculation has been used for many years and is a relatively simple process to operate. Necessary equipment is commercially available, and costs and energy consumption are relatively low. The process can be applied to almost any aqueous waste stream containing precipitable and/or suspended material.

Flocculation must be carried out in conjunction with a solid–liquid separation process, usually sedimentation. Often, it is preceded by precipitation. Flocculation is judged to be a viable candidate process for hazardous aqueous-waste treatment, particularly where suspended solids and/or heavy-metal removal is an objective. It may be used in conjunction with sedimentation as a pretreatment step prior to a subsequent process such as activated carbon adsorption. In most instances the applicability of the technique, the flocculating chemicals to be used, and the chemical dose can be judged based upon experience and simple laboratory treatability tests.

Ion Exchange.   Ion exchange is a proven process with a long history of use. It will remove dissolved salts, primarily inorganics, from aqueous solutions. For many applications, particularly where product recovery is possible, ion exchange is a relatively economical process. Also, it is characterized by low energy requirements.

Ion exchange is judged to have some potential for the application of interest in situations where it is necessary to remove dissolved inorganic species. However, other competing processes—precipitation, flocculation, and sedimentation—are more broadly applicable to mixed-waste streams containing suspended solids and a spectrum of organic and inorganic species. These competing processes also usually are more economical. Thus, the use of ion exchange probably would be limited to situations where a polishing step was required to remove an inorganic constituent that could not be reduced to satisfactory levels by preceding treatment processes. Therefore, ion exchange is believed to have some potential for current application.

Resin Adsorption.   Because of selectivity, rapid adsorption kinetics, and chemical regenerability, resins have a wide range of potential applications. The primary disadvantage is high initial cost—although resins recently have been quoted to be $11 to $33 per kilogram ($5 to $15 per pound). It is not economically competitive with carbon for high-volume, high-concentration, mixed-constituent wastes, but benefits may be gained by sequential resin and carbon adsorption.

Energy requirements heavily depend on whether solute recovery from the wash media is practiced. Without solute recovery energy costs account for 5% of operating costs; however, with solute recovery using distillation, energy costs could account for 50% of operating costs.

As with activated carbon, the only major environmental impacts relate to the regeneration process. If not reused, spent regenerant requires disposal, frequently by incineration or land disposal.

Resin sorption has been judged to be a viable candidate for treatment of hazardous aqueous organic wastes. The technology, however, has not been as well defined as carbon adsorption.

Reverse Osmosis. Reverse osmosis is a relatively new process that has been reduced to practice for some applications. A number of competitive suppliers of reverse osmosis systems exist. Energy requirements for commercially available systems are about 7.61×10^6 to 9.51×10^6 J/m^3 of product water (8–10 kWh/1000 gal). Reverse osmosis is a relatively costly process but is capable of producing high-purity water. The principal application is concentration of dilute solutions of inorganic and some organic solutes.

The state of development of the process is such that it is necessary to conduct extensive bench- and pilot-scale testing prior to almost any potential application to ascertain feasibility.

Reverse osmosis in its present state of development is judged to have limited potential for application. Its use probably will be limited to polishing operations subsequent to other more conventional processes.

Solvent Extraction. Solvent extraction is judged to have minimal potential for the application of interest. Broad spectrum sorbents such as activated carbon are expected to be more effective in treating dilute waste streams containing a diversity of organic compounds. Carbon adsorption also will be more economical unless a valuable product can be recovered, which is unlikely in most cases expected to be encountered.

Stripping. Air stripping is judged to have minimal potential for application. The process is difficult to optimize for hazardous aqueous-waste streams containing a spectrum of volatile and nonvolatile compounds. Air stripping does have appeal as pretreatment prior to another process, such as adsorption, to extend the life of the sorbent by removing sorbable organic constituents. However, air-pollution-control requirements are likely to be severe, thus making the economics less attractive. Note that some air stripping of volatile components results in safety hazards or air-quality problems. This is expected to be most severe for biological treatment processes using aeration devices.

Steam stripping has merit for wastes containing high concentrations of highly volatile compounds. It is a proven process for some applications but will require laboratory and bench-scale investigations prior to application to waste streams containing multiple organic compounds. Both energy requirements and costs are relatively high. By-product recovery to offset costs from the types of hazardous-waste streams under consideration is unlikely.

Steam stripping is judged to have greatest potential as a pretreatment step to reduce the load of volatile compounds to a subsequent treatment process.

Ultrafiltration.   Ultrafiltration is a commercially used process with several industrial applications. It is characterized by high capital and operating costs. Energy costs could run as high as 30% of direct operating costs.

Ultrafiltration is judged to have limited potential for the application of interest. Its use probably would be limited to relatively low-volume streams containing substantial quantities of high-molecular-weight solutes or suspended materials. Pilot testing is a prerequisite to use.

The current use of these treatment technologies in several industries is shown in Table 3-6. The technology usage by industry is grouped as having wide use, limited use, or potential use. A widely used technology indicates that based on the industry data-collection portfolios and plant visits, the technology is reported as being used at a significant number of plants within the industry and thus is a common technology for that particular industry. The limited use category indicates that a minority of plants in the database are using that particular technology. The potential use category indicates that either a full-scale or a pilot-scale demonstration has been carried out, or that the technology has been used in a related industry and has potential application in the industry indicated.

The performance of the technologies in removing classical pollutants from industrial wastewaters is documented in the literature. The performance data on removal of toxic pollutants, except for some metals and phenols, is more limited. Lack of long-term data precludes the determination of reliable quantitative performance evaluation of these technologies for removal of toxic pollutants. It should also be recognized that in several cases the performance data shown in the data sheets for the removal of toxic pollutants represent incidental rather than intentional control. For example, the removal of organic toxics may be observed for the removal of inorganic toxic materials, such as heavy metals. Such removal is of interest in determining the overall pollutant removal capabilities of the technology.

Table 3-6. Treatment Technology Matrix by Industry

Industry	Activated Carbon Adsorption	Chemical Oxidation	Chemical Precipitation	Chemical Reduction	Coagulation and Flocculation	Distillation	Electrodialysis	Evaporation	Filtration	Flotation	Flow Equalization	Ion Exchange	Neutralization	Oil Separation	Polymeric Adsorption	Reverse Osmosis	Screening	Sedimentation	Stripping	Solvent Extraction	Ultrafiltration
Auto and other laundries	*	o								o			o			o	o				
Coal mining		o	o	o				*				o					o				
Inorganic chemicals manufacturing			o								o		o					o			
Iron and steel manufacturing				o	o								o	o				o			
Leather tanning and finishing	*	*			*				*										o		
Metal finishing		o			o			o	o	o		o		o			o				
Aluminum forming		*	o										o	o				o			
Battery manufacturing			o	o									o	o				o			
Coil coating			o	o										o				o			
Electrical and electronic comp.	*	*	o									*	o					o			*
Foundries		*	o															o			
Coil equipment and supplies											o	o	o			o		o			
Porcelain enameling			o					*		*		*	o			*		o			
Explosives manufacturing			o					o					o					o			
Gum and wood chemicals											o			o				o			

(continued)

Table 3-6. (continued)

Industry	Activated Carbon Adsorption	Chemical Oxidation	Chemical Precipitation	Chemical Reduction	Coagulation and Flocculation	Distillation	Electrodialysis	Evaporation	Filtration	Flotation	Flow Equalization	Ion Exchange	Neutralization	Oil Separation	Polymeric Adsorption	Reverse Osmosis	Screening	Sedimentation	Stripping	Solvent Extraction	Ultrafiltration
Pharmaceutical manufacturing					○						○		○					○	○		
Nonferrous metals manufacturing			○		○							*	○					○			
Ore mining and dressing			○		○								○					○			
Organic chemicals manufacturing																					
Paint and ink formulation					○													○			
Petroleum refining										○									○		
Pulp and paper mills	*				○					○	○		○		*		○	○			
Rubber processing					○						○		○	○				○			
Soap and detergent manufacturing														○							
Steam electric power plants			○		○						○		○					○			
Textile mills		○			○								○				○	○			
Timber products processing								○					○	○				○			

○ Wide use.
* Potential application.

106

TOXIC MATERIALS IN POTW SLUDGES

In an effort to determine the source and occurrence of priority toxic pollutants in sewerage and in POTWs, the U.S. EPA initiated a program of sampling and analysis to study the occurrence of 129 priority toxic pollutants in POTWs [2,3]. Results of the study were published in draft form in November 1981.

The study sampled 50 POTWs, all providing a minimum of secondary treatment. Most plants were subjected to a minimum of six days of 24-h sampling of influents, effluents, and sludge streams. Each sample was analyzed for conventional, selected nonconventional, and priority pollutants. The overall database generated during the study was derived from 1532 individual samples collected at the 50 POTWs.

Although the study was geared extensively toward effluent discharges, we have attempted to extract from the text pertinent information from the database dealing with the occurrence of priority toxic pollutants in sludge streams. Table 3-7 presents occurrence data and concentration ranges for the POTW raw sludge samples (raw sludges include thickened sludges) and gives the total number of samples analyzed, the percentage of times the pollutant was detected, and the minimum and maximum concentrations of detected pollutants. The table shows that 67 priority pollutants were detected in the raw sludges, with 12 of the 15 most frequently occurring and most highly concentrated toxic pollutants in raw sludges being metals.

The data for raw sludges in the study are relatively extensive, but the data made available for treated sludge streams were definitely lacking. The focus of this discussion had initially been to provide removal efficiencies for priority toxic pollutants in POTW sludge streams after some method of sludge treatment (i.e., aerobic digestion, heat treatment). However, of the 50 POTWs sampled only 7 included data on pollutant concentrations employing various sludge treatment processes. Individual plant characteristics are provided in Table 3-8. The pretreatment and treatment data for the seven plants are shown in Tables 3-9 to 3-17.

Although removal efficiencies could not be calculated from the data, and a sufficient amount is lacking to reliably make any type of conclusion regarding the occurrence of priority pollutants in treated sludge streams, the available data does suggest that significant amounts of pollutants are being concentrated in treated sludges in some treatment plants. Consequently, questions must be raised regarding proper disposal of the treated sludges. POTWs produce significant quantities of sludge annually, and if what the data suggests is true, large quantities of priority toxic pollutants are being disposed of either properly or improperly. If the data revealed are indeed representative of priority pollutant concentrations in treated sludge streams, proper disposal of these wastes should immediately be addressed.

(Text continues on page 130)

Table 3-7. Priority Pollutants Occurring in POTW Raw Sludges

Parameter	Number of Samples Analyzed	Samples Detected (%)	Minimum Value Detected[a]	Maximum Value Detected[a]
Lead	458	99	31	170,000
Zinc	459	99	70	5,180,000
Copper	456	98	100	180,000
Cyanide	449	96	10	286,000
Bis(2-ethylhexyl)phthalate	422	95	2	47,000
Toluene	419	95	1	427,300
Chromium	458	94	110	160,000
Arsenic	458	93	1	6,000
Silver	458	93	6	24,000
Antimony	452	88	1	2,000
Cadmium	458	87	4	95,000
Nickel	458	86	12	84,000
Selenium	453	77	1	1,567
Methylene chloride	419	70	1	10,500
Mercury	455	67	110	690,000
Ethylbenzene	419	63	1	4,200
1,2-*trans*-dichloroethylene	419	61	1	96,000
Benzene	419	58	1	953
Trichloroethylene	418	56	1	32,700
Phenanthrene	422	49	1	35,000
Pyrene	422	49	1	33,000
Phenol	423	48	5	3,700
Anthracene	422	43	1	35,000
Fluoranthene	422	42	1	37,000
Di-*n*-butyl benzene	422	41	2	6,900
Tetrachloroethylene	419	41	1	2,400
Butyl benzyl phthalate	422	39	2	45,000

1,1-dichloroethane	419	35	1	630
Naphthalene	422	32	1	8,100
Chrysene	421	29	1	15,000
Beryllium	459	27	2	310
1,2-benzanthracene	422	24	1	15,000
Chloroform	419	23	1	366
1,1,1-trichloroethane	419	18	1	10,910
1,4-dichlorobenzene	422	18	2	7,300
1,1,2,2-tetrachloroethane	419	14	1	3,040
1,2,4-trichlorobenzene	422	14	2	15,000
Pentachlorophenol	423	13	10	10,500
Thallium	456	12	1	251
3,4-benzofluoranthene	422	11	1	2,400
1,3-dichlorobenzene	422	10	14	4,100
Di-n-octyl phthalate	422	9	4	1,024
Dichlorodifluoromethane	419	9	2	4,300
Diethyl phthalate	422	9	1	786
11,12-benzofluoranthene	423	9	1	6,200
Vinyl chloride	418	8	8	62,000
1,2-dichloroethane	419	8	1	10,010
Chloroethane	419	7	5	71,000

(continued)

Table 3-7. *(continued)*

Parameter	Number of Samples Analyzed	Samples Detected (%)	Minimum Value Detected[a]	Maximum Value Detected[a]
1,2-dichloropropane	419	7	1	44
Dichlorobromomethane	419	6	3	260
Fluorene	422	6	1	6,300
Methyl chloride	419	6	1	6,100
Acenapthene	422	5	6	12,000
Benzo(a)pyrene	422	5	1	92,000
Dimethyl phthalate	422	5	3	650
Methyl bromide	419	5	33	30,000
Trichlorofluoromethane	419	5	2	353
Carbon tetrachloride	419	4	5	3,030
1,1,2-trichloroethane	419	4	1	2,100
Chlorodibromomethane	419	3	10	75
1,1-dichloroethylene	419	3	1	14,000
Acrylonitrile	419	2	4	290
Hexachlorobenzene	422	2	28	780
Ideno(1,2,3-c,d)pyrene	422	2	17	102
2-chlorophenol	423	2	11	72
2,4-dichlorophenol	423	2	14	3,800
Parachlorometa cresol	423	1	12	35

Note: Pollutants reported as not detected or less than detection limit were treated as not detected for purposes of this table.
[a] All units are $\mu g/L$.

110

Table 3-8. POTW Characteristics

Plant:	4	5	6	7	20	30	60
Biological treatment process	AS	AS	AS	AS	AS	AS	AS
Secondary design flow, MGD	120	25	7.4	66	50	24.0	5.0
Avg. daily flow, MGD (historical previous 12 months)	80–85	26	7	50	20–25	20.0	3.5
Avg. daily flow, MGD^a (actual)	83.7	21.7	7.1	48.6	22.8	19.7	4.0
% Combined/separate sewers	50/50	0/100	0/100	10/90	0/100	95/5	5/95
BOD, mg/L inf/eff	152/22	138/13	263/18	169/29	238/42	308/23	557/17
TSS, mg/L inf/eff	164/43	147/12	632/27	135/18	205/69	55/7	442/33
% Indust. contribution	7.2	8.0	12.3	16.9	30.0	31.9	12.7
Major indust. contributors	Food and kindred prod.; transport. equip.	Transport. equip.; paper and allied prod.	Food and kindred prod.; chemicals and allied prod.; rubber and plastic prod.; fabricated metal prod.	Fabricated metal prod.; chemicals and kindred prod. and allied prod.; leather prod.; primary metals ind.	Fabricated metal prod.	Chemicals and allied prod.; textile mill prod.	Leather prod.

Note: AS = activated sludge.

^{a}MGD = million gallons per day; actual recorded value through total treatment (secondary or tertiary) during sampling period.

Table 3-9. Summary of Analytical Data (Plant 4)

Parameter	Untreated Combined Sludge	Digested Combined Sludge
BOD (mg/L)	18,300	5,500
Total susp. solids (mg/L)	59,667	41,917
COD (mg/L)	59,783	34,083
Oil and grease (mg/L)	10,128	9,048
Ammonia nitrogen (mg/L)	207	523
Phenols, total	677	1,158
Solids, settleable (mg/L)	NR	NR
Solids, total (mg/L)	63,583	45,917
Solids, diss., total (mg/L)	1,790	659
Solids, sus., total vol. (mg/L)	29,933	15,067
Solids, volatile, total (mg/L)	34,683	21,533
Solids, diss., volatile (mg/L)	1,058	405
TOC (mg/L)	14,538	11,602
Benzene	40	20
Carbon tetrachloride	270	ND
Chlorobenzene	ND	252
Chloroform	1	ND
Dichlorobromomethane	84	ND
1,2-dichloroethane	ND	65
1,1-dichloroethylene	2,347	3
1,2-*trans*-dichloroethylene	54,993	9,800
1,2-dichloropropane	4	7
Ethylene	1,467	910
Methylene chloride	142	16
1,1,2-tetrachloroethane	441	ND
1,1,2,2-tetrachloroethane	43	194
Tetrachloroethylene	958	10
Toluene	984	1,847
1,1,1-trichloroethane	507	37
Trichloroethylene	467	120
Vinyl chloride	ND	27
2,4-dimethylphenol	ND	ND
2,4-dinitrophenol	ND	200
2-nitrophenol	ND	37
Pentachlorophenol	23	ND
Phenol	103	70
Acenaphthene	ND	65
Bis(2-ethylhexyl)phthalate	8,188	8,437
Butyl benzyl phthalate	2,650	4,400
Di-*n*-butyl phthalate	4,270	33
1,2-dichlorobenzene	282	116
1,3-dichlorobenzene	252	10

(continued)

Table 3-9. (continued)

Parameter[a]	Untreated Combined Sludge	Digested Combined Sludge
1,4-dichlorobenzene	1,128	425
Diethyl phthalate	ND	ND
Fluoranthene	114	51
Isophorone	ND	ND
Naphthalene	640	445
Acenaphthylene	ND	28
Anthracene	602	375
1,2-benzanthracene	16	59
3,4-benzofluoranthene	23	25
11,12-benzofluoranthene	23	25
Chrysene	16	59
Dimethyl phthalate	ND	ND
Fluorene	19	38
Ideno(1,2,3-c,d)pyrene	ND	ND
Phenanthrene	602	375
Pyrene	121	90
Alpha—BHC (mg/L)	1,000	1,000
Gamma-BHC (mg/L)	1,000	1,000
Dieldrin (mg/L)	1,000	1,000
Heptachlor (mg/L)	1,000	1,000
Antimony	116	116
Arsenic	403	333
Beryllium	25	30
Cadmium	518	660
Chromium	17,500	25,833
Copper	10,800	15,333
Cyanide	193	407
Lead	41,000	49,333
Mercury (mg/L)	360,000	463,333
Nickel	2,567	3,083
Selenium	20	23
Silver	1,967	2,378
Zinc	143,333	171,667
Aluminum	NR	NR
Barium	NR	NR
Calcium (mg/L)	NR	NR
Iron	NR	NR
Magnesium (mg/L)	NR	NR
Manganese	NR	NR
Sodium (mg/L)	NR	NR

Notes: Pollutants not listed were never detected.   ND = not detected; NR = not run.
[a] Units are µg/L except where indicated otherwise.

Table 3-10. Summary of Analytical Data (Plant 5)

Parameter[a]	Combined Sludge	Digested Sludge
BOD (mg/L)	15,671	3,182
Total susp. solids (mg/L)	26,433	17,438
COD (mg/L)	35,283	12,712
Oil and grease (mg/L)	4,188	2,746
Ammonia nitrogen (mg/L)	35	419
Phenols, total	99	258
Solids, settleable (mg/L)	992	860
Solids, total (mg/L)	29,864	18,407
Solids, diss., total (mg/L)	3,431	1,016
Solids, sus., total vol. (mg/L)	19,182	8,362
Solids, volatile, total (mg/L)	657	161
Solids, diss., volatile (mg/L)	2,426	359
TOC (mg/L)	7,527	1,275
Benzene	42	5
Chlorobenzene	ND	1
Chlorodibromomethane	3	1
2-chloroethyl vinyl ether	ND	0
Chloroform	2	6
Dichlorobromomethane	ND	ND
1,1-dichloroethylene	22	ND
1,2-trans-dichloroethylene	1,541	22
1,2-dichloropropane	8	ND
1,3-dichloropropylene	ND	ND
Ethylbenzene	166	73
Methyl chloride	45	9
Methylene chloride	101	11
1,1,2,2-tetrachloroethane	40	45
Tetrachloroethylene	14	ND
1,1,1-trichloroethane	ND	2
Trichloroethylene	163	2
Toluene	199	124
Vinyl chloride	3,292	117
2,4-dichlorophenol	ND	4
2,4-dimethylphenol	ND	ND
Pentachlorophenol	ND	5
Phenol	27	48

(continued)

Table 3-10. *(continued)*

Parameter[a]	Combined Sludge	Digested Sludge
Acenaphthene	20	13
Anthracene	92	102
Bis(2-ethylhexyl)phthalate	3,598	4,170
Butyl benzyl phthalate	450	152
Di-*n*-butyl phthalate	ND	ND
1,2-dichlorobenzene	3	ND
1,4-dichlorobenzene	15	6
Diethyl phthalate	ND	ND
Di-*n*-octyl phthalate	47	48
Fluoranthene	ND	10
Naphthalene	118	32
Phenanthrene	92	102
Pyrene	10	9
Beta-BHC (mg/L)	1,000	1,000
Gamma-BHC (mg/L)	1,000	1,000
Antimony	55	52
Arsenic	78	69
Beryllium	6	6
Cadmium	385	312
Chromium	8,317	7,767
Copper	3,000	2,767
Cyanide	221	28
Lead	8,967	9,167
Nickel	1,077	1,240
Selenium	53	41
Silver	1,103	800
Zinc	23,833	24,167
Aluminum	NR	NR
Barium	NR	NR
Calcium (mg/L)	NR	NR
Iron	NR	NR
Magnesium (mg/L)	NR	NR
Manganese	NR	NR
Sodium (mg/L)	NR	NR

Notes: Pollutants not listed were never detected.   ND = not detected; NR = not run.

[a] Units are μg/L except where indicated otherwise.

Table 3-11. Summary of Analytical Data (Plant 6)

Parameter[a]	Primary Sludge	Digested Sludge
BOD (mg/L)	18,857	16,856
Total susp. solids (mg/L)	51,342	47,758
COD (mg/L)	55,386	51,787
Oil and grease (mg/L)	4,988	4,948
Ammonia nitrogen (mg/L)	82	585
Phenols, total	4,672	7,420
Solids, settleable (mg/L)	1,825	1,775
Solids, total (mg/L)	53,717	49,232
Solids, diss., total (mg/L)	4,618	1,140
Solids, sus., total vol. (mg/L)	41,017	37,250
Solids, volatile, total (mg/L)	41,398	37,662
Solids, diss., volatile (mg/L)	382	534
TOC (mg/L)	24,008	25,625
Benzene	12	11
Carbon tetrachloride	10	ND
Chlorobenzene	0	0
Chloroethane	167	1,800
Chloroform	ND	ND
Dichlorobromomethane	ND	ND
1,1-dichloroethylene	212	ND
1,2-dichloroethylene	ND	ND
1,1-dichloroethylene	ND	ND
1,2-trans-dichloroethylene	878	116
Ethylbenzene	317	392
Methyl chloride	ND	ND
Methylene chloride	64	28
1,1,2,2-tetrachloroethane	ND	1
Tetrachloroethylene	58	52
Toluene	475	423
1,1,1-trichloroethane	33	ND
1,1,2-trichloroethane	ND	ND
Trichloroethylene	32	4
Trichlorofluoromethane	ND	ND
Vinyl chloride	34,000	33,800
2,4-dichlorophenol	ND	ND
Parachlorometa cresol	ND	ND
Pentachlorophenol	ND	ND
Phenol	882	1,833

(continued)

Table 3-11. (continued)

Parameter[a]	Primary Sludge	Digested Sludge
Acenaphthene	ND	ND
Anthracene	1,505	994
Bis(2-ethylhexyl)phthalate	5,200	4,086
1,2-benzanthracene	565	62
1,12-benzoperylene	ND	ND
Butyl benzyl phthalate	1,500	ND
Chrysene	565	62
Di-n-butyl phthalate	543	ND
1,3-dichlorobenzene	ND	ND
1,4-dichlorobenzene	ND	ND
Diethyl phthalate	ND	ND
2,4-dinitrotoluene	ND	ND
Fluorene	ND	ND
Naphthalene	ND	ND
Phenanthrene	1,505	994
Antimony	393	350
Arsenic	310	212
Beryllium	11	9
Cadmium	82,500	79,833
Chromium	74,333	73,833
Copper	51,333	46,000
Cyanide	22,123	17,597
Lead	16,167	9,667
Mercury (mg/L)	601,667	486,667
Nickel	17,000	17,500
Selenium	162	152
Silver	823	802
Zinc	386,667	365,000
Aluminum	NR	NR
Barium (mg/L)	NR	NR
Calcium	NR	NR
Iron (mg/L)	NR	NR
Magnesium	NR	NR
Manganese (mg/L)	NR	NR
Sodium	NR	NR

Notes: Pollutants not listed were never detected.   ND = not detected; NR = not run.
[a] Units are μg/L except where indicated otherwise.

Table 3-12. Summary of Analytical Data (Plant 7)

Parameter[a]	Combined Sludge	Heat Trt. Sludge	Heat Trt. Decant
BOD (mg/L)	27,484	29,013	13,964
Total susp. solids (mg/L)	35,057	26,313	2,877
COD (mg/L)	52,081	57,922	20,110
Oil and grease (mg/L)	7,983	9,608	533
Ammonia nitrogen (mg/L)	257	516	433
Phenols, total	1,133	3,512	4,122
Solids, settleable (mg/L)	670	343	33
Solids, total (mg/L)	41,248	34,823	14,116
Solids, diss., total (mg/L)	4,262	9,578	11,239
Solids, sus., total vol. (mg/L)	22,103	15,217	1,615
Solids, volatile, total (mg/L)	27,934	23,131	11,496
Solids, diss., volatile (mg/L)	2,243	7,581	9,881
TOC (mg/L)	9,050	8,433	8,767
Benzene	95	507	22
Chlorobenzene	6	1	0
Chloroform	7	ND	5
Dichlorobromomethane	32	ND	5
1,1-dichloroethylene	352	ND	5
1,2-trans-dichloroethylene	1,517	283	9
1,2-dichloropropane	ND	8	5
1,3-dichloropropylene	ND	55	5
Ethylbenzene	2,100	460	18
Methyl chloride	8	1	38
1,1,2,2-tetrachloroethane	26	ND	5
Tetrachloroethylene	1	15	5
Toluene	4,615	2,342	63
1,1,1-trichloroethane	ND	ND	5
Trichloroethylene	2	7	5
Pentachlorophenol	1,000	ND	250
Phenol	173	1,717	907
Anthracene	827	407	100
1,2-benzanthracene	153	25	50
Bis(2-ethylhexyl)phthlate	11,257	10,117	1,498
Butyl benzyl phthalate	1,162	735	100
Chrysene	153	25	50
Di-n-butyl phthalate	318	265	100
1,2-dichlorobenzene	233	50	17
1,3-dichlorobenzene	35	ND	100

(continued)

Table 3-12. *(continued)*

Parameter[a]	Combined Sludge	Heat Trt. Sludge	Heat Trt. Decant
1,4-dichlorobenzene	28	10	100
Diethyl phthalate	ND	ND	100
Fluoranthene	143	13	50
Naphthalene	180	16	2
Phenanthrene	827	407	100
Pyrene	160	14	50
Delta—BHC (mg/L)	1,000	1,000	1,000
Gamma—BHC (mg/L)	1,000	1,000	167
Heptachlor (mg/L)	1,000	1,000	333
Heptachlor epoxide (mg/L)	1,000	1,000	83
Antimony	1,403	1,047	54
Arsenic	332	207	56
Beryllium	10	10	1
Cadmium	498	313	139
Chromium	72,667	56,000	9,569
Copper	45,833	35,333	5,701
Cyanide	2,503	278	49
Lead	44,167	6,133	1,701
Mercury	205,000	140,000	1,000
Nickel	27,333	20,667	9,888
Selenium	153	93	8
Silver	177	185	27
Thallium	10	10	52
Zinc	128,333	98,833	32,602
Aluminum	NR	NR	15,523
Barium	NR	NR	935
Boron	NR	NR	1,055
Calcium (mg/L)	NR	NR	308
Cobalt	NR	NR	190
Iron	NR	NR	46,081
Magnesium (mg/L)	NR	NR	70
Manganese	NR	NR	1,104
Molybdenum	NR	NR	154
Sodium (mg/L)	NR	NR	164
Tin	NR	NR	391
Titanium	NR	NR	171
Vanadium	NR	NR	662
Yttrium	NR	NR	18

Notes: Pollutants not listed were never detected.   ND = not detected; NR = not run.
[a] Units are μg/L except where indicated otherwise.

Table 3-13. Summary of Analytical Data (Plant 8)

Parameter[a]	Combined Sludge	Heat Trt. Sludge	Heat Trt. Decant
BOD (mg/L)	27,200	55,833	9,378
Total susp. solids (mg/L)	76,755	189,100	273
COD (mg/L)	118,333	290,667	16,333
Oil and grease (mg/L)	24,550	62,375	38
Ammonia nitrogen (mg/L)	653	709	900
Phenols, total	2,413	7,150	3,200
Solids, settleable (mg/L)	810	940	1
Solids, total (mg/L)	84,439	194,996	10,590
Solids, diss., total (mg/L)	3,306	7,550	10,317
Solids, sus., total vol. (mg/L)	44,733	102,933	172
Solids, volatile, total (mg/L)	53,549	109,432	8,942
Solids, diss., volatile (mg/L)	2,343	5,733	8,774
TOC (mg/L)	13,833	17,333	6,767
Acrylonitrile	25	165	19
Benzene	16	773	22
Chlorobenzene	5	33	1
Chloroethane	2,000	5	15
Chloroform	5	5	1
Dichlorodifluoromethane	494	1,893	30
1,1-dichloroethane	1	5	1
1,1-dichloroethylene	5	5	1
1,2-trans-dichloroethylene	1,259	805	2
1,2-dichloropropane	2	3	1
Ethylbenzene	359	1,266	7
Methyl bromide	5	5	6
Methyl chloride	5	125	34
Methylene chloride	5	5	19
Tetrachloroethylene	3	240	2
1,1,2,2-tetrachloroethylene	5	13	1
Toluene	7,635	41,575	1,825
1,1,1-trichloroethane	5	13	2
1,1,2-trichloroethane	3	5	2
Trichloroethylene	5	185	2
Vinyl chloride	250	5	33
2,4-dichlorophenol	ND	ND	2
Pentachlorophenol	823	1,300	60
Phenol	610	1,238	334
Acenaphthene	1,150	ND	10
Anthracene	1,565	690	3
1,2-benzanthracene	750	119	3

(continued)

Table 3-13. *(continued)*

Parameter[a]	Combined Sludge	Heat Trt. Sludge	Heat Trt. Decant
Benzo(a)pyrene	ND	11	25
Bis(2-ethylhexyl)phthlate	10,500	20,200	6
Butyl benzyl phthalate	17,775	33,425	3
2-chloronaphthalene	400	ND	3
Chrysene	750	119	3
Di-*n*-butyl phthalate	1,045	1,688	3
1,2-dichlorobenzene	258	ND	2
1,3-dichlorobenzene	475	ND	2
1,4-dichlorobenzene	325	ND	2
Diethyl phthalate	ND	ND	4
Dimethyl phthalate	160	ND	6
Fluoranthene	600	577	3
Hexachlorobenzene	195	ND	9
Hexachlorobutadiene	675	ND	8
Naphthalene	1,159	686	2
Phenanthrene	1,565	690	3
Pyrene	734	768	3
Antimony	1,015	1,825	500
Arsenic	695	1,463	500
Beryllium	22	44	2
Cadmium	450	780	2
Chromium	101,250	150,000	1,155
Copper	120,500	202,500	212
Cyanide	3,040	1,526	44
Lead	98,750	95,250	192
Mercury (mg/L)	172,500	50,500	225
Nickel	60,000	92,000	6,945
Selenium	170	328	500
Silver	160	160	2
Zinc	715,000	1,335,000	7,495
Aluminum	ND	ND	1,323
Barium	ND	ND	373
Calcium (mg/L)	ND	ND	146
Iron	ND	ND	19,300
Magnesium (mg/L)	ND	ND	33
Manganese	ND	ND	843
Sodium (mg/L)	ND	ND	97

Notes: Pollutants not listed were never detected. ND = not detected; NR = not run.
[a] Units are μg/L except where indicated otherwise.

Table 3-14. Summary of Analytical Data (Plant 20)

Parameter[a]	Primary Sludge	Secondary Sludge	Nitrification Sludge
BOD (mg/L)	15,833	7,543	8,350
Total susp. solids (mg/L)	50,500	NR	NR
COD (mg/L)	41,067	9,813	9,960
Oil and grease (mg/L)	5,337	1,823	270
Ammonia nitrogen (mg/L)	70	27	8
Phenols, total	720	418	393
Solids, settleable (mg/L)	998	983	870
Solids, total (mg/L)	58,947	8,827	8,499
Solids, diss., total (mg/L)	NR	NR	NR
Solids, sus., total vol. (mg/L)	30,000	6,333	NR
Solids, volatile, total (mg/L)	35,750	6,011	6,089
Solids, diss., volatile (mg/L)	NR	NR	NR
TOC (mg/L)	1,304	583	335
Benzene	11	ND	ND
Bromoform	ND	ND	ND
Carbon tetrachloride	ND	ND	ND
Chlorodibromomethane	ND	ND	ND
Chloroform	ND	ND	ND
Dichlorobromomethane	ND	ND	ND
Dichlorodifluoromethane	ND	ND	ND
1,1-dichloroethane	ND	ND	ND
1,2-trans-dichloroethylene	507	ND	ND
Ethylbenzene	61	ND	ND
Methyl bromide	ND	ND	ND
Methylene chloride	ND	ND	ND
Methyl chloride	ND	ND	ND
Tetrachloroethylene	ND	ND	ND
Toluene	197	53	125
1,1,1-trichloroethane	ND	ND	ND
Trichloroethylene	309	ND	ND
Vinyl chloride	ND	ND	ND
2,4-dichlorophenol	ND	ND	ND
2,4-dimethylphenol	ND	ND	ND
Phenol	53	60	23
2,4,6-trichlorophenol	ND	ND	ND
Acenaphthylene	ND	ND	305
Anthracene	37	3	700
1,2-benzanthracene	12	8	54
1,2-benzanthracene	ND	ND	70

(continued)

Table 3-14. *(continued)*

Parameter[a]	Primary Sludge	Secondary Sludge	Nitrification Sludge
11,12-benzofluoranthene	ND	ND	20
1,12-benzoperylene	ND	ND	65
Bis(2-ethylhexyl)phthalate	2,357	671	425
Butyl benzyl phthalate	827	ND	ND
Chrysene	ND	ND	70
1,2-dichlorobenzene	88	ND	ND
Diethyl phthalate	200	ND	ND
Fluoranthene	10	ND	365
Fluorene	ND	ND	250
Isophorone	ND	ND	ND
Naphthalene	ND	ND	170
Phenanthrene	37	3	700
Pyrene	10	ND	310
1,2,4-trichlorobenzene	308	7	ND
Alpha—BHC (mg/L)	ND	ND	ND
Gamma—BHC (mg/L)	ND	ND	ND
4,4′—DDD (mg/L)	ND	ND	ND
Heptachlor epoxide (mg/L)	ND	ND	ND
Antimony	186	10	50
Arsenic	258	78	115
Cadmium	86	46	16
Chromium	3,433	998	1,010
Copper	14,433	3,500	2,450
Cyanide	38,050	2,917	1,650
Lead	6,467	24,167	19,500
Mercury (mg/L)	127,852	24,167	19,500
Nickel	3,200	832	775
Selenium	134	100	100
Silver	773	247	185
Zinc	27,617	8,750	5,950
Aluminum	NR	NR	NR
Barium	NR	NR	NR
Calcium (mg/L)	NR	NR	NR
Iron	NR	NR	NR
Magnesium (mg/L)	NR	NR	NR
Manganese	NR	NR	NR
Sodium	NR	NR	NR

Notes: Pollutants not listed were never detected. ND = not detected; NR = not run.
[a] Units are μg/L except where indicated otherwise.

Table 3-15. Summary of Analytical Data (Plant 30)

Parameter[a]	Primary Sludge	Secondary Sludge	Combined Sludge (Ht. Trt.)
BOD (mg/L)	359	4,832	15,601
Total susp. solids (mg/L)	760	8,977	23,468
COD (mg/L)	778	11,200	23,500
Oil and grease (mg/L)	31	760	1,533
Ammonia nitrogen (mg/L)	15	17	1,277
Phenols, total	656	234	430
Solids, settleable (mg/L)	171	915	307
Solids, sus., total vol (mg/L)	232	6,217	3,677
Solids, total (mg/L)	1,211	11,233	30,325
Solids, volatile, total (mg/L)	388	7,532	7,798
TOC (mg/L)	132	1,855	7,183
Benzene	1	ND	24
Bromoform	ND	ND	ND
Chlorobenzene	ND	17	15
Chlorodibromomethane	ND	ND	ND
Chloroethane	ND	ND	ND
2-chloroethyl vinyl ether	ND	ND	ND
Chloroform	ND	ND	ND
Dichlorobromomethane	ND	ND	ND
Dichlorodifluoromethane	ND	ND	ND
1,2-dichloroethane	96	1,577	53
1,2-trans-dichloroethylene	1	ND	ND
Ethylbenzene	15	3	5
Methyl bromide	ND	ND	ND
Methyl chloride	ND	ND	ND
Methylene chloride	1	7	2
Tetrachloroethylene	1	ND	ND
Toluene	601	391	41
1,12-trichloroethane	ND	ND	ND
Trichloroethylene	1	ND	ND
Vinyl chloride	ND	ND	ND
Pentachlorophenol	ND	ND	ND
Phenol	2	112	7,643
Acenaphthylene	ND	ND	843
Anthracene	0	1	20
3,4-benzofluoranthene	ND	1	ND
11,12-benzofluoranthene	ND	1	ND
Bis(2-ethylhexyl)phthalate	199	432	528

(continued)

Table 3-15. *(continued)*

Parameter[a]	Primary Sludge	Secondary Sludge	Combined Sludge (Ht. Trt.)
Butyl benzyl phthalate	ND	ND	75
Chrysene	ND	1	ND
Di-*n*-butyl phthalate	ND	ND	ND
1,2-dichlorobenzene	ND	ND	26
1,4-dichlorobenzene	ND	ND	30
Di-*n*-octyl phthalate	ND	ND	0
1,2-diphenylhydrazine	ND	ND	ND
Fluoranthene	3	1	5
Naphthalene	ND	1	32
Phenanthrene	0	1	20
Pyrene	5	3	1
Aldrin (mg/L)	ND	ND	ND
Alpha—BHC (mg/L)	ND	ND	ND
Delta—BHC (mg/L)	ND	ND	ND
Gamma—BHC (mg/L)	ND	ND	ND
Heptachlor (mg/L)	ND	ND	ND
Antimony	2	17	21
Arsenic	10	99	823
Beryllium	24	2	11
Cadmium	10	26	61
Chromium	133	5,800	11,900
Copper	290	3,800	12,133
Cyanide	921	2,837	15,433
Lead	262	1,557	8,833
Mercury (mg/L)	8,000	15,333	22,667
Nickel	50	307	890
Selenium	2	1	9
Silver	57	110	283
Zinc	2,700	69,667	94,667
Aluminum	NR	NR	NR
Barium	NR	NR	NR
Calcium (mg/L)	NR	NR	NR
Iron	NR	NR	NR
Magnesium (mg/L)	NR	NR	NR
Manganese	NR	NR	NR
Sodium (mg/L)	NR	NR	NR

Notes: Pollutants not listed were never detected.   ND = not detected; NR = not run.
[a] Units are μg/L except where indicated otherwise.

Table 3-16. Summary of Analytical Data (Plant 60)

Parameter[a]	Thickened Combined Sludge	Heat Treated Combined Sludge
BOD (mg/L)	10,375	39,500
Total susp. solids (mg/L)	43,000	186,600
COD (mg/L)	28,000	30,000
Oil and grease (mg/L)	1,325	14,050
Ammonia nitrogen (mg/L)	122	95
Phenols, total	5,175	3,475
Solids, sus., total vol (mg/L)	28,000	103,750
TOC (mg/L)	10,700	13,544
Benzene	2	38
Chlorobenzene	10	7
Chloroethane	7	100
Chloroform	0	4
Dichlorodifluoromethane	100	4
1,1-dichlorethane	403	2
1,2-dichloroethane	22	25
1,1-dichlorethylene	25	22
1,2-trans-dichloroethylene	138	8
Ethylbenzene	51	103
Methyl chloride	227	111
Tetrachloroethylene	25	127
Toluene	2,402,537	214
1,1,1-trichlorethane	12	25
Trichloroethylene	25	21
Trichlorofluoromethane	25	24
2,4-dimethylphenol	50	50
2,4-dinitrophenol	600	600
4-nitrophenol	300	300
Parachlorometa cresol	100	100
Pentachlorophenol	200	300
Phenol	2,000	1,825
Acenaphthene	20	123
Bis(2-ethylhexyl)phthlate	2,865	2,410,380
Butyl benzyl phthalate	45	20
Di-n-butyl phthalate	32	153
1,2-dichlorobenzene	378	1,725
1,3-dichlorobenzene	152	45
1,4-dichlorobenzene	650	454
Fluoranthene	178	221
Naphthalene	1,168	1,870
Nitrobenzene	100	100

(continued)

Table 3-16. *(continued)*

Parameter[a]	Thickened Combined Sludge	Heat Treated Combined Sludge
Acenaphthylene	13	10
Anthracene	10	10
Dimethyl phthalate	20	20
Phenanthrene	432	1,007
Pyrene	140	174
Aldrin (mg/L)	ND	ND
Alpha—BHC (mg/L)	ND	ND
Beta—BHC (mg/L)	ND	ND
Delta—BHC (mg/L)	ND	ND
Gamma—BHC (mg/L)	ND	ND
4,4′—DDE	ND	ND
Alpha-endosulfan (mg/L)	ND	ND
Beta-endosulfan (mg/L)	ND	ND
Heptachlor epoxide (mg/L)	ND	ND
Antimony	51	NR
Arsenic	52	NR
Beryllium	15	NR
Cadmium	241	13
Chromium	66,667	44,700
Copper	6,167	144
Cyanide	3,125	8
Lead	18,875	237
Mercury (mg/L)	284,750	1,833
Nickel	2,667	7
Selenium	24	NR
Silver	207	NR
Thallium	34	NR
Zinc	67,333	157
Aluminum	NR	1,177
Barium	NR	NR
Boron	NR	NR
Calcium (mg/L)	NR	364,333
Iron	NR	3,642
Magnesium (mg/L)	NR	25,610
Manganese	NR	191
Sodium (mg/L)	NR	1,295,000
Tin	NR	NR
Titanium	NR	500
Vanadium	NR	NR

Notes: Pollutants not listed were never detected.   ND = not detected; NR = not run.
[a] Units are μg/L except where indicated otherwise.

Table 3-17. Chattanooga 6-Day and 30-Day Studies Combined Data

Parameter[a]	Primary Sludge	Vacuum Filter Filtrate	Digester Supernatant
BOD (mg/L)	10,906	72	1,673
Total susp. solids (mg/L)	25,110	409	1,535
COD (mg/L)	40,361	323	2,718
Oil and grease (mg/L)	3,678	14	153
Ammonia nitrogen (mg/L)	26	71	87
Phenols, total	1,242	224	362
Solids, settleable (mg/L)	973	1	200
Solids, total (mg/L)	28,097	1,768	3,182
Solids, diss., total (mg/L)	1,783	1,310	5,226
Solids, sus., total vol (mg/L)	14,697	98	937
Solids, volatile, total (mg/L)	14,003	301	1,326
Solids, diss., volatile (mg/L)	759	159	378
TOC (mg/L)	3,372	139	438
Benzene	16	2	10
Carbon tetrachloride	ND	NR	NR
Chlorobenzene	6	ND	35
Chlorodibromomethane	ND	ND	ND
Chloroethane	ND	ND	0
Chloroform	62	ND	0
Dichlorobromomethane	0	ND	ND
1,1-dichloroethane	17	ND	2
1,2-dichloroethane	4	NR	NR
1,1-dichloroethylene	ND	NR	NR
1,2-*trans*-dichloroethylene	13	ND	16
Dichlorofluoromethane	1	NR	NR
1,2-dichloropropane	ND	NR	NR
1,3-dichloropropylene	ND	NR	NR
Ethylbenzene	120	1	231
Methyl chloride	ND	NR	NR
Methylene chloride	29	10	13
1,1,2,2-tetrachloroethane	0	NR	NR
Tetrachloroethylene	103	NR	NR
Toluene	779	7	140
1,1,1-trichloroethane	80	0	0
1,1,2-trichloroethane	53	NR	NR
Trichloroethylene	329	NR	NR
Trichlorofluoromethane	1	0	0
Vinyl chloride	0	ND	2
Parachlorometa cresol	ND	ND	1
2,4,6-trichlorophenol	ND	1	ND
Acenaphthylene	4	5	ND

(continued)

Table 3-17 *(continued)*

Parameter[a]	Primary Sludge	Vacuum Filter Filtrate	Digester Supernatant
Anthracene	97	2	5
1,2-benzanthracene	41	ND	1
Benzo(a)pyrene	26	NR	NR
11,12-benzofluoranthene	26	NR	NR
Bis(2-ethylhexyl)phthalate	1,529	13	57
Butyl benzyl phthalate	100	ND	5
2-chlorophenol	ND	1	ND
Chrysene	47	0	0
Di-*n*-butyl phthalate	73	1	7
2,4-dichlorophenol	ND	1	1
Diethyl phthalate	31	10	10
Dimethyl phthalate	ND	NR	NR
Di-*n*-octyl phthalate	4	3	12
Fluorene	46	1	4
Ideno(1,2,3-*c,d*)pyrene	4	NR	NR
Isophrone	ND	NR	NR
Naphthalene	606	14	63
1*n*-nitrosodiphenylamine	ND	ND	3
n-nitrosodi-*n*-propylamine	ND	NR	NR
Phenanthrene	189	2	9
Acenapthene	13	ND	1
2,4-dimethylphenol	ND	3	ND
2,4-dinitrophenol	ND	NR	NR
2-nitrophenol	ND	NR	NR
4-nitrophenol	ND	NR	NR
Pentachlorophenol	1	2	ND
Phenol	211	52	38
Pyrene	127	1	2
Benzidine	ND	NR	NR
Gamma—BHC	ND	NR	NR
Bis(2-chloroethyl)ether	ND	NR	NR
Bis(2-chloroethyloxy)methane	ND	1	1
Bis(2-chloroisopropyl)ether	ND	NR	NR
2-chloronaphthalene	ND	NR	NR
4-chlorophenyl phenyl ether	3	NR	NR
1,2-dichlorobenzene	53	0	7
1,3-dichlorobenzene	22	1	21
1,4-dichlorobenzene	66	0	17
2,6-dinitritoluene	2	NR	NR
1,2-diphenylhydrazine	ND	2	1
Alpha-endosulfan	ND	NR	NR
Fluoranthene	194	1	4
Alpha—HDC	2	NR	NR

(continued)

Table 3-17. (continued)

Parameter[a]	Primary Sludge	Vacuum Filter Filtrate	Digester Supernatant
Hexachlorobenzene	ND	NR	NR
Hexachloroethane	1	NR	NR
1,2,4-trichlorobenzene	1,136	2	81
Antimony	92	1	5
Arsenic	666	6	65
Beryllium	71	1	3
Cadmium	175	2	22
Chromium	13,225	120	843
Copper	7,453	88	727
Cyanide	8,134	460	33
Lead	5,821	75	568
Mercury	34,501	533	6,217
Nickel	4,076	57	350
Selenium	7	1	4
Silver	625	5	76

Notes: Pollutants not listed were never detected.   ND = not detected; NR = not run.
[a] Units are μg/L except where indicated otherwise.

The U.S. EPA also initiated an additional study to document the priority toxic pollutant variability of the influent waste stream to a single POTW for a period of 30 consecutive days to determine the occurrence of these pollutants. Results of this 30-day study were then compared to a similar EPA study on the same POTW but for a period of only six consecutive sampling days. The combined studies provided a 36-day database for the POTW, and are shown in Table 3-17.

TREATMENT OF INDUSTRIAL HAZARDOUS-WASTE SLUDGES AND RESIDUES

Previous sections have dealt primarily with treatment processes and systems for treatment of aqueous and dilute hazardous-waste streams. Most waste-treatment systems, and particularly those whose principal function is the removal of pollutant concentrations, also produce residues or sludges that may also be hazardous and require further processing. Many industrial production processes also generate hazardous residues, including distillation bottoms, filter cake, and various sludges and slurries of process by-products. This section discusses several of the more predominant techniques for treating these residual materials.

The principal means of control for hazardous solid residues in the United States today is land disposal. Residues that are primarily organic in composition are most commonly destroyed in high-temperature incinerators. These techniques are discussed more thoroughly in subsequent chapters.

Three technologies for industrial hazardous-waste sludge and residue treatment are covered: wet oxidation, land treatment, and chemical fixation/stabilization/encapsulation. Emphasis is placed upon these processes because they are generally applicable to a reasonably wide range of waste streams and hazardous materials. Many other waste-specific processes are used or have been studied for specific residues, such as ammonium-carbonate treatment for metal removal from metal-finishing sludges, and solvent-extraction calcination and cementation for inorganic residues. Discussing the many variants of these treatment systems is beyond the scope of this chapter.

In addition to material pertaining to industrial hazardous-waste sludge and residue treatment, the chapter includes a few examples of various manufacturing processes and the methods by which the subsequent waste streams were treated.

The listing of such waste as hazardous follows 40 CFR 261.3(c)(2), the Definition of Hazardous Waste, which states: "Any solid waste generated from the treatment, storage, or disposal of a hazardous waste, including any sludge, spill residue, ash emission control dust or leachate (but not including precipitation run-off) is a hazardous waste." However, the EPA allows for exemptions from the hazardous classification if it can be demonstrated that the waste does not meet the criteria under which it was listed as hazardous. A review process was undertaken to evaluate the "delisting petitions" submitted to the EPA over the last two years to determine the type of waste-treatment processes employed. Of the approximately 50 delisting petitions reviewed, several are presented here. The petitions selected contain the most complete data on the concentrations of priority pollutants in waste sludge streams before and after some method of treatment. Also included in this section are descriptions of the treatment processes used by different manufacturing processes as well as the manufacturing processes themselves.

Wet Oxidation

Wet air oxidation (WAO) is a commercially proven technology to destroy organics in wastewater and sludges. In conventional WAO, waste is pumped into the system by a high-pressure pump and mixed with air from an air compressor. The waste is passed through a heat exchanger and then into a reactor where atmospheric oxygen reacts with the organic matter in the waste, sometimes in the presence of catalysts. The oxidation is accomplished by a temperature rise, the gas and liquid phases are separated, and the liquid is circulated through the heat exchanger before discharge. Gas and liquid are both exhausted through control valves. System pressure is controlled to maintain the reaction temperature as changes occur in feed characteristics (i.e., organic content, heat value, temperature). The mass of water in the system serves as a heat sink to prevent a runaway reaction that might be caused by a high influx of concentrated organics.

In the process, sulfur is converted to SO_4, nitrogen to NH_3, chlorine to HCl, and organics to water and CO_2. Although the process has been primarily employed for treatment of municipal and nonhazardous industrial wastewaters and sludges, data are emerging on hazardous-waste-control applications.

Design Criteria.

organic feed concentration: 5.0–30 g/L

temperature: 150–350 °C

pressure: 2000–4000 psi

residence time: 15–90 min

catalysts: required to improve performance for halogenated organics (Cu, HBr/HNO$_3$, and various proprietary co-catalyst systems)

flow rate: 10–350 g/min

Performance Data. Table 3-18 summarizes the results of a variety of reported laboratory– and pilot-scale wet oxidation studies. Note that the data in this table are reported as percentage destruction of the original compound and not as percentage total organic destroyed. Care must therefore be taken in interpreting these data because many compounds exhibiting high-destruction efficiencies may, in fact, be primarily converted to other organic compounds that may be equally hazardous. Total organic destruction efficiencies are in excess of 99%. Consequently, detailed effluent analysis is necessary. In some cases wet oxidation effluents may have to be directed to subsequent adsorption or biological treatment processes to effect complete control of hazardous materials. Table 3-18 lists reported destructive efficiencies for wet air oxidation.

In general, studies to date indicate that

1. Aliphatic compounds with multiple halogen atoms can be destroyed within conventional wet oxidation conditions.
2. Aromatic hydrocarbons are easily oxidized (e.g., toluene, acenapthene, or pyrene).
3. Halogenated aromatic compounds can be oxidized provided there is at least one nonhalogen functional group present on the aromatic ring.
4. Halogenated aromatic compounds such as 1,2 dichlorobenzene or Aroclor 1254, a PCB, are resistant under conventional conditions.

**Table 3-18. Destruction Efficiencies for Hazardous Compounds
by Wet Air Oxidation**

Compound	Temp. (°C)	Catalyst[a]	Percent Destroyed	Reaction time (min)
Acenaphthene	320	0	99.96	60
	275	0	99.99	60
Acetic acid	165	+	36.0	90
Acetonitrile	200	+	97.0	60
Acrolein	320	0	99.96	60
	275	0	99.05	60
Acrylonitrile	320	0	99.91	60
	275	0	99.00	60
	275	+	99.50	60
Aroclor 1016	195	+	31.0	60
Aroclor 1254 (PCB)	320	0	63.0	60
	320	+	2.0	60
	250	+	95.0	120
Butylphthalate	165	+	99.999	60
Carbon tetrachloride	275	0	99.99	60
	275	+	99.99	60
Chloroform	275	0	99.92	60
	275	+	99.94	60
2-chlorophenol	320	0	99.86	60
	275	0	94.96	60
	275	+	99.88	60
	232	+	97.5	60
	232	0	99.2	60
	204	0	94.7	60
2,4-dichloroaniline	275	+	99.8	60
1,2-dichlorobenzene	275	0	2.98	60
	275	+	32.2	60
	320	+	69.11	60
1,2-dichloroethane	275	0	99.8	60
	275	+	99.9	60
2,4-dichlorophenol	320	0	99.88	60
	275	0	99.74	60
	165	+	65.0	15
2,4-dimethylphenol	370	0	99.99	60
	275	0	99.99	60
1,2-dimethylphthalate	204	0	98.0	30
	232	0	99.0	30
	260	+	99.0	30
Dioxin	200	+	99.0	240
1,2-diphenylhydrazine	320	0	99.98	60
	275	0	99.88	60

(continued)

Table 3-18. *(continued)*

Compound	Temp. (°C)	Catalyst[a]	Percent Destroyed	Reaction time (min)
Ethylenedibromide	165	+	94.0	60
Formic acid	275	0	99.3	60
Glycolic acid	165	+	99.0	30
Hexachlorobutadiene	250	+	99.997	60
Hexachlorocyclopentadiene	250	0	90.0	60
Hexachlorocyclopentadiene	250	0	90.0	60
	275	0	98.2	60
	300	0	99.85	60
Kepone	280	+	93.0	60
	250	+	93.0	360
Malathion	200	0	99.87	60
	250	0	99.85	60
	300	0	99.97	60
	165	+	99.0	60
4-nitrophenol	320	0	99.96	60
	275	0	99.60	60
	204	+	89.5	60
	260	+	95.0	60
N-nitrosodimethylamine	275	0	99.56	60
	275	+	99.38	60
Orthoxylene	200	+	99.83	60
Pentachlorophenol	320	0	99.88	60
	275	0	81.96	60
	275	+	97.30	60
	165	+	99.0	30
Phenol	320	0	99.97	60
	275	0	99.77	60
	275	+	99.93	60
	232	+	97.2	60
Potassium thiocyanate	275	0	99.98	60
Pyrene	275	0	99.995	60
	275	+	99.997	60
Sodium cyanide	275	0	99.96	60
Toluene	275	0	99.73	60
	275	+	99.96	60
2,4,6-trichloroaninine	280	+	99.97	60

Sources: Miller, et al., 1980 (May), Destruction of toxic chemicals by catalyzed wet oxidation, *Proceedings of the 35th Industrial Waste Conference,* Purdue University, W. Lafayette, Ind.

Bailrod, et al., 1979 (May), Wet oxidation of organic substances in *Proceedings of the 34th Industrial Waste Conference,* Purdue University, W. Lafayette, Ind.

Wet Oxidation and Ozonation of Specific Organics Pollutants 1982 (April), EPA Draft Report, Michigan Technological University.

Randal and Knopp, 1980, Detoxification of specific organic substances by wet oxidation. *J. Water Pollution Control Federation.*

Catalyzed Wet Oxidation of Hazardous Waste, 1981, EPA Draft Report, I. T. Enviroscience.

[a] 0 indicates no catalyst used; + indicates catalyst used.

5. Halogenated condensed ring compounds, such as the pesticides, aldrin, deldrin, or endrin, are expected to resist conventional wet oxidation.
6. DDT, although destroyed, results in intractable oil formation on conventional wet oxidation.
7. Heterocyclic compounds containing O, N, or S atoms provide a point of attack for oxidation and are *expected* to be destroyed by wet oxidation.

Stabilization/Fixation/Encapsulation

Stabilization, fixation, and encapsulation processes may be used to treat industrial hazardous-waste sludges and residues. These processes represent an alternative to ponding, lagooning, ocean disposal, mine disposal, or conventional landfilling. Many of the approaches used in solidifying or stabilizing hazardous industrial wastes originated in the area of radioactive waste management. The objective of these treatments is the production of a material of better physical handling leachability, and landfilling characteristics than the wastes from which it is derived.

Several stabilization/solidification/encapsulation methods now available or under development have as their goal the safe ultimate disposal of hazardous waste. Ultimate disposal implies the final disposition of persistent, nondegradable, cumulative, and/or harmful waste. The four primary goals of treating hazardous waste for ultimate disposal are

1. To improve the handling and physical characteristics of the waste
2. To decrease the surface area across which transfer or loss of contained pollutants can occur
3. To limit the solubility of any pollutants continued in the waste
4. To detoxify contained pollutants

These goals can be met in many ways, but not all techniques attempt to meet all the goals. Thus individual treatment techniques may solve one particular set of problems but be completely unsatisfactory for others. Process selection involves weighing advantages and disadvantages for the particular situation.

Representative Processes. The seven major categories of industrial waste stabilization/fixation/encapsulation are

1. Cement-based processes
2. Pozzolanic processes (not containing cement)
3. Thermoplastic techniques (including bitumen, paraffin, and polyethylene incorporation)

4. Organic polymer techniques (including urea-formaldehyde, unsaturated polyesters)
5. Surface encapsulation techniques (jacketing)
6. Self-cementing techniques (for high-calcium-sulfate sludges)
7. Glassification and production of synthetic minerals or ceramics

Cement-based Processes.   Common (Portland) cement is produced by firing a charge of limestone and clay or other silicate mixtures at high temperatures. The resulting clinker is ground to a fine powder to produce a cement consisting of about 50% tricalcium and 25% dicalcium silicates (also present are about 10% tricalcium aluminate and 10% calcium aluminoferrite). The cementation process is brought about by adding water to the anhydrous cement powder. This first produces a colloidal calcium-silicate-hydrate gel of indefinite composition and structure. Hardening of the cement is a lengthy process brought about by the interlacing of thin, densely packed, silicate fibrils growing from the individual cement particles. This fibrillar matrix incorporates the added aggregates and/or waste into a monolithic, rock-like mass.

Most hazardous wastes slurried in water can be mixed directly with cement, and the suspended solids will be incorporated into the rigid matrices of the hardened concrete. This process is especially effective for waste with high levels of toxic metals, because at the pH of the cement mixture most multivalent cations are converted into insoluble hydroxides or carbonates. Metal ions may also be incorporated into the crystal structure of the cement minerals that are formed. Materials in the waste, such as sulfides, asbestos, latex, and solid-plastic wastes, may actually increase the strength and stability of the waste concrete.

The presence of certain inorganic compounds and organic impurities in the hazardous waste and mixing water can act as interfering agents, deleterious to the setting and curing of the waste-bearing concrete.

Additives have been developed to improve the physical characteristics and to decrease leaching losses from the resulting solidified sludge. Some of these are clay, vermiculite, and soluble silicate. Many additives are proprietary.

Pozzolanic Processes (Not Containing Cement).   Techniques based on lime products usually depend on the reaction of lime with a fine-grained siliceous (pozzolanic) material and water to produce a concrete-like solid (sometimes referred to as a pozzolanic concrete). The most common pozzolanic materials used in waste treatment are fly ash, ground blast-furnace slag, and cement-kiln dust. All of these materials are themselves waste products with little or no commercial value at this time. The use of these waste products to consolidate another waste is often advantageous to the processor, who can treat two waste products at the same time. For example, the production of a pozzolanic reaction with power-plant fly ash permits the flue-gas cleaning sludge to be combined with the normal fly ash output and lime (along with other additives) to produce an easily handled solid.

Certain treatment systems fall in the category of cement-pozzolanic processes and have been used for some time outside the United States. In this case both cement and lime-siliceous materials are combined to give the best and most economical containment for the specific waste being treated.

Thermoplastic Techniques (Including Bitumen, Paraffin, and Polyethylene). The use of thermoplastic solidification systems in radioactive waste disposal has led to the development of waste-containment systems that can be adapted to industrial waste. When the radioactive waste with bitumen or other thermoplastic material is processed, the waste is dried, heated, and dispersed through a heated plastic matrix. The mixture is then cooled to solidify the mass, and it is usually buried in a secondary containment system such as a steel drum. Variations of this treatment system can use thermoplastic organic materials such as paraffin or polyethylene.

In many cases the types of waste rule out the use of any organic-based treatment systems. Organic chemicals that are solvents for the matrix obviously cannot be used directly in the treatment system. Strongly oxidizing salts such as nitrates, chlorates, or perchlorates will react with the organic matrix materials and cause slow deterioration. At the elevated temperatures necessary for processing, the matrix-oxidizer mixtures are extremely flammable.

Leach or extraction testing undertaken on anhydrous salts embedded in bitumen as a matrix indicates that rehydration of the embedded compound can occur when the sample is soaked in water, and this can cause the asphalt or bitumen to swell and split apart, greatly increasing the surface area and rate of waste loss. Some salts (such as sodium sulfate) will naturally dehydrate at the temperatures required to make the bitumen plastic; so these easily dehydrated compounds must be avoided in thermoplastic stabilization.

Organic Polymer Techniques (Including Urea-Formaldehyde, Unsaturated Polyesters). Organic polymer techniques were developed to solidify waste for transportation. The most thoroughly tested organic polymer solidification technique is the urea-formaldehyde (UF) system. The polymer is generally formed in a batch process where the wet or dry wastes are blended with a prepolymer in a waste receptable (steel drum) or in a specially designed mixer. When these two components are thoroughly mixed, a catalyst is added and mixing is continued until the catalyst is thoroughly dispersed. Mixing is terminated before the polymer has formed, and the resin-waste mixture is transferred to a waste container if necessary. The polymerized material does not chemically combine with the waste; it forms a spongy mass that traps the solid particles. Any liquid associated with the waste will remain after polymerization. The polymer mass must often be dried before disposal.

Several organic polymer systems are available that are not based on UF resins. Dow Industrial Division is developing a vinyl esterstyrene system (Binder 101) for use with radioactive waste. Testing of this material is currently underway in the Nuclear Regulatory Commission Research Programs.

The Polymeric Material Section at Washington State University has developed a polyester resin system that is being used in waste solidification. This system is currently in a pilot-plant stage in the processing of hazardous wastes.

Surface Encapsulation Techniques (Jacketing).   Many waste treatment systems depend on binding particles of waste material together. To the extent to which the binder coats the waste particles, the waste is encapsulated. However, the systems addressed under surface encapsulation are those in which a waste that has been pressed or bonded together is enclosed in a coating or jacket of inert material. A number of systems for coating solidified industrial wastes have been examined for EPA by TRW Corporation. In most cases, coated materials have suffered from lack of adhesion between coatings and bound wastes, and lack of long-term integrity in the coating materials. After investigating many alternative binding and coating systems, TRW Corporation has produced a system that overcomes most of these problems.

The TRW surface encapsulation system requires that the waste material be thoroughly dried. The dried wastes are stirred into an acetone solution of modified 1,2-polybutadiene for 5 minutes. The mixture is allowed to set for 2 hours. The optimum amount of binder is 3–4% of the fixed material on a dryweight basis. The coated material is placed in a mold, subjected to slight mechanical pressure, and heated to between 120 °C and 200 °C (250 °F and 400 °F) to produce fusion. The agglomerated material is a hard, tough, solid block. A polyethylene jacket 3.5 mm (1/4 in.) thick is fused over the solid block and adheres to the polybutadiene binder. In a 360- to 450-kg (800- to 1000-lb) block, the polyethylene would amount to 4 percent of the fused waste on a weight basis.

Another process has been developed for encapsulation of whole 55 gallon (208 L) metal drums containing hazardous waste. This process involves the overpacking of leaking or damaged drums in 85-gal (322 L) polyethylene (PE) encapsulates and field welding (with PE) PE lids on them. These encapsulates were developed to render drummed materials safe and secure for transportation and eventual landfilling. The encapsulates have demonstrated good mechanical strength and are presently undergoing field leaching tests.

Self-Cementing Techniques (High-Calcium-Sulfate Sludges). Some industrial wastes such as flue-gas cleaning or desulfurization sludges contain large amounts of calcium sulfate and calcium sulfite. A technology has been developed to treat these types of wastes so that they become self-cementing. Usually a small portion (8–10% by weight) of the dewatered waste sulfate/sulfite sludge is calcined under carefully controlled conditions to produce a partially dehydrated cementitious calcium sulfate or sulfite. This calcined waste is then reintroduced into the bulk of the waste sludge along with other proprietary additives. Fly ash is often added to adjust the moisture content. The finished product is a hard, plasterlike material with good handling characteristics and low permeability.

Glassification and Production of Synthetic Minerals or Ceramics. Where material is extremely dangerous or radioactive, it is possible to combine the waste with silica and either fuse the mixture in glass or form a synthetic silicate mineral. Glasses or crystalline silicates are only very slowly leached by naturally occurring waters, so these waste products are generally considered safe materials for disposal without secondary containment. No work using glassification of industrial wastes is now going on.

Design Criteria. There are over 40 vendors of different stabilization/fixation/encapsulation processes in the United States. Since most of these processes are proprietary, the exact operational principles (mix–waste ratios, pretreatments, and processing equipment) are not available for discussion here. However, in general, the selection of any particular technique for hazardous-waste treatment must include careful consideration of the containment required, the cost of processing, the increase of bulk of materials, and the changes in the handling characteristics. The design and location of any placement area or landfill that eventually receives the treated wastes are also major considerations when deciding on the degree of containment and the physical properties required.

Selection of the best treatment system requires detailed knowledge of the constituents and characteristics of the waste to be treated, the amount of waste to be handled, and the location and environment of the waste disposal site. The waste should be fully characterized, including information such as the process producing the waste and the methods of transportation, storage, and pretreatment (if any).

Consideration should be given to waste and process compatibility to avoid detrimental effects such as heat generation, release of toxic materials, toxic gases, fire, explosion, and so forth. Compatibility considerations of selected waste categories are summarized in Table 3-19.

Table 3-19. Compatibility of Waste Categories with Solidification/Stabilization Techniques

Waste Component	Treatment Type						
	Cement Based	Lime Based	Thermoplastic	Organic Polymer (UF)[a]	Surface Encapsulation	Self-Cementing	Classification and Synthetic Mineral Formation
Organics							
1. Organic solvents and oils	Many impede setting, may escape as vapor	Many impede setting, may escape as vapor	Organics may vaporize on heating	May retard set of polymers	Must first be absorbed on solid matrix	Fire danger on heating	Wastes decompose at high temperatures
2. Solid organics (e.g., plastics, resins, tars)	Good–often increases durability	Good–often increases durability	Possible use as binding agent	May retard set of polymers	Compatible–many encapsulation materials are plastic	Fire danger on heating	Wastes decompose at high temperatures
Inorganics							
1. Acid wastes	Cement will neutralize acids	Compatible	Can be neutralized before incorporation	Compatible	Can be neutralized before incorporation	May be neutralized to form sulfate salts	Can be neutralized and incorporated
2. Oxidizers	Compatible	Compatible	May cause matrix breakdown, fire	May cause matrix breakdown	May cause deterioration of encapsulating materials	Compatible if sulfates are present	High temperatures may cause undesirable reactions

	1	2	3	4	5	6	7
3. Sulfates	May retard setting and cause spalling unless special cement is used	Compatible	May dehydrate and rehydrate, causing splitting	Compatible	Compatible	Compatible	Compatible in many cases
4. Halides	Easily leached from cement, may retard setting	May retard setting, most are easily leached	May dehydrate	Compatible	Compatible	Compatible if sulfates are also present	Compatible in many cases
5. Heavy metals	Compatible	Compatible	Compatible	Acid pH solubilizes metal hydroxides	Compatible	Compatible if sulfates are present	Compatible in many cases
6. Radioactive materials	Compatible	Compatible	Compatible	Compatible	Compatible	Compatible if sulfates are present	Compatible

Source: Guide to the Disposal of Chemically Stabilized and Solidified Waste, U.S. EPA, Report #SW-872, September, 1980.

[a] Urea-formaldehyde resin.

Performance Testing Data: Physical Testing.   Performance testing for stabilized–fixed or encapsulated waste generally consists of various physical stability and leachability evaluations.

Treated wastes are subjected to physical tests to (a) determine particle size, distribution, porosity, permeability, and wet and dry density; (b) evaluate bulk-handling properties; (c) predict the reaction of a material to applied stresses in embankments, landfills, and so on and (d) evaluate durability under freeze–thaw conditions, and so on.

Some typical results of physical testing of stabilized and untreated industrial waste are listed in Table 3-20. The data show that stabilization processes generally increase density and strength and decrease permeability. Note that many of the treated samples lack durability.

Chemical leach testing of wastes is used to examine or predict the chemical stability of treated wastes when they are in contact with aqueous solutions that might be encountered in a landfill. The procedures demonstrate the degree of immobilization of contaminants produced by the treatment process. Many techniques for leach testing are available. Unfortunately, no single leach testing system can duplicate the variable conditions that may be encountered by landfilled treated wastes.

Most test procedures are conducted at temperatures (20–25 °C) and pressures normally occurring in the laboratory. The major variables encountered in comparing different leaching procedures are the nature of the leaching solution, waste-to-leaching solution ratios, number of elutions of leaching solution used, contact time of waste and leaching solution, surface area of waste exposed, and agitation technique employed.

There is no uniform opinion as to how each of these variables should be treated in a testing procedure. Results of leaching tests are commonly reported by vendors of waste-treatment systems. The protocols of leaching tests employed vary widely—from a one hour, unstirred, distilled-water leaching test on undisturbed treated waste samples to extended, repeated leaching of ground samples by aggressive leaching solutions. Some vendors report results of field tests. Table 3-21 lists typical types of leaching tests and the results form the major vendors of waste-treatment systems.

Appendix 3-3 summarizes the current practices of several industrial operations regarding the treatment and disposal of hazardous-waste sludges.

TREATMENT OF UNUSUALLY TOXIC WASTES

Treatment of unusually toxic wastes, such as those containing concentrations of PCBs and CDDs, is laden with problems far beyond the scope of normal chemical/physical/biological (CPB) treatment. Increased public

(Text continues on page 148)

Table 3-20. Typical Results from Physical Testing of Stabilized and Untreated Industrial Wastes

Type of Waste and Treatment	Unit Weight		Unconfined Compressive Strength (lb/in³)	Permeability (cm/s)	Durability (test cycles to failure)	
	Bulk (lb/ft³)	Dry (lb/ft³)			Wet/Dry	Freeze/Thaw
Nickel-cadmium battery waste						
untreated	—	43.9	—	5.7×10^{-6}	—	—
lime-based pozzolan product	104.0	86.2	169.0	1.9×10^{-6}	9	—
patented additives, soil-like material	93.2	47.3	7.96	1.9×10^{-4}	1	1
Chlorine production waste						
untreated	—	64.0	—	1.0×10^{-4}	—	—
lime-based pozzolan product	103.0	88.6	133.0	8.5×10^{-7}	—	—
patented additives, soil-like material	106.0	81.3	21.6	3.6×10^{-5}	2	1
Calcium fluoride waste						
untreated	—	46.8	—	3.5×10^{-5}	—	—
lime-based pozzolan product	85.9	66.6	26.2	3.8×10^{-5}	1	1
patented additives, soil-like material	86.2	52.8	25.6	8.7×10^{-6}	1	2
Electroplating waste						
untreated	—	28.1	—	3.1×10^{-5}	—	—
lime-based pozzolan product	100.0	77.4	77.3	4.0×10^{-7}	5	—
patented additives, soil-like material	87.1	47.4	32.4	1.1×10^{-5}	2	—
organic resin, rubber-like material	75.4	52.7	747.0	1.1×10^{-4a}	1	1
plastic encapsulation	73.6	73.6	1540.0	Impervious	NF(0.00)[b]	12 NF(0.00)[b]

(continued)

Table 3-20. (continued)

Type of Waste and Treatment	Unit Weight Bulk (lb/ft³)	Dry (lb/ft³)	Unconfined Compressive Strength (lb/in³)	Permeability (cm/s)	Durability (test cycles to failure) Wet/Dry	Freeze/Thaw
Flue-gas cleaning waste						
untreated	—	58.8	—	3.6×10^{-5}	—	—
lime-based pozzolan product	100.0	80.9	100.0	2.0×10^{-6}	3	2
patented additives, soil-like material	77.0	43.4	23.7	1.6×10^{-4}	1	1
cement-based, concrete-like product	101.0	94.9	2570.0	7.9×10^{-4a}	NF(15.80)[b]	10

Source: Guide to the Disposal of Chemical Stabilized and Solidified Waste, U.S. EPA, Report #SW-872, September, 1980.
Note: — = not tested.
[a] Value questionable because flow restriction caused by sample support may have influenced flow through sample.
[b] NF indicates no failure in 12 cycles; figure in parentheses is the percent weight lost after 12 cycles.

Table 3-21. Results of Physical Property and Leaching Tests by Sludge Stabilization Vendors

Vendor or Data Source	Solidification Basis	Physical Property Tests and Results	Typical Permeabilities (cm/s)	Leaching Tests and Results
Atcor Washington, Inc. (Div. Chem. Nuclear Systems, Inc.), Peekskill, N.Y. 10566	Cement	Numerous results available. Product is monolithic cement structure with no freewater	—	Numerous results available. Leaching rate found acceptable for shallow land burial
Chemfix, Inc., Kenner, La. 70063	Cement and soluble silicates	Treated material varies from soil-like to concrete-like monolith with high bearing capacity	10^{-5}–10^{-6}	Extensive leaching tests have been run on a variety of processed materials. Tests run include cyclic leaching tests, saturation extraction tests, and nonequilibrium extraction systems
Bravo Lime Co., Pittsburgh, Pa. 15222 ("Calcilox")	Cementitious product from basic, glassy blast-furnace slag	Treated material dry with clay-like consistency (like compacted clayish soil)	—	Field leaching test results[a] with FGD sludges available. Typically leaching rates were reduced 1–2 orders of magnitude compared to untreated material
Environmental Technology Corp., Pittsburgh, Pa. 15220	Lime plus ion-exchange medium and binder	Product resembles clay in appearance and properties. Strengths average about 9.6×10 N/m^2	10^{-6}	Unique open-trench leach test gave results (1 of 10 sludges) after 1 month 1–5,000 mg/L dissolved solids 5–800 mg/L SO^4. and less than 0.01 mg/L of Ni, Mn, Cr, Fe, and Zn

(continued)

Table 3-21. (continued)

Vendor or Data Source	Solidification Basis	Physical Property Tests and Results	Typical Permeabilities (cm/s)	Leaching Tests and Results
I.U. Conversion Systems, Horsham, Pa. 19044 ("POZ-O-TEC")	Fly ash and lime-based (pozzolanic)	Unconfined compressive strength (FGD sludges) 1.2×10^{-3} N/m²	10^{-5}–10^{-7} after 2–4 weeks	The company maintains that forced leachate tests are not valid for their product. Different tests have shown a reduction in concentration of most species in leachate of 1/2 to 1/200 that form un-treated wastes
Ontario Liquid Waste Disposal, Ltd. (Canadian Waste Technology), Markham, Ontario, Canada L3R-1G6	Silicate compounds	For cost reasons, end product is usually low strength; but strengths up to 21×10^6 N/m² are possible	—	Heavy-metal concentrations in leachate are commonly below 1 mg/L from acid, metal-bearing, treated wastes
Polymeric Materials Section, Dept. of Material Science, Washington State Univ., Pullman, Wash. 99164	Polyester resin	40% resin product has compressive strength of 15×10^6 N/m²	—	Leachability after dissolution of surface materials is practically negligible
Sludge Fixation Technology Inc., Orchard Park, N.Y. 14127 ("Terra-Crete")	Self-cementing process for FGD sludges (calcination)	Strengths from 9.6×10^4 to 5.7×10^5 N/m² are possible, depending upon proportions	10^{-6}–10^{-7}	Data on leaching of treated lead-rich FGD sludges shows less than 0.01 mg/L in leach liquid
Stabatrol Corporation, Norristown, Pa. 19401 ("Terra-Tite")	Cementitious additives	Unfined compressive strengths to 4.8×10^{-5} N/m²	10^{-7}	"insignificantly low"

Stablex Corporation, Radnor, Pa. 19087 ("Seal-o-safe")	Cement and pozzolanic materials	Strengths typical for grouts used as fillers and soil stabilizers, but much less than concrete	10^{-7}	Grinding product to powder and immersing in water for 3 h gave very little (1 mg/L) loss of materials
TJK, Inc., North Hollywood, Calif. 91605	Cement	Strengths vary widely but $5-10 \times 10^5$ N/m² are not uncommon with 20% (w/v) additives	—	Grinding product to powder and immersing in water for 6 h showed only low levels of pollutants were lost (or about 1 mg/L)
Todd Shipyards Corp., Galveston, Tx. 77553 ("Safe-T-Set")	Organic polymer (not UF)	—	—	Nine tests reported: Escape of radioactive material, temp. cycle. several immersion tests, off-gas tests, and biological and radiation stability
TRW Systems Group, Redondo Beach, Calif. 90278	Inorganic cements and polybutadiene resins	Tests include mechanical testing, bulk density, surface hardness, compressive strength	—	Leaching tests at pH 3.8 to 4.0
Werner and Pfleiderer Corp., Waldwick, N.J. 07463	Bitumen encapsulation	Plastic solids usually placed in steel drums	Negligible	Leach rates average 100 times less than comparable cement-treated wastes

Source: EPA-600/2-79-056 and company literature.
Note: These data are provided by the vendor companies for illustrative purposes only and are not verified.
[a]FGD = flue gas desulfurization.

awareness and the possibility of harm from even the most minor of mishaps require complex, fail-safe systems to isolate the community and the workers from the wastes until thorough treatment has been accomplished. These problems, the availability of inexpensive landfills, and the small volumes of wastes generated have severely restricted efforts to find effective treatment or disposal strategies.

Recently, however, the passage of the Comprehensive Environmental Response, Compensation, and Liability Act, coupled with the greater realization that a landfill will ultimately leak, is leading more and more wise waste managers to examine permanent methods of disposal for their highly toxic wastes. This examination has led to consideration of more exotic techniques (e.g., plasma arc decomposition and freeze crystallization) as well as the novel application of more conventional technologies (e.g., solvent extraction and distillation).

To date, much of the research into the treatment of highly toxic wastes has centered on wastes with two types of constituents: polychlorinated biphenyls (PCBs) and chlorinated dibenzo-p-dioxins (CDDs). Although this emphasis is largely attributable to toxicity and the volumes of wastes generated, the influence of regulatory pressures cannot be overlooked. Logically speaking, if one were to consider all wastes with carcinogenicity equivalent to PCBs as highly toxic, the chlorofluoromethyl ethers and some of the chlorinated solvents should have been equally well researched. Nevertheless, these other wastes have not received the same attention in the open literature, even though they have been studied. Because there is a great deal of interest in the treatment of PCBs and CDDs and also a need to bound this section, the following discussion is generally confined to the treatment of these two types of wastes.

Current Techniques

Numerous technologies have been applied to the treatment of CDDs and PCBs. In some cases these techniques are merely pretreatment strategies that serve to concentrate the toxic compounds for more effective destruction. A classic example is solvent extraction. These techniques are discussed in the following sections unless they are major areas of active research and development, like supercritical fluids.

When evaluating these methods, one must keep in mind the stability and toxicity of the end products. Methods that produce more dangerous compounds or concentrate the compounds in a less treatable media are no bargain and have not been considered for application. With these caveats in mind, it is appropriate to now discuss the specific technologies.

Solvent Extraction. This approach has been widely applied to the pretreatment of CDDs and PCBs. The purpose of solvent extraction in this context is to concentrate the compounds in a solvent that will not interfere with subsequent attempts to destroy or react the CDDs or PCBs. For CDDs, investigators have found that some common solvents are quite effective under proper conditions at extracting tetrachlorodibenzo-*p*-dioxin (TCDD) from liquids, soils, or sludges [6,7]. Likewise, a number of processes have been developed [8] that use polyethylene glycols to extract PCBs from transformer oil and water. Common solvents should also work quite well in extracting PCBs [9] and could be used as pretreatment prior to incineration. Although any application requires a rather specific consideration of matrix effects, target compound polarity and suitability of the solvent for subsequent treatment have been demonstrated to have broad applicability to most compounds.

Ultraviolet Irradiation. Photolysis for the destruction of CDDs has been an active area of research for several years [7]. Investigators have concluded that there are three requirements for CDD photolysis:

1. Dissolution in a light-transmitting film
2. Presence of an organic hydrogen donor
3. Ultraviolet light

These principles have been reduced to practice in Verona, Mo., where 99.9% TCDD reduction (to the 0.1–0.5-ppm level) was achieved [6]. Based on other work by Dow Chemical Company, we can reasonably assume that the process achieves ring disruption rather than dehalogenation.

Similar work has not been published in the United States for PCBs. However, it is reasonable to assume that similar success could be achieved if the proper wavelengths and hydrogen donors were identified.

Ozonolysis. Although this technique has not been demonstrated in any full-scale facility, its application to both CDDs and PCBs is quite promising. This assessment is based upon examining the results of several laboratory tests and the success of ozonation in destroying cyanides and pesticides. With respect to CDDs, one study demonstrated that if dioxins were suspended as an aerosol combined with carbon tetrachloride, 97% degradation of TCDD was possible. In another experiment ozone in conjunction with UV radiation has been shown effective for the destruction of polychlorinated phenols and pesticides. In both cases the key requirements were to concentrate the CDDs in a medium where they were susceptible to attack and to provide a free radical for reaction with the dioxin molecule.

PCBs have also been destroyed by using ozone. The Royal Military College of Canada has demonstrated 95% destruction of PCBs in the presence of a 1:1 weight ratio of ozone [8]. Although the associated power requirement would make the process very expensive, its applicability to the destruction of PCBs in soils and sludges does hold promise. Furthermore, if an effective concentrating and activating solvent can be identified, improved economics may result. The additional advantage of ozonolysis for CDDs and PCBs is that fairly innocuous by-products should be produced.

Polymerization.   This process has been demonstrated by a number of companies [8] to be effective for the destruction of PCBs. In general, the processes (which have the additional advantage of being movable to different locations) involve the mixing of the PCB-contaminated media with a reagent containing sodium or calcium. The resultant mixture (after elevated temperature reaction) contains innocuous salts and polymerized biphenyl molecules. The mixture also has a considerable heat value and is an excellent fuel. The reagents used in the various processes are considered confidential, with the exception of the sodium glycolate used by the Franklin Institute in Philadelphia. It is clear that they all are "super bases" that attack the chlorine atoms rather than the ring structures. Similar polymerization techniques have not been applied to CDDs.

Dechlorination.   Dechlorination has been used to treat several pesticides. It is usually conducted in the presence of a catalyst, such as nickel borohydride. Its major potential application with respect to CDDs is the removal of some chlorines from wastes containing the terra and higher chlorinated isomers as a pretreatment prior to incineration. This step would minimize TCDD formation as a product of incomplete combustion (PIC). Dechlorination does not seem useful in and of itself because the process is not totally effective in removing chlorine. A potential use for PCB treatment is discussed under microbial degradation.

Microbial Degradation.   Aerobic degradation has been investigated for both PCBs and CDDs [7,10]. Although TCDD and Aroclors 1221 and 1232 exhibited significant degradation, this occurred only at low concentrations and at slow rates. The long-term degradation of monochlorobiphenyls is well established in natural systems.

The promise of microbial degradation for PCBs is in conjunction with other technologies or separately as an in situ cleanup technique [11]. Methods such as dechlorination might be used to reduce the PCBs to the mono- and bichlorinated species. In that form they would be susceptible to attack by specially acclimated microbes.

Inappropriate Techniques.   Activated carbon adsorption, chlorolysis, and ammination are demonstrated techniques for the concentration and/or destruction of CDDs and/or PCBs [7,12]. Unfortunately, in every case, residuals produced are more toxic and/or more concentrated and consequently more difficult to handle. These techniques are currently considered inappropriate for high-toxicity waste treatment.

Research Trends in Hazardous-Waste Treatment

Research into the treatment of toxic and hazardous wastes continues to be a high-level priority of the EPA. In addition, promulgation of regulations associated with the Resource Conservation and Recovery Act have encouraged the establishment of a strong research community in the private sector. Between these two groups, rapid advances are being made in the state-of-the-art of hazardous waste treatment. This advancement is facilitated by the fact that, while EPA's effort is geared toward the development of new technologies, commercial hazardous waste treaters have been investigating the treatment of specific problem wastes. Taken together, these approaches are resulting in the examination of a large number of alternatives.

Regardless of the approach taken to development of treatment information, no change has occurred in the basic approach of bench/pilot/full-scale demonstrations. The technologies described herein have all been demonstrated in bench-scale units and are being further examined based on the results of those tests. In general, treatment research in the United States and Canada is focusing on the applications of one or more of the following types:

1. Production process applications
2. Concentration of natural degradation phenomena
3. Laboratory chemical purification strategies

Examples of each approach appear in the following sections.

Solvent Extraction.   The identification of appropriate solvents for waste treatment is a major area of research. This is true for both metals and organic compounds. With respect to metals, several studies are underway to identify solvents useful for the recovery of metals from electroplating sludges. Typically, these processes are initiated with resolubilization in sulfuric acid or ammonia leach, similar to approaches used in nonferrous smelting [13]. One sample application is the selective recovery of copper from a mixed-metal hydroxide sludge by using LIX 64N at a pH of 2. When coupled with electrowinning for copper extraction, sulfuric acid is also generated. This significantly enhances the economics of the process and provides a very effective technique for copper reclamation. A similar approach for the removal of toxic metals is discussed in the section on hydrometallurgy.

Solvent extraction for removing organic chlorides also has a long history of application, especially in the organic chemical industry. In addition, specific examples were cited in the section on the treatment of unusually toxic wastes. Its other major application is for the cracking of polymers in solids and tars to facilitate subsequent treatment. The number of applications of this approach is equivalent to the number of occasions for its use. In many cases the process is merely a scaled-up version of analytical laboratory approaches to emulsion breaking and chemical synthesis.

Hydrometallurgy. An active area of research is the combined application of solvent extraction, electrowinning, chemical precipitation, and filtration to the recovery of pure metals from hydroxide sludges. The major driving force behind this effort is the massive volume and metals currently being land disposed. It is hoped that central treatment facilities can be established to facilitate economical treatment of wastes and to guarantee consistent quality feedstocks for effective treatment. One example of the application of this approach is the work of Mehta [14] on a hydroxide sludge containing zinc, cadmium, copper, chromium, and nickel. The following are steps taken in one of Mehta's projects.

1. The copper was solvent extracted with LIX 64N at a pH of 2 followed by electrowinning of the copper at appropriate potential, current density, and temperature.
2. The aqueous phase from the solvent extraction process (contains nickel, trivalent chromium, and cadmium) was then treated with an oxidizing agent (at pH = 2) to oxidize the chromium for subsequent precipitation as lead chromate. The lead chromate, after sufficient product is accumulated, may easily be refined to produce fairly pure chromium.
3. The remaining solution was solvent extracted by LIX 64N at a pH of 4 to 5 to separate out the nickel. The nickel was then electrowinned from the solvent layer.
4. The remaining aqueous solution was then electrowinned to produce pure cadmium or precipitated for sale as a zinc-cadmium slurry to an appropriate refinery.

The attraction of these approaches to treatment is that they can be used to either selectively remove a single toxic compound (like cadmium) prior to disposal or to actually produce a useful product. Mehta's tests have demonstrated that acid extraction at a pH of approximately 1 will solubilize over 95% of the metals in a sludge, thereby making them available for subsequent treatment.

Supercritical Fluids (SCFs).   SCFs are fluids existing at or above the lowest temperature at which condensation may occur. Above the critical temperature certain fluids exhibit characteristics that enhance their solvent properties. Organic materials, which are only slightly soluble in particular solvents at room temperature, become completely miscible with the solvent when under supercritical conditions. The exceptional solvent properties result from the rapid mass transfer ability and the very low density that characterizes an SCF. Major advantages of SCFs are short residence times with no char formation.

The major applications of SCFs to date have been for regeneration of activated carbon and for hydrocarbon extraction from coal. During the late 1970s, supercritical carbon dioxide was investigated through an EPA grant. It was found to be an excellent solvent for a wide variety of organics. Supercritical toluene and *p*-cresol were effective in dissolving the hydrogen-rich components of liquified coal. More recently, the MODAR Division of O'Donnel and Rich Enterprises has been conducting bench-scale studies involving the oxidation of a variety of organics in solution with supercritical water. Organics investigated include DDT, chlorinated solvents, PCBs, and hexachlorocyclopentadiene. Rapid and complete oxidation was observed for the five runs as is shown in Tables 3-22 and 3-23 [15].

MODAR (the process vendor) is now participating in a joint venture with the State of New York to dispose of TCDD contaminated soils from a landfill in a 50 gal/day unit. The main hindrance to the wide scale application of the technology at this time appears to be the cost effective maintenance of the high temperatures (350 °C) and pressures (200 atm) required for this process. Even if these conditions are expensive to maintain, however, application to unusually toxic wastes may be possible.

Ultraviolet Irradiation/Chemical Oxidation.   As was discussed in the previous section, combined UV-chemical oxidant treatment shows considerable promise for treating chlorinated organics in solution. The key to successful treatment appears to be provision of sufficient energy of short enough wavelengths so that absorption for molecule energy activation can occur. Ozone, peroxide, or other oxidants can then cost effectively bring about degradation. The biggest factors in the resurgence of interest in UV irradiation are the development of improved lamps with tighter and lower bands of radiation and designs that deliver higher irradiating surface to waste volume ratios. Since water is transparent to ultraviolet light, this process appears very promising for water-containing compounds requiring chemical oxidation to render them nonhazardous (e.g., pesticides, PCBs, and cyanides). Unfortunately, additional demonstrations are probably some time from fruition.

Table 3-22. Composition of Feed Mixtures for Runs A–E

	wt (%)	wt (% Cl)
Run A		
DDT	4.32	2.133
Run B		
1,1,1-trichloroethane	1.01	0.806
1,2-ethylene dichloride	1.01	0.739
1,1,2,2-tetrachloroethylene	1.01	0.866
o-chlorotoluene	1.01	0.282
1,2,4-trichlorobenzene	1.01	0.591
Biphenyl	1.01	—
o-oxylene	5.44	—
Run C		
Hexachlorocyclohexane	0.69	0.497
DDT	1.00	0.493
4,4'-dichlorobiphenyl	1.57	0.495
Hexachlorocyclopentadiene	0.65	0.505
Run D		
PCB 1242	0.34	0.14
PCB 1254	2.41	1.30
Transformer oil	29.26	—
Run E		
4,4'-dichlorobiphenyl	3.02	0.96

Source: P. B. Reichardt, et al. 1981 (Jan.), Kinetic study of biodegradation of biphenyl and its mono-chlorinated analogies by a mixed marine community, *Environmental Science & Technology.*

Microbial Degradation.   Recently conducted laboratory tests [10] have demonstrated that many priority pollutants are biodegradable in acclimated microbial cultures. A summary of some of the more interesting results for acclimated cultures is given in Table 3-24.

But beyond these tests, many specific organisms have already been identified in natural and municipal systems that are effective for the destruction of hazardous compounds [16, 17]. Defining performance requirements for applications to specified wastes and developing populations appropriate for seeding complex waste streams yet remain to be accomplished. Both areas continue to be investigated by researchers.

Table 3-23. Results of Oxidation of Organic Chlorides

	A	*B*	*C*	*D*	*E*
			Run Numbers[a]		
Carbon analysis					
Organic carbon in (ppm)	26,700.	25,700.	24,500.	38,500.	33,400.
Organic carbon out (ppm)	2.0	1.0	6.4	9.4	
Destruction efficiency (%)	99.993	99.996	99.975	99.991	99.97
Combustion efficiency (%)	100.	100.	100.	100.	100.
Gas composition					
O_2	25.58	32.84	37.10	10.55	19.00
CO	59.02	51.03	46.86	70.89	70.20
CH_2	—	—	—	—	—
H_2	—	—	—	—	—
CO_2	—	—	—	—	—
Chloride analysis					
Organic chloride in (ppm)	876.	1266.	748.	775.	481.
Organic chloride out (ppm)	0.23	0.037	0.028	0.032	0.36
Organic chloride conversion (%)	99.997	99.997	99.996	99.996	99.993

[a]Residence time was 1.1 min for all runs.

Table 3-24. Some Biodegradation Results

Compound	Initial Concentration (mg/L)	Performance Summary
Phenol	10,000	D
2-chlorophenol	10,000	D
2,4-dichlorophenol	10,000	D
2,4,6-trichlorophenol	10,000	D
Pentachlorophenol	10,000	D
2,4-dinitrophenol	10,000	D
4,6-dinitro-*o*-cresol	10,000	N
Chlorobenzene	10,000	D
1,2,4-trichlorobenzene	10	N
Toluene	10	D
2,4-dinitrotoluene	10	N
Anthracene	5	D
Pyrene	5	D
Bis(2-chloroethyl)ether	10	D

D = Significant degradation with rapid adaptation; N = Insignificant degradation.

APPENDIX 3-1
POLLUTANT REMOVABILITY/TREATABILITY

Treatment Process	Range of Removal (%)	Range of Effluent Conc. (g/L)
Antimony		
Activated carbon adsorption	0–33	1.3–590
granular		
ozone		
Chemical precipitation with sedimentation		
lime	40–99	ND–180
polymer unspecified	28	18
sodium carbonate	84	15–57
Coagulation and flocculation	51	10–120
Filtration	0–92	BDL–700
Flotation	4–95[a]	ND–2,300
Reverse osmosis	10–60	1–200
Sedimentation	0–86	2–1,000
Activated sludge	33–90	BDL–670
Lagoons (aerated)	82–99	ND–30
Arsenic		
Activated carbon adsorption (granular)	0–99	1–42
Chemical oxidation (ozone)	0–48	BDL–43
Chemical precipitation with sedimentation		
lime	20–99	ND–110
alum	0	62–62
barium chloride	33	2
FeCl(3)	75	1
sodium carbonate	96	10–68
combined precipitants	25–99	ND–3
Chemical precipitation with filtration	33	4.0
Coagulation and flocculation	40–92	BDL–62
Filtration	0–99	BDL–100
Flotation	8–99	ND–18
Reverse osmosis	57–99	1–49
Sedimentation	0–99	BDL–230
Activated sludge	20–98	BDL–160
Lagoons (aerated)	99	ND–20
Asbestos (Total)		
Chemical precipitation with sedimentation		
lime	95	6.1E6–2.3E9[a]
barium chloride	75	5.7E8[a]

Pollutant Removability/Treatability

Treatment Process	Range of Removal (%)	Range of Effluent Conc. (g/L)
Filtration	90	4.7E8[a]
Sedimentation	75–99	1.9E7–3.3E10[a]
Asbestos, Chrysotile		
Sedimentation	95–99	3.3E5–1.5E4[a]
Beryllium		
Chemical precipitation with sedimentation		
polymer/unspecified	33	4
sodium carbonate	75	1–11
combined precipitants	80–99	ND–10
Chemical precipitation with filtration	0	1.0
Filtration	0–7	0.02–10
Reverse osmosis	85	0.5–5
Sedimentation	0–98	0.9–20
Activated sludge		BDL–BDL
Lagoons (aerated)	50	1
Cadmium		
Activated carbon adsorption (granular)	76–95	1.5–40
Chemical precipitation with sedimentation		
lime	0–99	ND–80
alum	38–88	12–47
polymer/unspecified	0–99	5–100
sodium carbonate	67–99	1–5
sodium hydroxide	22–99	ND–930
combined precipitants	11–99	ND–80
Chemical precipitation with filtration	0–99	ND–19
Coagulation and flocculation	99	BDL–10
Filtration	0–99	ND–97
Flotation	0–99[b]	BDL–72
Oil separation	98	BDL–200
Reverse osmosis	0–60	0.5–48
Sedimentation	0–99	BDL–200
Ultrafiltration	67–93	BDL–200
Activated sludge	0–99	BDL–13
Lagoons (aerated)	97	2
Chromium		
Activated carbon adsorption		
granular	10–95	4–260

Pollutant Removability/Treatability

Treatment Process	Range of Removal (%)	Range of Effluent Conc. (g/L)
powdered	3–97	24–110
Chemical precipitation with sedimentation		
lime	47–99	ND–18,000
alum	13–93	31–170,000
barium chloride	50	25
polymer/unspecified	74–99	5–790,000
sodium carbonate	99	27–430
sodium hydroxide	53–99	18–3,000
combined precipitants	33–99	ND–2,500
Chemical precipitation with filtration	50–99	5.0–2,200
Chemical reduction	18–99	90–130,000
Coagulation and flocculation	72–99	17–1,300
Filtration	0–99	4–320
Flotation	20–99	2–620
Neutralization	99	40
Oil separation	82–98	9–240
Reverse osmosis	3–95	1–900
Sedimentation	0–99	BDL–30,000
Ultrafiltration	67	68–2,900
Activated sludge	5–98	BDL–20,000
Lagoons		
aerated	0–99	9–1,100
non-aerated	99	ND
Rotating biological contactors		
Copper		
Activated carbon adsorption		
granular	13–85	4–360
powdered	61–96	5.5–29
Chemical oxidation		
chlorine	14	320
ozone	12	88–590
Chemical precipitation with sedimentation		
lime	29–100	ND–220
alum	30–99	ND–270,000
barium chloride	50	20
polymer/unspecified	0–99	4–2,200,000
sodium carbonate	83	48–1,300
sodium hydroxide	36–98	1–5,900
combined precipitants	43–99	4–400
Chemical precipitation with filtration	14–99	16–1,700

Pollutant Removability/Treatability

Treatment Process	Range of Removal (%)	Range of Effluent Conc. (g/L)
Chemical reduction	21–89	BDL–15
Coagulation and flocculation	35–99	10–80
Filtration	0–99	4–4,500
Flotation	9–98	5–960
Neutralization	98	30
Oil separation	93–99	BDL–450
Reverse osmosis	29–93	9–28,000
Sedimentation	0–99	ND–660
Ultrafiltration	44–90	BDL–1,100
Activated sludge	2–99	BDL–170
Lagoons (aerated)	26–94	7–110
Cyanides		
Activated carbon adsorption		
granular	1–90	2–52
powdered	50–69	20–45
Chemical oxidation		
chlorine	86–99	2–130
ozone	33–98	2–95
Chemical precipitation with sedimentation		
lime	67–99	ND–5,500
alum	60	2–120
polymer/unspecified	0–99	BDL–5,200
Chemical precipitation with filtration		
Coagulation and flocculation	60	BDL–5
Filtration	0–99	3–260
Flotation	0–62	10–2,300
Oil separation	13	BDL–13
Reverse osmosis	23–91	1–2,200
Sedimentation	20–90	BDL–4,500
Stripping (air)	60	51,000
Activated sludge	0–99	ND–38,000
Lagoons (aerated)	91–99	ND–150
Trickling filters	79	16
Lead		
Activated carbon adsorption (granular)	2–72	18–79
Chemical oxidation (ozone)	29	22–900
Chemical precipitation with sedimentation		
lime	0–99	ND–580
alum	0–96	28–140,000

Pollutant Removability/Treatability

Treatment Process	Range of Removal (%)	Range of Effluent Conc. (g/L)
polymer/unspecified	26–99	BDL–1,000
sodium carbonate	94–99	15–1,900
sodium hydroxide	99	ND
combined precipitants	0–99	ND–230
Chemical precipitation with filtration	81–99	ND–68
Chemical reduction	25–67	62–120,000
Coagulation and flocculation	0–99	BDL–100
Filtration	3–99	BDL–2,100
Flotation	9–99	ND–1,000
Oil separation	97–99	BDL–600
Reverse osmosis	0–96	1–520
Sedimentation	0–99	ND–6,800
Ultrafiltration	44–95	BDL–1,000
Activated sludge	10–99	ND–120
Lagoons		
aerated	80–99	ND–80
non-aerated	99	ND
Mercury		
Chemical precipitation with sedimentation		
lime	69–96	0.1–13
alum	6–93	1.7–4,000
barium chloride	87	0.5
polymer/unspecified	0–7	1–26
combined precipitants	0–97	0.2–980
Chemical precipitation with filtration	67	0.1
Coagulation and flocculation	10	0.3–1
Filtration	0–89	0.1–2,900
Flotation	33–88	BDL–2
Oil separation	80	BDL–2
Reverse osmosis	60	0.1–1
Sedimentation	0–97	BDL–84
Ultrafiltration	0–33	0.4–2
Activated sludge	33–94	ND–0.9
Lagoons (aerated)	99	0.1–1.6
Nickel		
Activated carbon adsorption (granular)	10–68	BDL–700
Chemical precipitation with sedimentation		
lime	6–99	ND–5,200

Pollutant Removability/Treatability

Treatment Process	Range of Removal (%)	Range of Effluent Conc. (g/L)
alum	0–98	1.0–51,000
polymer/unspecified	8–99	5–6,400
sodium carbonate	96	18–640
sodium hydroxide	99	ND–210
combined precipitants	0–99	ND–10,000
Chemical precipitation with filtration	47–99	ND–1,700
Coagulation and flocculation	82–99	BDL–2,500
Filtration	0–99	BDL–700
Flotation	29–99	ND–270
Neutralization	99	20
Oil separation	96	BDL–500
Reverse osmosis	9–89	1–210
Sedimentation	0–99	BDL–2,000
Ultrafiltration	32	10–500
Activated sludge	0–99	ND–400
Lagoons (aerated)	0–50	5–40
Selenium		
Activated carbon adsorption (granular)	0–50	1–50
Chemical oxidation (ozone)	99	ND–1
Chemical precipitation with sedimentation		
lime	0–99	ND–87
sodium carbonate	96	4–280
combined precipitants	0–13	2–7
Filtration	0–10	BDL–100
Reverse osmosis	67–85	1–6.1
Sedimentation	0–98	2–32
Lagoons		
aerated	50–99	ND–200
non-aerated	44	18
Silver		
Activated carbon adsorption (granular)	12–36	1.7–100
Chemical precipitation with sedimentation		
lime	20–99	ND–90
alum	21	7–170
polymer/unspecified	0–67	12–35
sodium carbonate	97	2–22
sodium hydroxide	6	11–64
combined precipitants	9–43	0.4–10

Pollutant Removability/Treatability

Treatment Process	Range of Removal (%)	Range of Effluent Conc. (g/L)
Chemical precipitation with filtration	40–78	4.5–34
Coagulation and flocculation	10	1–72
Filtration	0–91	BDL–100
Flotation	45	BDL–66
Reverse osmosis	0–92	0.2–78
Sedimentation	0–96	1–100
Activated sludge	3–99	BDL–95
Thallium		
Chemical precipitation with sedimentation		
lime	58–75	1–20
sodium carbonate	99	41
Chemical precipitation with filtration	0	50
Chemical reduction	94	1
Reverse osmosis	50–99	1–4
Sedimentation	0–83	2–100
Activated sludge		BDL–BDL
Lagoons (aerated)	7–80	13–58
Zinc		
Activated carbon adsorption		
granular	5–99	1–6,000
powdered	26–98	45–140
Chemical oxidation (ozone)	96	90–46
Chemical precipitation with sedimentation		
lime	25–99	2–26,000
alum	11–88	110–350,000
barium chloride	50	3
polymer/unspecified	10–99	ND–1,800,000
sodium carbonate	84–99	330–11,000
sodium hydroxide	80–99	44–560
combined precipitants	14–99	2–17,000
Chemical precipitation with filtration	82–99	10–60
Chemical reduction	87	21–1,500
Coagulation and flocculation	11–98	BDL–5,700
Filtration	0–99	16–18,000
Flotation	12–99	ND–53,000
Neutralization	99	30
Oil separation	88–97	BDL–680
Reverse osmosis	31–99	2–8,600

Pollutant Removability/Treatability

Treatment Process	Range of Removal (%)	Range of Effluent Conc. (g/L)
Sedimentation	0–99	BDL–100,000
Ultrafiltration	75–98	BDL–420
Activated sludge	0–92	BDL–38,000
Lagoons		
aerated	34–99	49–510
non-aerated	86	100–120
Bis(chloromethyl)ether		
Activated sludge	99	ND
Dimethyl phthalate		
Filtration	99–99	ND–BDL
Reverse osmosis	18–84	1–170
Sedimentation	98–99	ND–93
Activated sludge	99–99	ND–200
Diethyl phthalate		
Chemical precipitation with sedimentation		
lime	36	BDL–73
sodium hydroxide	63–95	ND–92
combined precipitants	99	ND–3.0
Chemical precipitation with filtration	25–99	ND–75
Filtration	50–99	ND–11,000
Flotation	99	ND
Oil separation	92	65
Reverse osmosis	63–99 [b]	BDL–1
Sedimentation	99–99	ND–44
Ultrafiltration	95 [b]	BDL
Activated sludge	20–99	ND–69
Di-n-butyl phthalate		
Activated carbon adsorption (granular)	0–99	BDL–11
Chemical oxidation (ozone)	78	2.7
Chemical precipitation with sedimentation		
lime	93 [b]	BDL–BDL
alum	93–99	ND–550
FeCl(3)	5	5.4
combined precipitants	97–99	ND–ND
Coagulation and flocculation	0	BDL–0.6
Filtration	0–96	ND–1,300
Flotation	0–99	ND–300
Oil separation	94	49

Pollutant Removability/Treatability

Treatment Process	Range of Removal (%)	Range of Effluent Conc. (g/L)
Reverse osmosis	38–99[b]	BDL–11
Sedimentation	83	BDL–36
Ultrafiltration	86	13
Activated sludge	84–99	ND–7
Lagoons (aerated)	99	ND
Trickling filters	25	6
Di-n-octyl phthalate		
Activated carbon adsorption (granular)	20	4
Coagulation and flocculation	90	ND
Filtration	50–99	ND–4
Flotation	61–99	ND–33
Activated sludge		5,000
Bis(2-ethylhexyl)phthalate		
Activated carbon adsorption		
granular	26–66	4.7–410
powdered	97	10
Chemical precipitation with sedimentation		
lime	41–99	ND–40
alum	0–99	ND–67
barium chloride	95	2.4
polymer/unspecified	50	BDL
sodium hydroxide	73–92	BDL–52
combined precipitants	96	BDL–10
Chemical precipitation with filtration	95	BDL–84
Coagulation and flocculation	16–91	BDL–44
Filtration	21–98	BDL–16,000
Flotation	10–98	30–1,100
Oil separation	91–94	44–130
Reverse osmosis	25–99[b]	BDL–31
Sedimentation	0–80	BDL–170
Ultrafiltration	99[b]	BDL
Activated sludge	15–99	ND–220
Lagoons		
aerated	26–99	ND–28
non-aerated	72–99	ND–11
Trickling filters	83	6
Butyl benzyl phthalate		
Activated carbon adsorption (granular)	53–99	BDL–17
Chemical oxidation (ozone)	98[b]–98[b]	*BDL*–BDL

Pollutant Removability/Treatability

Treatment Process	Range of Removal (%)	Range of Effluent Conc. (g/L)
Chemical precipitation with sedimentation (lime)	99[b]	BDL–BDL
Coagulation and flocculation	94	BDL
Filtration	52–99	ND–10
Flotation	97–99	ND–42
Reverse osmosis	20–98	BDL–0.8
Sedimentation	95	ND–BDL
N-nitrosodiphenylamine		
Chemical precipitation with sedimentation (polymer/unspecified)	99	ND
Filtration	99	ND–0.4
Flotation	66	620
Sedimentation	99	ND
Activated sludge	69–99	ND–1.6
N-nitrosodi-n-propylamine		
Filtration	99	ND–BDL
Benzidine		
Activated sludge	0	4
Phenol		
Activated carbon adsorption		
granular	0–98	BDL–49
powdered	85	10
Chemical precipitation with sedimentation		
lime	69[b]	BDL
alum	0–99	ND–140
combined precipitants	33–96	10–47
Coagulation and flocculation	75–91	BDL–3
Filtration	22–99	ND–34,000
Flotation	0–80	9–2,400
Oil separation	99	ND–820
Reverse osmosis	80	0.2–0.7
Sedimentation	33–99	BDL–670
Solvent extraction	3–96	2,220–4,600,000
Activated sludge	8–99	ND–1,400
Lagoons (aerated)	ND–24	25–99
2-chlorophenol		
Filtration	0	2–17
Oil separation	99	ND
Sedimentation	99	BDL
Activated sludge	92–99	ND–100

Pollutant Removability/Treatability

Treatment Process	Range of Removal (%)	Range of Effluent Conc. (g/L)
2,4-dichlorophenol		
Chemical precipitation with sedimentation (sodium carbonate)	99	ND
Filtration	63–99	ND–2
Sedimentation	98	10–48
2,4,6-trichlorophenol		
Filtration	80	69
Oil separation	99	ND
Ultrafiltration	99	ND
Activated sludge	37–99	ND–4,300
Lagoons (aerated)	99	ND
Pentachlorophenol		
Activated carbon adsorption (granular)	59–99	BDL–49
Chemical precipitation with sedimentation (alum)	99	100
Filtration	99	ND–12
Flotation	19	5–30
Sedimentation	55	12–24
Activated sludge	67–99	ND–3,100
Lagoons (aerated)	99	ND–ND
2-nitrophenol		
Sedimentation	99	ND
Activated sludge	99–99	ND–BDL
4-nitrophenol		
Activated sludge	99	ND
Lagoons (aerated)	23	10
2,4-dinitrophenol		
Chemical precipitation with sedimentation (polymer/unspecified)	99	ND
Activated sludge	99	ND
2,4-dimethylphenol		
Chemical precipitation with sedimentation (lime)	48–88[b]	BDL–11
Flotation	99	ND–28
Sedimentation	99	ND
Activated sludge	99	ND–8
Total phenols		
Activated carbon adsorption		

Pollutant Removability/Treatability

Treatment Process	Range of Removal (%)	Range of Effluent Conc. (g/L)
granular	38–97	5–4,300
powdered	99–99	10–60
Chemical oxidation (ozone)	50–90	10–130
Chemical precipitation with sedimentation		
lime	0–99	60–1,600
alum	8–56	28–220,000
FeCl(3)	33	20
sodium hydroxide	90	60–70
combined precipitants	48–99	ND–1,300
Chemical precipitation with filtration	99	ND
Coagulation and flocculation	0–26	20–340
Filtration	0–60	10–64,000
Flotation	3–94	10–23,000
Oil separation	0–43	20–1,600
Reverse osmosis	5–81	12–20
Sedimentation	0–96	5–84
Ultrafiltration	27	16
Activated sludge	11–99	7–18
Lagoons		
aerated	33–99	3–20
non-aerated	40	30–50
Trickling filters	97	1,000
p-Chloro-m-cresol		
Activated carbon adsorption (granular)	17	BDL–BDL
Chemical precipitation with lime	44	62
Activated sludge	94–99	ND–4
Cresol		
Solvent extraction	74–98	800–330,000
Benzene		
Activated carbon adsorption (granular)	64–90	BDL–210
Chemical oxidation (ozone)	80[b]	BDL
Chemical precipitation with sedimentation		
lime	99	ND–1.0
alum	99	ND–310
polymer/unspecified	35	720
sodium hydroxide	99	2.0
combined precipitants	50–90	46–3,800

Pollutant Removability/Treatability

Treatment Process	Range of Removal (%)	Range of Effluent Conc. (g/L)
Reverse osmosis	30–80	0.4–1.4
Sedimentation	64–99	ND–2,400
Solvent extraction	75–97	7,000–35,000
Ultrafiltration	99[b]	ND
Activated sludge	75–99	ND–64
Lagoons (aerated)	74–95	5–10
Chlorobenzene		
Activated carbon adsorption (granular)	98	BDL
Activated sludge	38–99	ND–100
1,2-dichlorobenzene		
Activated carbon adsorption (granular)	99	BDL–5.4
Chemical precipitation with sedimentation (alum)	99	ND–0.05
Coagulation and flocculation	99	BDL–13
Flotation	76	18–260
Oil separation	99	ND
Activated sludge	69–99	ND–69
Lagoons (aerated)	99	ND
1,3-dichlorobenzene		
Oil separation	99	ND
Activated sludge	99	ND
1,4-dichlorobenzene		
Activated sludge	79–99	ND–21
Lagoons (aerated)	99	ND
1,2,4-trichlorobenzene		
Activated carbon adsorption (granular)	99	ND–94
Chemical precipitation with sedimentation (alum)	91	150–150
Coagulation and flocculation	91	150
Activated sludge	49–99	ND–920
Hexachlorobenzene		
Activated sludge	97–99	ND–0.8
Ethylbenzene		
Activated carbon adsorption (granular)	50	BDL–1.3
Chemical precipitation with sedimentation alum	70–99	ND–38,000

Pollutant Removability/Treatability

Treatment Process	Range of Removal (%)	Range of Effluent Conc. (g/L)
Oil separation	18	330–630
Reverse osmosis	0–64	4–6
Sedimentation	17–99	BDL–1,100
Stripping (steam)	54–87	130,000
Activated sludge	69–99	0.9–250
Lagoons (aerated)	91–97	5–1,000
Chloroform		
Activated carbon adsorption (granular)	64–99	ND–18
Chemical precipitation with sedimentation		
lime	99	ND–BDL
alum	40–99	ND–550
sodium carbonate	33	2.0
sodium hydroxide	91	5
combined precipitants	94	BDL–4,700
Filtration	50	BDL–500
Flotation	20–99	ND–24
Reverse osmosis	0–93 [b]	BDL–31
Sedimentation	0–74	2–230
Stripping (steam)	49–99	ND–65,000
Activated sludge	9–99	ND–58
Lagoons (aerated)	99	ND–1,000
Carbon tetrachloride		
Chemical precipitation with sedimentation		
alum	94	65–180
polymer/unspecified	99	ND
combined precipitants	99	ND
Coagulation and flocculation		
Filtration	89–99	ND–55
Flotation	75	BDL–210
Sedimentation	99	ND
Activated sludge	98	BDL–0.1
Chloroethane		
Activated carbon adsorption (granular)	0–99	ND–240,000
1,1-dichloroethane		
Activated carbon adsorption (granular)	42–99	ND–45,000
Filtration	0–99	ND–180
Sedimentation	0	2
Activated sludge	99–99	ND–290

Pollutant Removability/Treatability

Treatment Process	Range of Removal (%)	Range of Effluent Conc. (g/L)
polymer/unspecified	81	130
combined precipitants	98–99	ND–22
Coagulation and flocculation	98	BDL–1.3
Flotation	3–99	ND–970
Oil separation	83	ND–BDL
Reverse osmosis	0–33	1–1
Sedimentation	64–99	ND–2,400
Solvent extraction	97	4,000
Activated sludge	77–99	ND–3,000
Lagoons (aerated)	99	ND
Nitrobenzene		
Chemical precipitation with sedimentation		
alum	68–99	ND–35
combined precipitants	99	5
Ultrafiltration	71 [b]	BDL
Toluene		
Activated carbon adsorption (granular)	23–99	BDL–630
Chemical oxidation (ozone)	31	0.9–1.2
Chemical precipitation with sedimentation		
lime	0–99	ND–5.0
alum	0–73	3–3,600
polymer/unspecified	39	1,900
FeCl(3)	0	1.1
combined precipitants	84–96	73–4,200
Coagulation and flocculation	74	BDL–14
Flotation	10–99	ND–2,100
Oil separation	83	ND–BDL
Reverse osmosis	12–15	0.7–29
Sedimentation	17–83	BDL–1,100
Solvent extraction	94–96	1,600–2,300
Activated sludge	17–99	ND–1,400
Lagoons (aerated)	95–99	ND–BDL
2,6-dinitrotoluene		
Activated sludge	99	ND–200
Styrene		
Solvent extraction	93	1,000
Xylenes		
Filtration	75	12,000
Flotation	95–99	ND–1,000

Pollutant Removability/Treatability

Treatment Process	Range of Removal (%)	Range of Effluent Conc. (g/L)
Oil separation	58	10
Solvent extraction	97–98	1,000–1,000
2-chloronaphthalene		
Flotation	0	17
Reverse osmosis		1.4
Benzo(a)anthracene		
Chemical precipitation with sedimentation (sodium hydroxide)	64	BDL
Oil separation	9	10
Benzo(b)fluoranthene		
Chemical oxidation (ozone)	90[b]	BDL
Oil separation	9	10
Reverse osmosis		2.1–12
Sedimentation	86–99	ND–BDL
Benzo(k)fluoranthene		
Activated carbon adsorption (granular)	90	BDL
Sedimentation	99	ND
Benzo(a)pyrene		
Activated carbon adsorption (granular)	95	BDL–0.8
Chemical oxidation (ozone)	95[b]	BDL
Chemical precipitation with sedimentation (lime)	91[b]	ND–BDL
Oil separation	23	10
Sedimentation	83–99	ND–10
Ideno(1,2,3-c,d)pyrene		
Activated sludge	99	ND
Benzo(ghi)perylene		
Sedimentation	99	ND
Acenaphthene		
Chemical precipitation with sedimentation (lime)	75[b]	BDL
Filtration	73–99	ND–10
Oil separation	99	6
Reverse osmosis	9–99[b]	BDL
Sedimentation	99	ND–53
Activated sludge	0–99	ND–2.0

Pollutant Removability/Treatability

Treatment Process	Range of Removal (%)	Range of Effluent Conc. (g/L)
Acenaphthylene		
Reverse osmosis		1–1.6
Sedimentation	99	ND–19
Anthracene		
Activated carbon adsorption (granular)	50–97	BDL–0.4
Chemical oxidation (ozone)	98	BDL–0.4
Chemical precipitation with sedimentation		
lime	99	ND–BDL
polymer/unspecified		BDL
FeCl(3)	50	0.1
Filtration	0–99	ND–32,000
Flotation	45–98	0.2–600
Oil separation	99	ND
Reverse osmosis	77–99	BDL–1
Sedimentation	0–92	BDL–51
Activated sludge	52–99	ND–500
Chrysene		
Chemical precipitation with sedimentation		
lime	99	ND–10
polymer/unspecified	99	ND
sodium hydroxide	64	BDL
Oil separation	9	10
Sedimentation	0	10–19
Fluoranthene		
Activated carbon adsorption (granular)	88–95	BDL–BDL
Chemical precipitation with sedimentation		
lime	98[b]	ND–BDL
polymer/unspecified	99	ND
Filtration	20–50	0.05–93
Reverse osmosis	63–98	BDL–7.4
Sedimentation	64–99	ND–33
Activated sludge	0	2
Fluorene		
Chemical oxidation (ozone)	50	0.1
Chemical precipitation with sedimentation		
lime	99[b]	ND–BDL

Pollutant Removability/Treatability

Treatment Process	Range of Removal (%)	Range of Effluent Conc. (g/L)
sodium hydroxide	88	BDL
Oil separation	99	ND
Sedimentation	40–99	ND–23
Activated sludge	99–99	ND–ND
Naphthalene		
Activated carbon adsorption		
granular	51	78
powdered	96	10
Chemical precipitation with sedimentation		
lime	97[b]	ND–BDL
sodium hydroxide	77	1.0
combined precipitants	33–70	BDL–10
Filtration	86	BDL–160
Flotation	33–99	ND–840
Reverse osmosis	99[b]	BDL
Sedimentation	41–99	ND–23
Activated sludge	0–99	ND–260
Lagoons		
aerated	99	ND
non-aerated	99	ND
Phenanthrene		
Activated carbon adsorption (granular)	97–99	BDL–BDL
Chemical precipitation with sedimentation (lime)	92–99	ND–BDL
Coagulation and flocculation		
Filtration	67	ND–32,000
Flotation	45–98	0.2–600
Oil separation	99	ND
Reverse osmosis	99[b]	BDL
Sedimentation	0	BDL–40
Activated sludge	52–99	ND–500
Pyrene		
Activated carbon adsorption (granular)	95–98	BDL–BDL
Chemical oxidation (ozone)	67	0.1
Chemical precipitation with sedimentation (polymer/unspecified)	99	ND
Filtration	0–10	0.09–3,200
Flotation	0	0.3–18
Oil separation	99	ND

Pollutant Removability/Treatability

Treatment Process	Range of Removal (%)	Range of Effluent Conc. (g/L)
Reverse osmosis	19–99[b]	BDL–3.9
Sedimentation	19–99	ND–21
Activated sludge	78	0.1–9
Aroclor 1016		
Filtration	16	480
Oil separation	98	BDL–8
Aroclor 1221		
Filtration	20	650
Oil separation	97	BDL–6
Aroclor 1232		
Filtration	16	480
Oil separation	98	BDL–8
Aroclor 1242		
Filtration	20	650
Oil separation	97	BDL–6
Aroclor 1248		
Filtration	16	480
Oil separation	98	BDL–8
Aroclor 1254		
Filtration	20	650
Oil separation	97	BDL–6
Aroclor 1260		
Filtration	16	480
Oil separation	BDL–8	98
Methyl chloride		
Sedimentation	84	BDL–39
Methylene chloride		
Activated carbon adsorption (granular)	0–99	1.8–940
Chemical precipitation with sedimentation		
lime	33	BDL–39
alum	94	ND–1,300
polymer/unspecified	93–93	BDL–130
FeCl(3)	92	14
sodium hydroxide	90	1–90
combined precipitants	13	900–9,800
Coagulation and flocculation	56	70–100
Filtration	5–99	ND–31,000
Flotation	0–84	2–6,000

Pollutant Removability/Treatability

Treatment Process	Range of Removal (%)	Range of Effluent Conc. (g/L)
1,2-dichloroethane		
Activated carbon adsorption (granular)	21–99	ND–760,000
Chemical precipitation with sedimentation (alum)	99	ND–90
Stripping (steam)	70–99	5.9–440,000
Solvent extraction	99	20,000
1,1,1-trichloroethane		
Activated carbon adsorption (granular)	99–99	ND–1.9
Chemical oxidation		
Chemical precipitation with sedimentation		
alum	55	10–170
sodium hydroxide	0–99	ND–1
Chemical precipitation with filtration	75	0.3–1.0
Filtration	86–99	ND–4,400
Flotation	22–99	ND–860
Sedimentation	19–96	2–2,500
Stripping (steam)	9	42,000
Activated sludge	99–99	ND–3.3
Lagoons (aerated)	96	22
1,1,2-trichloroethane		
Activated carbon adsorption (granular)	99–99	ND–ND
Chemical precipitation with sedimentation		
sodium hydroxide	50	1.0
combined precipitants	99	ND–ND
Stripping (steam)	98–99	ND–200
Solvent extraction	90	16,420
1,1,2,2-tetrachloroethane		
Stripping (steam)	99–99	100–78,000
Solvent extraction	92	4,200
Activated sludge	99	ND–BDL
Vinyl chloride		
Activated carbon adsorption (granular)	52	1,100–9,600
1,2-dichloropropene		
Activated carbon adsorption (granular)	65–99	ND–BDL

Pollutant Removability/Treatability

Treatment Process	Range of Removal (%)	Range of Effluent Conc. (g/L)
Chemical precipitation with sedimentation (combined precipitants)		
Activated sludge	99–99	ND–ND
1,3-dichloropropane		
activated sludge	99	ND–0.89
Dichlorobromomethane		
Coagulation and flocculation	75	BDL
Flotation	99	ND
Activated sludge	99	ND–1.5
Chlorodibromomethane		
Chemical precipitation with sedimentation		
lime	99	ND
combined precipitants	99	ND–ND
Sedimentation	99	ND–1
Trichlorofluoromethane		
Flotation	99	ND
Stripping (steam)	24	34,000
Activated sludge	96	1.7–89
Trichloroethylene		
Activated carbon adsorption (granular)	58–99	BDL–5
Chemical precipitation with sedimentation		
lime		
alum	10–99	17
combined precipitants	99	ND–300
Filtration	0–99	ND–2,000
Flotation	86	130
Sedimentation	21–93	33–3,000
Stripping (steam)	99–99	ND–ND
Activated sludge	56–99	ND–84
1,1-dichloroethylene		
Activated carbon adsorption (granular)	99	ND–1.4
Chemical precipitation with sedimentation (alum)	99	ND–22
Filtration	52–57	ND–130
Sedimentation	87	40–70
1,2-trans-Dichloroethylene		
Activated carbon adsorption		

Pollutant Removability/Treatability

Treatment Process	Range of Removal (%)	Range of Effluent Conc. (g/L)
(granular)	96–98	1.1–140
Chemical precipitation with sedimentation (alum)	27	5–190
Sedimentation	38–44	5–19
Stripping (steam)	9–99	15,000–1,300,000
Tetrachloroethylene		
Activated carbon adsorption (granular)	68	BDL–32
Chemical oxidation (ozone)	99	ND
Chemical precipitation with sedimentation		
alum	99	ND–700
sodium hydroxide	99	ND
combined precipitants	95–99	ND–7
Filtration	0–0	ND–210
Flotation	0–99	ND–1,000
Oil separation	13–99	ND–71
Sedimentation	21–99	ND–3,000
Stripping (steam)	37–99	ND–6,800
Ultrafiltration	93	200
Activated sludge	55–99	ND–40
Lagoons (aerated)	99	ND
Alpha-endosulfan		
Oil separation	99	ND
Alpha-BHC		
Coagulation and flocculation	91	BDL
Filtration	77	1.9–6
Oil separation	86	ND–BDL
Ultrafiltration	79[b]	BDL
Beta-BHC		
Filtration	21	55
Ultrafiltration	50[b]	BDL
Activated sludge	99	ND
Gamma-BHC		
Filtration	64	BDL
4,4'-DDE		
Ultrafiltration	64[b]	BDL
4,4'-DDT		
Chemical precipitation with sedimentation (alum)	52	1
Coagulation and flocculation	76	BDL

Pollutant Removability/Treatability

Treatment Process	Range of Removal (%)	Range of Effluent Conc. (g/L)
Endrin aldehyde		
Oil separation	99	ND
Heptachlor		
Chemical precipitation with sedimentation (alum)	29	1.0
Coagulation and flocculation	64	BDL
Activated sludge	75	1.6
Chlordane		
Filtration	37	24
Acrolein		
Filtration	99	ND
Isophorone		
Chemical precipitation with sedimentation (lime)	6	560
Flotation	99	ND
Sedimentation	35–99	ND–110
Activated sludge		BDL–BDL

Notes: BLD = below detection limit; ND = not detected.
[a]Range of effluent conc. (fibers/L).
[b]Approximate value.

APPENDIX 3-2
CHEMICAL TREATABILITY SUMMARY

An extensive amount of information on the treatability of hundreds of chemical compounds was collected during the course of a program being conducted by Touhill, Shuckrow, and Associates, Inc. (TSA) for the EPA. The information was assembled in an appendix to the TSA report entitled "Concentration Technologies for Hazardous Aqueous Waste Treatment." To provide a quick reference on the treatability of the 505 chemical compounds listed in the report appendix, we have prepared a concise summary and present it here. The full appendix and its references span several hundred pages, so those readers interested in additional details should consult the full report.

This summary will be most useful to those desiring a 'first-cut'' idea of how well certain treatment processes can deal with the chemical compounds listed.

Compounds are arranged in alphabetical order according to the chemical classification (in order) alcohol, aliphatic, amine, aromatic, ether, halocarbon (halogenated aliphatics), metal, PCB, pesticide, phenol, phthalate, and polynuclear aromatic. The roman numeral in the "Process-Treatability" column refers to the following unit processes.

Process	Appendix 3-2 Process Code No.
Biological	I
Coagulation/precipitation	II
Reverse osmosis	III
Ultrafiltration	IV
Stripping	V
Solvent extraction	VII
Carbon adsorption	IX
Resin adsorption	X
Miscellaneous adsorbents	XII

For each compound a summary statement describing its treatability is given with information on treatability by more than one concentration technology provided for the majority of compounds.

Many compounds are known by several names. Attempts were made to use preferred or generic names according to *The Merck Index*. However, in some cases it was necessary to use names that were used in the reference documents. Users of the table are advised to check for compounds under several potential compound names. Additionally, some of the chemical classifications overlap; hence, users also should check related classifications.

This information was collected in partial fulfillment of Contract No. 68-03-2766 by Touhill, Shuckrow, and Associates, Inc. under the sponsorship of the EPA.

Chemical Treatability Summary

Chemical		Process–Treatability	Ref.
A. Alcohols			
Ally alcohol	IX	22% reduction @ 100 mg/L	29
n-amyl alcohol	I	toxic @ 350 mg/L	51
(1-pentanol)	IX	72% reduction @ 1000 mg/L	29
Borneol	I	90% reduction	65
Butanol	I	70–100% reduction	9, 41, 43, 51, 65
	IX	53–100% reduction @ 0.1 to 1000 mg/L	14, 29, 63
	X	100% reduction @ μg/L	14
Sec-butanol	I	98% reduction	65
Tert-butanol	I	98% reduction	43, 65
	IX	30% reduction @ 1000 mg/L	29
1,4-butanediol	I	99% reduction	65
Cyclohexanol	I	96% reduction	65
	IX	100% reduction @ 100 μg/L	14
	X	100% reduction @ 100 μg/L	14
Decanol	IX	100% reduction @ 100 μg/L	14
	X	100% reduction @ 100 μg/L	14
Dimethylcyclohexanol	I	92% reduction	65
1,2-ethanediol	I	depressed performance @ 484 mg/L	64
Ethanol	I	70–100% reduction @ up to 1000 mg/L	9, 43, 64
	III	20–100% reduction @ 1000 mg/L–dependent upon membrane	13, 24
	VII	7% reduction @ 286 mg/L	23
	IX	10% reduction @ 1000 mg/L	14
Ethylbutanol	I	75–100% reduction	9, 41, 43
2-ethylbutanol	IX	86% reduction @ 1000 mg/L	29
2-ethylhexanol	I	75–85% reduction	41
	IX	98% reduction @ 700 mg/L	29
2-ethyl-l-hexanol	IX	100% reduction @ 100 μg/L	14
	X	100% reduction @ μg/L	14
Furfuryl alcohol	I	97% reduction	65
m-heptanol	IX	100% reduction @ 100 μg/L	14
	X	100% reduction @ 100 μg/L	14
1-hexanol	I	70–100% reduction	9, 41
m-hexanol	IX	96% reduction @ 1000 μg/L	29
Isobutanol	IX	42% reduction @ 1000 μg/L	29
Isopropanol	I	70–100% reduction	9, 41, 43, 65
	IX	13% reduction @ 1000 μg/L	29

Chemical Treatability Summary

Chemical		Process–Treatability	Ref.
Methanol	I	30–85% reduction	41, 60, 9, 43, 64
	III	0–40% reduction @ 1000 mg/L– dependent upon membrane	13, 24
	IX	4–33% reduction @ 15–1000 mg/L	29, 63
4-methylcyclohexanol	I	94% reduction	65
Octanol	I	30–75% reduction	9, 43
	IX	100% reduction @ 100 μg/L	14
	X	100% reduction @ 100 μg/L	14
Pentanol	IX	100% reduction @ 100 μg/L	14
	X	100% reduction @ 100 μg/L	14
Pentarythritol	I	No toxic effect	21
Phenyl methyl carbinol	I	85–95% reduction	43
Propanol	IX	100% reduction @ 100 μg/L 19% reduction @ 1000 μg/L	14, 29
	X	100% reduction @ 100 μg/L	14
i-propanol	III	20–100% reduction @ 1000 mg/L– deplendent upon membrane	13, 24
m-propanol	I	99% reduction	65
B. Aliphatics			
Acetaldehyde	I	30–95% reduction	9, 41, 60
	IX	12% reduction @ 1000 mg/L	29
Acetic acid	III	20–80% reduction @ 1000 mg/L– dependent upon membrane	13, 24
	IX	24% reduction @ 1000 mg/L	29
Acetone	I	50–100% reduction	9, 56, 64
	III	15–100% reduction @ 1000 mg/L– dependent upon membrane	13, 24
	IX	22% reduction @ 1000 mg/L	29
Acetone cyanohydrin	IX	30–60% reduction @ 100–1000 mg/L	63
Acetonitrile	I	Inhibitory @ 500 mg/L	52, 64
Acetylglycine	I	Readily oxidized @ 500 mg/L	54
Acrolein	VII	Extractable with xylene	22
	IX	30% reduction @ 1000 mg/L	22, 29
Acrylic acid	I	50–95% reduction	9, 41, 43
	IX	64% reduction @ 1000 mg/L	22, 29
Acrylonitrile	I	70–100% reduction	22, 41
	V	Could be flash evaporated	22
	VII	Extractable with ethyl ether	22

Chemical Treatability Summary

Chemical		Process–Treatability	Ref.
Adipic acid	I	Readily oxidixed @ 1000 mg/L	52
Alanine	I	Readily degraded @ 500 mg/L	64
Ammonium oxalate	I	92% reduction	65
Amyl acetate	IX	88% reduction @ 985 mg/L	29
Butanedinitrile	I	Toxic @ 500 mg/L; also reported to be readily but slowly oxidized	52, 54
Butanenitrile	I	Toxic @ 500 mg/L; also reported to be readily but slowly oxidized	52, 54
Butyl acetate	IX	85% reduction @ 1000 mg/L	29
Butyl acrylate	IX	96% reduction @ 1000 mg/L	29
Butylene oxide	I	Degraded very slowly	54
Butyraldehyde	IX	53% reduction @ 1000 mg/L	29
Butyric acid	I	50–95% reduction; rapidly oxidized	9, 41, 52, 54
	IX	60% reduction @ 1000 mg/L 100% reduction @ 100 μg/L	14, 29
	X	100% reduction @ 100 μg/L	14
Calcium gluconate	I	Rapidly oxidized	64
Caproic acid	IX	90–98% reduction @ 0.1–1000 mg/L	14, 29
	X	50% reduction @ 1000 mg/L	14
Caprolactam	I	94% reduction	65
Citric acid	I	Biodegradable; depressed O_2 consumption	64
Crotonaldehyde	I	90–100% reduction	9, 41, 43
	IX	46% reduction @ 1000 mg/L	29
Cyclohexanolone	I	92% reduction	65
Cyclohexanone	I	96% reduction	65
	IX	67% reduction @ 1000 mg/L	29
Cyclopentanone	I	96% reduction	65
Cystine	I	Completed inhibited O_2 consumption @ 1000 mg/L	64
L-cystine	I	Slowly oxidized @ 1000 mg/L	54
Decanoic acid	IX	100% reduction @ 100 μg/L	14
	X	100% reduction @ 100 μg/L	14
Dicyclopentadiene	IX	Found to vaporize	68
Diethylene glycol	I	95% reduction	65
	IX	26% reduction @ 1000 mg/L	29

Chemical Treatability Summary

Chemical		Process–Treatability	Ref.
Diisobutyl ketone	IX	100% reduction @ 300 mg/L	29
Diisopropyl methyl-phosphonate	IX	98% reduction @ 2680 µg/L	68
Dimethyl sulfoxide	III	63–88% reduction @ 250–1000 mg/L	13
Dipropylene glycol	IX	16% reduction @ 1000 mg/L	29
2,3-dithiabutane	I	100% reduction @ 120 µg/L	60
Dodecane	IX	100% reductgion @ 100 µg/L	14
	X	25% reduction @ 100 µg/L	14
Dulcitol	I	Slightly inhibitory @ 1700 mg/L	20
Erucic acid	I	Oxidized @ 500 mg/L	54
Ethyl acetate	I	90–100% reduction	9, 41, 43
	IX	50% reduction @ 1000 mg/L	29
Ethyl acrylate	I	90–100% reduction	9, 41, 43
	IX	78% reduction @ 1015 mg/L	29
Ethylene glycol	I	97% reduction	61
	IX	7% reduction @ 1000 mg/L	29
2-Ethylhexylacrylate	I	90–100% reduction	9, 41, 43
Formaldehyde	I	Conflicting data; removable and inhibitory @ 720–3000 mg/L	21, 64
	III	20–80% reduction @ 1000 mg/L–dependent upon membrane	13, 24
	IX	9% reduction @ 1000 mg/L	29
Formamide	I	Slowly oxidized @ 500 mg/L	54
Formic acid	I	Rapidly oxidized @ 720 mg/L	54
	IX	24% reduction @ 1000 mg/L	29
Glutamic acid	I	Readily oxidized	64
Glycerol	III	20–100% reduction @ 1000 mg/L–dependent upon membrane	13, 24
Glycerine	I	Readily oxidized @ 720 mg/L	64
Glycine	I	Rapidly oxidized @ 720 mg/L	64
Heptane	I	90–100% reduction	9, 41, 43, 52
Heptanoic acid	IX	10% reduction @ 100 µg/L	14
	X	50% reduction @ 100 µg/L	14
Hexadecane	IX	100% reduction @ 100 µg/L	14
	X	25% reduction @ 100 µg/L	14
Hexylene glycol	IX	61% reduction @ 1000 mg/L	29
Hydracrylonitrile	I	0–10% reduction	9

Chemical Treatability Summary

Chemical		Process–Treatability	Ref.
Isobutyl acetate	IX	82% reduction @ 1000 mg/L	29
Isophorone	I	93% reduction	65
	VII	Extractable with ethyl ether	22
Isoprene	IX	86% reduction @ 500–1000 mg/L	63
Isopropyl acetate	IX	68% reduction @ 1000 mg/L	29
Lactic acid	I	Rapidly oxidized @ 720 mg/L	7
Lauric acid	I	Slowly oxidized @ 500 mg/L	54
	IX	100% reduction @ 100 μg/L	14
	X	100% reduction @ 100 μg/L	14
L-malic acid	I	Rapidly oxidized @ 500 mg/L	54
DL-malic acid	I	Oxidized after 10–16 hr. lag period	54
Malonic acid	I	Inhibitory @ 500 mg/L	54
Methyl acetate	III	40–80% reduction @ 1000 mg/L– dependent upon membrane	13, 24
	IX	26% reduction @ 1030 mg/L	29
Methyl butyl ketone	IX	81% reduction @ 988 mg/L	29
Methyl decanoate	IX	100% reduction @ 100 μg/L	14
	X	100% reduction @ 100 μg/L	14
Methyl dodecanoate	IX	100% reduction @ 100 μg/L	14
	X	100% reduction @ 100 μg/L	14
Methyl ethyl ketone	VII	69–88% reduction @ 12,200 mg/L	23
	IX	47% reduction @ 1000 mg/L	29
Methyl hexadecanoate	IX	100% reduction @ 100 μg/L	14
	X	100% reduction @ 100 μg/L	14
Methyl isoamyl ketone	IX	85% reduction @ 986 mg/L	29
Methyl isobutyl ketone	IX	85% reduction @ 1000 mg/L	29
Methyl octadecanoate	X	100% reduction @ 100 μg/L	14
Methyl propyl ketone	IX	70% reduction @ 1000 mg/L	29
Myristic acid	IX	100% reduction @ 100 μg/L	14
	X	100% reduction @ 100 μg/L	14
Nitrilotriacetate	I	90% reduction @ 500 mg/L– after acclimation	69
Octadecane	IX	100% reduction @ 100 μg/L	14
	X	25% reduction @ 100 μg/L	14
Oleic acid	I	Inhibitory	20

Chemical Treatability Summary

Chemical		Process–Treatability	Ref.
Oxalic acid	I	Inhibitory @ 250 mg/L	64
Pentane	I	Inhibitory @ 500 mg/L	52
Pentanedinitrile	I	Slowly oxidized or toxic @ 500 mg/L	52, 54
Pentanitrile	I	Toxic @ 500 mg/L	52
Propanedinitrile	I	Toxic @ 500 mg/L	52
Propanenitrile	I	Toxic @ 500 mg/L	52
B-propiriolactone	I	Inhibitory @ 500 mg/L	55
Propionaldehyde	IX	28% reduction @ 1000 mg/L	29
Propionic acid	IX	100% reduction @ 100 μg/L 33% reduction @ 1000 mg/L	14, 29
	X	100% reduction @ 100 μg/L	14
Propyl acetate	IX	75% reduction @ 1000 mg/L	29
Propylene glycol	IX	12% reduction @ 1000 mg/L	29
Proplylene oxide	IX	26% reduction @ 1000 mg/L	29
Pyruvic acid	IX	100% reduction @ 100 μg/L	14
	X	100% reduction @ 100 μg/L	14
Sodium alkyl sulfate	I	Readily degraded	36
Sodium lauryl sulfate	I	Rapidly oxidized	36
Sodium-n-oleyl-n- methyl taurate	I	Readily oxidized	36
Sodium sulfo methyl myristate	I	Readily oxidized	36
Tannic acid	I	Inhibitory	20
Tetradecane	IX	100% reduction @ 100 μg/L	14
	X	50% reduction @ 100 μg/L	14
Tetraethylene glycol	IX	58% reduction @ 1000 mg/L	29
Thioglycolic acid	I	Inhibitory	64
Thiouracil	I	Very slowly oxidized @ 500 mg/L	55
Thiourea	I	Inhibitory @ 500 mg/L	20
Triethylene glycol	I	98% reduction	65
	IX	52% reduction @ 1000 mg/L	29
Urea	I	Inhibitory @ 1200 mg/L	64
Urethane	I	Inhibitory	55
Valeric acid	IX	80–100% reduction @ 0.1–1000 mg/L	14, 29
	X	50% reduction @ 100 μg/L	14
Vinyl acetate	IX	64% reduction @ 1000 μg/L	29

Chemical Treatability Summary

Chemical		Process–Treatability	Ref.
C. Amines			
Acetanilide	I	94% reduction	65
Allylamine	IX	31% reduction @ 1000 mg/L	29
p-aminoacetanilide	I	93% reduction	65
m-aminobenzoic acid	I	98% reduction	65
o-aminobenzoic acid	I	98% reduction	65
p-aminobenzoic acid	I	96% reduction	65
m-aminotoluene	I	98% reduction	65
o-aminotoluene	I	98% reduction	65
p-aminotoluene	I	98% reduction	65
Aniline	I	Inconsistent data; 100% reduction and inhibitory reported @ 500 mg/L	55, 65, 74
	III	3–100% reduction @ 1000 mg/L– dependent on membrane	13, 24
	IX	75–100% reduction @ 0.1–1000 mg/L	14, 29
	X	100% reduction @ 100 μg/L	14
Benzamide	I	Initially inhibitory; slowly degraded @ 500 mg/L	54
Benzidine	I	Inhibitory @ 500 mg/L; not reduced @ 1.6 μg/L	55, 65
Benzylamine	I	Inhibitory @ 500 mg/L	55
Butanamide	I	Slowly oxidized @ 500 mg/L	54
Butylamine	IX	52–100% reduction @ 0.1–1000 mg/L	14, 29
	X	100% reduction @ 100 μg/L	14
m-chloroaniline	I	97% reduction	65
o-chloroaniline	I	97% reduction	65
p-chloroaniline	I	96% reduction	65
Cyclohexylamine	IX	100% reduction @ 100 μg/L	14
	X	100% reduction @ 100 μg/L	14
Dibutylamine	IX	100% reduction @ 100 μg/L	14
	X	100% reduction @ 100 μg/L	14
Di-*n*-butylamine	IX	87% reduction @ 1000 mg/L	29
Diethanolamine	I	97% reduction	65
	IX	28% reduction @ 996 mg/L	29
Diethylenetriamine	IX	29% reduction @ 1000 mg/L	29
Dihexylamine	IX	100% reduction @ 100 μg/L	14
	X	100% reduction @ 100 μg/L	14

Chemical Treatability Summary

Chemical		Process–Treatability	Ref.
Diisopropanolamine	IX	46% reduction @ 1000 mg/L	29
Dimethylamine	IX	100% reduction @ 100 μg/L	14
	X	100% reduction @ 100 μg/L	14
2,3-dimethylaniline	I	96% reduction	65
2,5-dimethylaniline	I	96% reduction	65
3,4-dimethylaniline	I	76% reduction	65
Dimethylnitrosamine	IX	Not adsorbed	25
Di-*n*-propylamine	IX	80% reduction @ 1000 mg/L	29
Ethylenediamine	I	98% reduction	65
	IX	11% reduction @ 1000 mg/L	29
N-ethylmorpholine	IX	47% reduction @ 1000 mg/L	29
2-fluorenamine	I	Slowly biodegraded @ 500 mg/L	55
Hexylamine	IX	100% reduction @ 100 μg/L	14
	X	100% reduction @ 100 μg/L	14
2-methyl-5-ethylpyridine	IX	89% reduction @ 1000 mg/L	29
N-methyl morpholine	IX	42% reduction @ 1000 mg/L	29
Monoethanolamine	IX	7% reduction @ 1012 mg/L	29
Monoisopropanolamine	IX	20% reduction @ 1000 mg/L	29
Morpholine	IX	100% reduction @ 100 μg/L	14
	X	100% reduction @ 100 μg/L	14
B-naphthylamine	IX	Adsorbed	25
o-nitroaniline	I	99.9% reduction @ 18.5 mg/L	47
p-nitroaniline	I	99.9% reduction @ 6.7 mg/L	47
Octylamine	IX	100% reduction @ 100 μg/L	14
	X	100% reduction @ 100 μg/L	14
Pentanamide	I	Slowly oxidized @ 500 mg/L	54
p-(phenylzoa) aniline	I	Inhibitory @ 500 mg/L	55
Phenylenediamine	I	Toxic @ 500 mg/L	57
m-phenylenediamine	I	60% reduction	65
o-phenylenediamine	I	33% reduction	65
p-phenylenediamine	I	80% reduction	65
Piperidine	IX	100% reduction @ 100 μg/L	14

Chemical Treatability Summary

Chemical		Process–Treatability	Ref.
Piperidine	X	100% reduction @ 100 μg/L	14
Pyridine	IX	53% reduction @ 1000 mg/L	29
Pyrrole	IX	100% reduction @ 100 μg/L	14
	X	100% reduction @ 100 μg/L	14
Thiocetamide	I	Inhibitory @ 100 mg/L	64
Tributylamine	IX	100% reduction @ 100 μg/L	14
	X	100% reduction @ 100 μg/L	14
2,4,6-trichloro-aniline	I	Readily degraded @ 500 mg/L	57, 74
Triethanolamine	IX	33% reduction @ 1000 mg/L	29
D. Aromatics			
Acetophenone	IX	50–92% reduction @ 0.1–1000 mg/L	14, 29
	X	100% reduction @ 100 μg/L	14
sec-amylbenzene	I	Toxic @ 500 mg/L	57
tert-amylbenzene	I	Toxic @ 500 mg/L	57
Benzaldehyde	I	Conflicting data; reported to be toxic also 99% reduction	20, 55, 65
	IX	50–99% reduction @ 0.1–1000 mg/L	14, 29, 63
	X	100% reduction @ 100 μg/L	14
Benzene	I	90–100% reduction @ up to 500 mg/L	9, 37, 41, 43
	V	95–99% reduction	11, 22
	VII	97% reduction @ 71–290 mg/L	23
	IX	60–95% reduction @ 1 μg/L to 1500 mg/L	6, 16, 22, 25, 29, 31, 63
Benzene sulfonate	I	Slowly oxidized @ 500 mg/L	55
Benzene, toluene, xylene (BTX)	X	99% reduction @ 20–30 mg/L	27
Benzenethiol	I	Inhibitory @ 500 mg/L	55
Benzidine	IX	Adsorbed	25
Benzil	IX	50% reduction @ 100 μg/L	14
	X	100% reduction @ 100 μg/L	14
Benzoic acid	I	95–100% reduction	41, 65
	IX	91–100% reduction @ 0.1–1000 mg/L	14, 29
	X	100% reduction @ 100 μg/L	14
Benzanitrile	I	Inhibitory @ 500 mg/L	52
3,4-benzpyrene	I	Inhibitory @ 500 mg/L	52

Chemical Treatability Summary

Chemical		Process–Treatability	Ref.
sec-butylbenzene	I	Toxic @ 500 mg/L	57
tert-butylbenzene	I	Toxic @ 500 mg/L	57
Chloranil	I	Inhibitory @ 10 mg/L	43
Chlorobenzene	I	100% reduction @ 200 mg/L	61
	III	97–100% reduction @ 360 mg/L	22
	V	Steam strippable	22, 53
	VII	99% reduction w/chloroform solvent	22
	IX	50–85% reduction @ 1–416 mg/L	16, 22, 53
1-chloro-2- nitrobenzene	IX	Adsorbed	16
Cumene	IX	100% reduction @ 100 μg/L	14
	X	100% reduction @ 100 μg/L	14
1,2,4,5-dibenzpyrene	*I	Inhibitory @ 500 mg/L	55
m-dichlorobenzene	I	100% reduction @ 200 mg/L	22, 61
	V	Air and steam strippable	74
	VII	Extractable	22
	IX	95–100% reduction @ 0.1–416 mg/L	14, 22
	X	100% reduction @ 100 μg/L	14
o-dichlorobenzene	I	100% reduction @ 200 mg/L	61
	V	Air and steam strippable	22
	VII	Extractable	22
	IX	95–100% reduction @ 0.1–1000 mg/L	14, 22
	X	100% reduction @ 100 μg/L	14
p-dichlorobenzene	I	100% reduction @ 200 mg/L	61
	V	Steam strippable	22
	VII	Extractable	22
7,10-dimethylbenza- cridine	I	Inhibitory @ 500 mg/L	55
Dinitrobenzene	III	7–81% reduction @ 30 mg/L– dependent upon membrane	8
3,5-dinitrobenzoic acid	I	50% reduction	65
2,4-dinitrophenyl- hydrazine	III	3–91% reduction @ 30 mg/L– dependent upon membrane	13
2,4-dinitrotoluene	I	90–100% reduction @ 0.39–188 mg/L	22, 65
	VII	Extractable	22
	IX	95% reduction @ 416 mg/L	—
2,6-dinitrotoluene	VII	Extractable	22
	IX	95% reduction @ 416 mg/L	22

Chemical Treatability Summary

Chemical		Process-Treatability	Ref.
Ethylbenzene	I	90–100% reduction @ 0.192–105 mg/L	9, 16, 37, 41, 43
	II	56% reduction @ 153 mg/L with alum	16
	V	80–93% reduction	11, 22, 53
	VII	97% reduction	22, 23
	IX	50–84% reduction @ 1–115 mg/L	16, 22, 29, 53
Hexachlorobenzene	I	No reduction @ 200 mg/L	61, 74
	III	52% reduction @ 638 mg/L	22
	V	Steam strippable	53
	VII	Extractable	22
	IX	95% reduction @ 416 mg/L	22
Hydroquinone	III	2–80% reduction @ 1000 mg/L	13, 24
	IX	83% reduction @ 1000 mg/L	29
Hydroxybenzene-carbonitrile	I	Toxic @ 500 mg/L	52
Isophorone	IX	97% reduction @ 1000 mg/L	22, 29
2-methylbenzene-carbonitrile	I	Toxic @ 500 mg/L	52
3-methylbenzene-carbonitrile	I	Toxic @ 500 mg/L	52
4-methylbenzene-carbonitrile	I	Toxic @ 500 mg/L	52
4,4-methylene bis-(2-chloroaniline)	IX	Adsorbed	25
Methylethylpyridine	I	10–30% reduction	9
m-nitrobenzaldehyde	I	94% reduction	65
Paraldehyde	I	30–50% reduction	9
	IX	74% reduction @ 1000 mg/L	29
Pentamethylbenzene	I	Inhibitory @ 500 mg/L	57
m-propylbenzene	I	Very slowly oxidized @ 37.5 mg/L	37
Pyridine	IX	47–86% reduction @ 500–1000 mg/L	29, 63
Sodium alkyl-benzene sulfonate	I	Slowly oxidized	36
Styrene	I	70–100% reduction	9, 43
	V	98–99% reduction	11
	VII	93% reduction	23
	IX	55–97% reduction @ 20–200 mg/L	16, 29, 63
Styrene oxide	IX	95% reduction @ 1000 mg/L	29

Chemical Treatability Summary

Chemical		Process–Treatability	Ref.
1,2,3,4-tetrachloro-benzene	I	74% reduction @ 200 mg/L	61
1,2,3,5-tetrachloro-benzene	I	80% reduction @ 200 mg/L	61
1,2,4,5-tetrachloro-benzene	I	@ 200 mg/L, 80% reduction; @ 500 mg/L, very slowly oxidized	57, 61
Toluene	I	48–100% reduction @ 8 μg/L to 500 mg/L; 500 mg/L was inhibitory	9, 37, 41, 43, 52, 55, 60
	V	73–92% reduction	11, 22
	VII	94–96% reduction @ 41–44 mg/L	22, 23
	IX	79–98% reduction @ 0.12–317 mg/L	6, 22, 29
m-toluidine	I	100% reduction	74
Toxaphene	IX	99% reduction @ 155 μg/L	61
1,2,3-trichloro-benzene	I	100% reduction @ 200 mg/L	61
	V	50% reduction	22, 53
	VII	Extractable	22
	IX	70–100% reduction @ 0.1–416 mg/L	14, 22, 53
	X	100% reduction @ 100 μg/L	14
1,2,4-trichloro-benzene	I	100% reduction @ mg/L	61
1,3,5-trichloro-benzene	I	100% reduction @ 200 mg/L	61, 74
2,4,6-trichlorophen-oxyacetic acid	I	50% reduction @ 53 mg/L	39
2,4,5-trichlorophen-oxypropionic acid	I	99% reduction @ 107.5 mg/L	39
2,4,6-trinitrotoluene (TNT)	IV	80–93% TOC reduction @ 200 mg/L TOC	10
	IX	Adsorbed	1, 34
	X	99% reduction @ 81–116 mg/L	1, 34
2,6,6-trinitrotoluene	I	50–84% reduction @ 100 mg/L	33
Xylene	I	92–95% reduction @ 20–200 μg/L	60
	VII	97% reduction	23
	IX	68–99% reduction @ 0.14–200 mg/L	6, 63
m-xylene	I	Inhibitory @ 500 mg/L	57
o-xylene	I	Inhibitory @ 500 mg/L	57
p-xylene	I	Inhibitory @ 500 mg/L	57

Chemical Treatability Summary

Chemical		Process–Treatability	Ref.
E. Ethers			
bis(2-chloroiso-propyl)ether	III	37–94% reduction @ 250 mg/L–dependent upon membrane	13
	IX	100% reduction	22
bis(chloroethyl)ether	VII	Extractable	22
	IX	50% reduction @ 94 μg/L	22
bis(chloroisopropyl) ether	VII	Extractable with ethyl ether and benzene	22
Butyl ether	IX	100% reduction @ 197 mg/L	29
Dichloroisopropyl ether	IX	100% reduction @ 1008 mg/L	29
Diethyl ether	III	9.5–90% reduction @ 1000 mg/L–dependent upon membrane	13
Diethylene glycol monobutyl ether		83% reduction @ 1000 mg/L	29
Diethylene glycol monoethyl ether	IX	44% reduction @ 1010 mg/L	29
Ethoxytriglycol	IX	70% reduction @ 1000 mg/L	29
Ethylene glycol monobutyl ether	IX	56% reduction @ 1000 mg/L	29
Ethylene glycol monoethyl ether	IX	31% reduction @ 1022 mg/L	29
Ethylene glycol monoethyl ether acetate	IX	66% reduction @ 100 mg/L	29
Ethylene glycol monohexyl ether	IX	87% reduction @ 975 mg/L	29
Ethylene glycol monomethyl ether	IX	14% reduction @ 1024 mg/L	29
Ethyl ether	III	20–100% reduction @ 1000 mg/L–dependent upon membrane	24
Isopropyl ether	I	70–95% reduction	9, 41, 43
	IX	80% reduction @ 1023 mg/L	29
F. Halocarbons			
Bromochloro-methane	IX	Adsorbed	16
Bromodichloro-methane	V	Air and steam strippable	22
	VII	Soluble in most organics	22

Chemical Treatability Summary

Chemical		Process–Treatability	Ref.
Bromodichloro- methane	IX X	Adsorbed Adsorbed @ 2 mg/L	16, 38 38
Bromoform	I IX X	100% reduction @ 0.4–1.9 μg/L 100% reduction @ 100 μg/L 100% reduction @ 100 μg/L	60 14, 38 14, 38
Bromomethane	V VII IX	Air strippable Soluble in most organics Adsorbed	22 22 22
Carbon tetra- chloride	I II IX X	100% reduction @ 177 μg/L 51% reduction @ 144 μg/L with alum Adsorbed Adsorbed	16 16 6, 16, 22 27
Chloral	V	Steam strippable @ 693 mg/L	15
Chloral hydrate	VII	49% reduction @ 15,200 mg/L	23
Chloroethane	V VII IX	90% reduction by air stripping Extractable with alcohol and aromatics Adsorbed	22 22 22
Chloroethylene	V VII IX	Air strippable Soluble in most organics Adsorbed	22 22 22
Chloroform	V IX X	Steam strippable @ 140 mg/L Adsorbed Adsorbed	15 16, 38 27, 38
Chloromethane	V VII	Air strippable Soluble in most organics	22 22
Dibromochloro- methane	V VII IX X	Air and steam strippable Extractable with organics, ethers, and alcohols Adsorbed Adsorbed	22 22 16, 22, 38 38
Dichlorodifluoro- methane	VII	Extractable with organics, ethers, and alcohols	22
Dichloroethane	IX	Adsorbed @ 12 μg/L	6, 16
1,1-dichloroethane	V VII IX X	90% reduction with air stripping Extractable with alcohols and aromatics Adsorbed Adsorbed	22 22 22, 38 38
1,2-dichloroethane	I V	Reduced Air and steam strippable	60 15, 22

Chemical Treatability Summary

Chemical		Process–Treatability	Ref.
1,2-dichloroethane	VII	Extractable with alcohol and aromatics	22
	IX	81% reduction @ 1000 mg/L	22
	X	Adsorbed	38
Dichloroethylene	VII	99% reduction @ 1500 ppm; kerosene and C_{10}-C_{12} effective solvents	15, 23
1,1-dichloroethylene	V	Air and steam strippable	15, 22
	VII	Extractable with alcohols, aromatics, and ethers	22
	IX	Adsorbed	22
1,2-dichloroethylene	IX	Adsorbed	38
	X	Adsorbed	38
1,2-trans-dichloro-ethylene	V	Air and steam strippable	22
	VII	Soluble in most organics	22
	IX	Adsorbed	22
Dichlorofluoro-methane	IX	Adsorbed	22
Dichloromethane	V	90% reduction with air stripping; steam strippable @ 800 mg/L	15, 22
	VII	Soluble in most organics	22
	IX	Adsorbed	22
1,2-dichloropropane	V	Air and steam strippable	22
	VII	Soluble in most organics	22
	IX	93% reduction @ 1000 mg/L	22
1,2-dichloropro-pylene	V	Air and steam strippable	22
	VII	Soluble in most organics	22
	IX	Adsorbed	22
Ethylene chloride	VII	Kerosene and C_{10}-C_{12} organics effective solvents	15
Ethylene chloro-hydrin	VII	21% reduction @ 1640 mg/L	23
Ethylene dichloride	V	99% reduction @ 8700 mg/L	15, 61
	VII	94–100% reduction @ 23–1804 mg/L with kerosene and C_{10}-C_{12} organics	15
	IX	81% reduction @ 1000 mg/L	15, 29
	X	Adsorbed	27
Hexachloro-butadiene	V	Air and steam strippable	22
	VII	Soluble in most organics	22
	IX	100% reduction @ 100 μg/L	14
	X	100% reduction @ 100 μg/L	14
Hexachlorocyclo-pentadiene	V	Polymerizes with heat	22

Chemical Treatability Summary

Chemical		*Process–Treatability*	*Ref.*
Hexachloroethane	VII	Extractable with aromatics, alcohols, and ethers	22
	IX	100% reduction @ 100 μg/L	14
	X	100% reduction @ 100 μg/L	14
Methylene chloride	I	80–88% reduction @ 10–430 μg/L	60
	IX	73% reduction @ 190 μg/L	6
Pentachloroethane	VII	100% reduction with kerosene solvent @ 10 mg/L	15
Perchloroethylene	V	Steam strippable @ 15 mg/L	15
	VII	Extractable with kerosene and C_{10}–C_{12} solvents	15
Propylene dichloride	IX	93% reduction @ 1000 mg/L	29
Tetrachloroethane	VII	Kerosene and C_{10}–C_{12} organics provided 95% reduction	15
	IX	100% reduction @ 100 μg/L	14
	X	100% reduction @ 100 μg/L	14
1,1,1,2-tetrachloro-ethane	V	Steam strippable @ 513 mg/L	15
1,1,2,2-tetrachloro-ethane	V	Difficult to steam strip	15
	VII	Extractable with aromatics, alcohols, and ethers	22
	IX	Adsorbed	22
Tetrachloroethylene	V	90% reduction by air and steam stripping	22
	VII	Soluble in most organics	22
	IX	Adsorbed	22, 38
	X	Adsorbed	38
Tetrachloromethane	V	90% reduction by air and steam stripping	22
	VII	Soluble in most organics	22
Tribromomethane	V	Air and steam strippable	22
	VII	Soluble in most organics	22
	IX	Adsorbed	16, 22
Trichloroacetic	III	25–49% reduction @ 250 mg/L–dependent upon membrane	13
Trichloroethane	VII	97–99% reduction with kerosene and C_{10}–C_{12} solvents	15
1,1,1-trichloro-ethane	I	90% reduction @ 8–790 μg/L	60
	V	Air and steam strippable	15, 22
	VII	Extractable with alcohols and aromatics	22
	IX	Adsorbed	22
	X	Adsorbed @ 551 μg/L	38

Chemical Treatability Summary

Chemical		Process–Treatability	Ref.
1,1,2-trichloro-ethane	I	99% reduction @ 1305 μg/L	47
	V	Air and steam strippable	15, 22
	VII	Extractable with aromatics, methanol, and ether	22
	IX	Adsorbed	22
Trichloroethylene	I	99% reduction @ 78–214 μg/L	16, 60
	II	40% reduction @ 103 μg/L with alum	16
	V	Air and steam strippable	15
	VII	75% reduction with kerosene and C_{10}–C_{12} solvents	15, 22
	IX	99% reduction @ 21 μg/L	6, 22
Trichlorofluoro-methane	VII	Extractable with alcohols and ethers	22
	IX	Adsorbed	22
Trichloromethane	V	Air and steam strippable	22
	VII	Soluble in most organics	22
1,2,3-trichloro-propane	IX	100% reduction @ 100 μg/L	14
	X	100% reduction @ 100 μg/L	14
Vinyl chloride	I	100% reduction @ 8 μg/L	60
Vinylidene chloride	VII	92% reduction with kerosene and C_{10}–C_{12} solvents @ 13 mg/L	15
G. Metals			
Antimony	II	28,62,65% reduction @ 600 μg/L with alum, lime ferric chloride coagulants	32
Arsenic	II	76–90% reduction @ 5 mg/L with ferric sulfate and lime coagulants	53, 58
	IX	No reduction @ 1.1 μg/L	53
	XII	96% reduction @ 25 mg/L with silicon alloy adsorbent	
Arsenic (As^{+5})	II	94–97% reduction @ 21–25 mg/L with alum and lime coagulants	22
Barium	I	Inhibitory @ 100 mg/L	20
	II	36–99% reduction @ 0.08–5 mg/L with lime, alum, ferric sulfate	32, 53, 58
	III	87–99% reduction @ 0.8–9.2 mg/L	13
	IX	No reduction @ 32 μg/L	53
Beryllium	II	98–99% reduction @ 100 μg/L with alum, lime and ferric chloride	22, 32
Bismuth	II	94–96% reduction @ 600 μg/L with alum, lime and ferric chloride	32
Cadmium	I	Inhibitory @ 1–10 mg/L	20, 22, 60

Chemical Treatability Summary

Chemical		Process–Treatability	Ref.
Cadmium	II	45–98% reduction @ 9 µg/L–5 mg/L with lime, ferric chloride and ferric sulfate	32, 53, 58
	III	90–99% reduction @ 0.1–1.0 mg/L	13
	VI	Foam fractionation with sodium dodecylbenzene sulfonate	22
	IX	6–37% reduction @ 1.8–29 µg/L	53, 66
	XII	96% reduction @ 25 mg/L with silicon alloy adsorbent	
Chromic acid	III	85% reduction @ 200 mg/L	19
Chromium	I	27–78% reduction @ 0.8–4 mg/L	2
	II	27–54% reduction @ 0.1–5 mg/L with lime	12, 53
	III	85–98% reduction @ 1–12 mg/L	13
	VI	Reduction possible using quaternary ammonium salts	22
	IX	37–43% reduction @ 41–84 µg/L	53
	XII	100% reduction @ 300 mg/L with high clay soil adsorbent	22
Chromium (Cr^{+3})	I	Complete removal	3
	II	98–99% reduction @ 0.7–5 mg/L with ferric sulfate, lime, and ferric chloride	32, 58
	IX	5–48% reduction @ 100 mg/L	63
Chromium (Cr^{+6})	I	Inhibitory @ 100 mg/L	20
	II	22–65% reduction @ 0.7–5 mg/L with ferric sulfate, lime and ferric chloride	32, 58
	IX	16–36% reduction @ 100 mg/L	63
Cobalt	I	Inhibitory @ 0.08 mg/L	50
	II	18–91% reduction @ 500–800 mg/L	32
Copper	I	7–77% reduction @ 0.2–10 mg/L; reported to be inhibitory @ 0.5 mg/L	2, 4, 50, 59
	II	67–98% reduction @ 0.2–15 mg/L with alum, lime, and ferric sulfate coagulants	12, 22, 30, 53, 58
	III	95–100% reduction @ 0.6–12 mg/L	13
	IV	82% reduction @ 0.44 mg/L	49
	VI	Foam fractionation with sodium dodecylbenzene sulfonate	22
	IX	8–96% reduction @ 0.05–100 mg/L	53, 63
	XII	96–100% reduction @ 300 mg/L silicon alloy and high clay soil adsorbents	22
Iron	I	62% reduction @ 0.6 mg/L soluble iron	46
	II	26–99% reduction @ 0.2–10 mg/L with lime and ferric chloride coagulants	12, 53, 58

Chemical Treatability Summary

Chemical		Process–Treatability	Ref.
Iron	III	100% reduction @ 12 mg/L	13
	IV	85% reduction @ 6.8 mg/L	49
	IX	45–86% reduction @ 40–207 μg/L	53
Iron (Fe^{+3})	I	Inhibitory @ 100 mg/L	20
Iron (Fe^{+3})	I	Inhibitory @ 100 mg/L	20
Lead	I	Inhibitory @ 10 mg/L	20, 50
	II	43–99% reduction @ 0.02–5 mg/L with lime, ferric sulfate, and alum coagulants	22, 32, 53, 58
	III	98–100% reduction @ 0.9–12 mg/L	13
	VI	Foam fractionation with sodium dodecylbenzene sulfate	22
	IX	13–93% reduction @ 100 mg/L; no reduction @ 5–22 μg/L	53, 63
	XII	96% reduction with silicon alloy adsorbent; redwood bark also tried	22
Manganese	I	Conflicting data; 10 mg/L inhibited while 12–50 mg/L also reported to stimulate	20, 50
	II	18–98% reduction @ 0.04–5 mg/L with lime and ferric sulfate coagulants	32, 58
	IV	89% reduction @ 4.9 mg/L	49
	IX	1–50% reduction @ 0.002–100 mg/L	53, 63
Mercury	I	Conflicting data; 51–58% reduction @ 5–10 mg/L and inhibitory @ any concentration	28, 44
	II	25–98% reduction @ 0.001–5 mg/L with lime and ferric chloride coagulants	32, 53, 58
	VII	99% reduction @ 2 mg/L with high molecular weight amines and quaternary salts	22
	IX	80–99% reduction @ 0.001–100 mg/L with GAC and PAC plus chelating agent	22, 53, 63, 72
	XII	99% reduction using silicon alloy adsorbent	22
Molybdenum	II	No reduction with alum and lime; 68% reduction with ferric chloride @ 600 μg/L	32
Nickel	I	0–42% reduction @ 0.3–10 mg/L	2, 4, 59, 67, 70
	II	10–100% reduction @ 0.9–5 mg/L with alum, lime and ferric sulfate	12, 22, 32, 58
	III	93–97% reduction @ 12 mg/L	13
	IX	4–52% reduction @ 100 mg/L	63

Notes: GAC = granular activated carbon; PAC = powdered activated carbon.

Chemical Treatability Summary

Chemical		Process–Treatability	Ref.
Selenium	II	0–80% reduction @ 0.002–100 mg/L with lime, alum, and ferric chloride coagulants	22, 32, 53
	IX	96% reduction @ 500 mg/L after GAC and lime precipitation	22
Silver	II	38–98% reduction @ 0.006–500 mg/L with lime alum, and ferric chloride coagulants	22, 32, 53
Strontium	I	No affect @ 5–50 μg/L	50
Thallium	II	30–60% reduction @ 500 μg/L with lime, alum, and ferric chloride coagulants	22, 32
	IX	84% reduction after GAC and lime precipitation	22
Tin	II	92–98% reduction @ 500 μg/L with lime, alum, and ferric chloride coagulants	32
Aroclor 1254	IX	94–99% reduction @ 11–160 μg/L	8, 14, 17, 31, 61
	X	100% reduction @ 100 μg/L	14, 17
Aroclor 1254 and 1260	X	23–60% reduction @ 1–25 μg/L	45
	XII	73% reduction with polyvinyl chloride (PVC) chips, 37% reduction with polyurethane foam adsorbent	45
PCBs (unspecified)	IX	100% reduction @ 1–400 μg/L	6
J. Pesticides			
Aldrin	I	Not significantly degraded	62
	III	100% reduction	13
	IX	98–100% reduction @ 8–100 μg/L	6, 8, 14, 31
	X	100% reduction @ 100 μg/L	14
Aminotriazole	I	Not significantly degraded	62
Atrazine	III	84–98% reduction	13
	X	100% reduction @ 100 μg/L	14
Captan	III	99–100% reduction @ 689 μg/L	13
Chlordane	I	Slightly degraded	62
	IX	97–100% reduction @ 12–1430 μg/L	6
Chlorinated pesticides (unspecified)	X	79% reduction @ 33–118 mg/L	49

Notes: GAC = granular activated carbon; PAC = powdered activated carbon.

Chemical Treatability Summary

Chemical		Process–Treatability	Ref.
2,4-D butyl ester	IX	100% reduction @ 100 μg/L	20
	X	100% reduction @ 100 μg/L	20
2,4-D and related herbicides	X	95% reduction @ 20–1500 μg/L	32
2,4-D-isoctylester	I	Biodegradable	121
DDD	IX	99.8% reduction @ 56 μg/L	8, 38, 66
DDE	III	100% reduction	13
	IX	97% reduction @ 38 μg/L	8, 31, 61
DDT	I	Not significantly degraded	62
	II	98% reduction @ 10 μg/L with alum coagulant	6
	III	100% reduction	13
	IX	99% reduction @ 10–100 μg/L	6, 8, 14, 31, 61
	X	100% reduction @ 100 μg/L	14
DDVP	I	Degraded	74
Diazinon	I	Not significantly degraded	62, 74
	III	88–98% reduction	13
Dieldrin	I	Not significantly degraded	62
	II	55% reduction @ 10 μg/L with alum coagulant	6
	III	100% reduction	13
	IX	75–100% reduction @ 19–60 μg/L	6, 8, 31, 61
Endrin	I	Not significantly degraded	62
	II	35% reduction @ 10 μg/L with alum coagulant	6
	IX	80–99% reduction @ 10–62 μg/L	6, 8, 31, 61
Endrin and heptachlor	X	97% reduction @ 0.1–2 mg/L	27
Ferbam	I	Biodegradable	62
Heptachlor	I	Slightly degraded @ 500 mg/L	62
	III	100% reduction	13
	IX	99% reduction @ 6–80 μg/L	6
Heptachlorepoxide	III	99.8% reduction	13
Herbicides (unspecified)	IX	90–99% TOC reduction	31
Herbicide orange	I	77% reduction @ 1380 mg/L	65
Kepone	IX	100% reduction @ 4000 μg/L	6

Chemical Treatability Summary

Chemical		Process–Treatability	Ref.
Lindane	I	Not significantly degraded	62
	II	10% reduction @ 10 μg/L with alum coagulant	6
	III	99% reduction	13
	IX	30–99% reduction @ 10 μg/L	6
Malathion	I	Not significantly degraded	62, 74
	III	99% reduction	13
	III	99% reduction	13
Maneb	I	Biodegradable	62
Methyl parathion	I	Not significantly degraded	62, 74
	III	99% reduction	13
Parathion	I	Not significantly degraded	62, 74
	II	5% reduction @ 10 μg/L with alum	6
	III	99% reduction	13
	IX	99% reduction @ 10 μg/L	6
Pentachlorophenol (also see phenols)	I	Not significantly degraded @ 75–150 mg/L	62
Propoxur	I	Biodegradable	74
Randox	III	72–99% reduction	13
Tetraethyl pyro-phosphate	I	Not significantly degraded	62
Thanite	I	Biodegradable	62
Toxaphene	IX	97–99% reduction @ 36–155 μg/L	6, 8, 31
	X	99% reduction @ 70–2600 μg/L	27
2,4,5-T ester	II	65% reduction @ 10 μg/L with alum coagulant	6
	IX	80–95% reduction @ 10 μg/L	6
2,4,5-trichloro-phenoxyacetic acid	I	Slightly degraded @ 150 mg/L– 99% reduction after 7.5 days aeration	39
Trifluralin	III	100% reduction	13
Ziram	I	Slightly degraded	62
Zireb	I	Slightly degraded	62
K. Phenols			
Bisphenol A	X	94% @ 900 mg/L when pH adjusted	18
Brine phenol	X	99% reduction of phenol @ 10–400 mg/L	26
Butyl phenol	IX	95% reduction @ 300 μg/L	6

Chemical Treatability Summary

Chemical		Process–Treatability	Ref.
4-chloro-3-methyl-phenol	I	Toxic @ 50–100 mg/L; inhibitory but slowly degradable @ 50 mg/L	22, 56
	VII	Extractable with benzene, alcohol, and nitrobenzene	22
	IX	100% reduction @ 100 μg/L	14
	X	100% reduction @ 100 μg/L	14
2-chloro-4-nitro-phenol	I	72% reduction	65
Chlorophenol	V	Steam strippable	22
m-chlorophenol	I	100% reduction @ 200 mg/L	61
	X	Adsorbed	61
2-chlorophenol	I	90–95% reduction @ 150–200 mg/L	22
	III	66% reduction	22
	VII	Extractable with diisopropyl ether, benzene, butylacetate, and nitrobenzene	22
o-chlorophenol	I	96–100% reduction @ 200 mg/L	61, 65
p-chlorophenol	I	96–100% reduction @ 200 mg/L	61, 65
Cresol	IX	96% reduction @ 230 μg/L	6
m-cresol	I	96% reduction	65
	VII	91% reduction @ 291 mg/L	23
o-cresol	I	95% reduction	65
	VII	90–99% reduction @ 307–890 mg/L	23
p-cresol	I	96% reduction	65
	VII	91% reduction @ 291 mg/L	23
2,4-diaminophenol	I	83% reduction	65
2,4-dibromophenol	X	Adsorbed	26
Dichlorophenol	X	Adsorbed	26
2,3-dichlorophenol	IX	100% reduction @ 100 μg/L	14
	X	100% reduction @ 100 μg/L	14
2,4-dichlorophenol	I	98–100% reduction @ 60–200 μg/L	22, 39, 65
	X	100% reduction @ 430 mg/L	61
	XII	Extractable with benzene, alcohol, and nitrobenzene	22
2,5-dichlorophenol	I	100% reduction @ 200 mg/L	61
2,6-dichlorophenol	I	99% reduction @ 64 mg/L	39
Dimethylphenol	IX	99% reduction @ 1220 μg/L	6
2,3-dimethylphenol	I	96% reduction	65
2,4-dimethylphenol	I	94% reduction	65

Chemical Treatability Summary

Chemical		Process–Treatability	Ref.
2,4-dimethylphenol	VII	Extractable with benzene and alcohol	22
2,5-dimethylphenol	I	94% reduction	65
2,6-dimethylphenol	I	94% reduction	65
3,4-dimethylphenol	I	98% reduction	65
3,5-dimethylphenol	I	89% reduction	65
	IX	100% reduction @ 100 μg/L	14
4,6-dinitro-2-methylphenol	VII	Extractable with benzene and acetone	22
2,4-dinitrophenol	I	85% reduction	65, 71
	VII	Extractable with benzene and alcohol	22
	IX	Adsorbed	16
B-naphthol	X	100% reduction @ 100 μg/L	14
m-nitrophenol	I	95% reduction	65
o-nitrophenol	I	97–98% reduction	47, 65
2-nitrophenol	VII	Extractable with benzene and alcohol	22
p-nitrophenol	I	95–99% reduction	47, 65
	X	99% reduction @ 700–1800 mg/L	18, 26
4-nitrophenol	III	Removable	22
	VII	Extractable with benzene and alcohol	22
Nonylphenol	IX	Adsorbed	16
Pentachlorophenol	I	26% reduction @ 200 mg/L	61, 74
	VII	Extractable with benzene, alcohol and nitrobenzene	22
	IX	100% reduction @ 10 mg/L	6, 16
	X	100% reduction @ 100 μg/L	14
Phenol	I	62–100% reduction @ 5–500 mg/L; reported to be inhibitory @ 500 mg/L	22, 47, 48, 52, 55, 59, 61, 73, 74
	III	–6–100% reduction @ 1–1000 mg/L–dependent upon membrane	13, 22, 24, 42
	IV	75% reduction @ 1–100 mg/L	42
	V	Steam strippable	22
	VII	4–98% reduction @ 67–8800 mg/L	22, 23
	IX	80–100% reduction @ 0.1–1200 mg/L	6, 14, 16, 22, 29, 31
	X	99% reduction @ 500–5000 mg/L	18, 26
p-phenylazophenol	I	Inhibitory @ 500 mg/L	55
Resorcinol	IX	100% reduction @ 100 μg/L	14
	X	100% reduction @ 100 μg/L	14

Chemical Treatability Summary

Chemical		Process-Treatability	Ref.
2,3,5-trichloro-phenol	I	100% reduction @ 200 mg/L	61, 74
2,4,5-trichloro-phenol	I	99% reduction @ 19 mg/L	39
2,4,6-trichloro-phenol	I	100% reduction @ 20–200 mg/L reported to be inhibitory @ 50–200 mg/L	22, 39, 56, 61
	VII	Extractable with benzene, alcohol, and nitrobenzene	22
	IX	100% reduction @ 100 μg/L	14
	X	100% reduction @ 0.1–510 mg/L	14, 61
Trimethylphenol	IX	92% reduction @ 130 μg/L	6
Xylenol	VII	96% reduction @ 227 mg/L	23
L. Phthalates			
Bis(2-ethylhexyl) phthalate	I	70–78% reduction @ 5 mg/L	22
	II	80–90% reduction @ 0.5–3.5 μg/L with aluminum sulfate coagulant	22
	VII	Extractable with ethyl ether and benzene	22
	IX	98% reduction @ 1300 μg/L	5, 22
Butylbenzyl phthalate	I	Biodegradable	22
	VII	Extractable with ethyl ether and benzene	22
Dibutyl phthalate	IX	100% reduction @ 100 μg/L	14
	X	100% reduction @ 100 μg/L	14
Di-*n*-butyl phthalate	I	Biodegradable @ 200 mg/L	22
	II	60–70% reduction @ 2.5–4.5 μg/L with aluminum sulfate	22
	VII	Extractable with ethyl ether and benzene	22
Diethyl phthalate	II	Biodegradable	22
	VII	Extractable with ethyl ether and benzene	22
Diethylhexyl phthalate	X	100% reduction @ 100 μg/L	14
Di(2-ethylhexyl) phthalate	I	50–70% reduction	9
Dimethyl phthalate	I	Degradable; 100% reduction @ 215 μg/L	16, 22
	II	15% reduction @ 183 μg/L with alum	16
	VII	Extractable with ethyl ether and benzene	22
	IX	100% reduction @ 100 μg/L	14
	X	100% reduction @ 100 μg/L	14
Di-*n*-octyl phthalate	I	Biodegradable @ 63 mg/L	22
	VII	Extractable with ethyl ether and benzene	22

Chemical Treatability Summary

Chemical		Process–Treatability	Ref.
Isophthalic acid	I	95% reduction	65
Phthalimide	I	96% reduction	65
Phthalic acid	I	97% reduction	65
M. Polynuclear aromatics			
Acenaphthalene	X	100% reduction @ 100 μg/L	14
Acenaphthene	II	Precipitated with alum	22
Acenaphthylene	II	Precipitated with alum	22
Anthracene	I	Toxic @ 500 mg/L	55
	VII	Extractable with toluene	22
Benzanthracene	I	Slowly opxidized @ 500 mg/L	55
	II	Separable by gravity or sand filtration	22
11,12-benzofluor-anthene	II	Separable by gravity or sand filtration	22
Benzoperylene	I	Biodegradable	22
1,12-benzoperylene	II	Separable by gravity or sand filtration	22
Benzo(a)pyrene	II	Separable by gravity or sand filtration	22
Biphenyl	IX	100% reduction @ 100 μg/L	14
	X	100% reduction @ 100 μg/L	14
D-chloramphenicol	I	86% reduction	65
2-chloronaphthalene	II	Precipitated with alum	22
Chrysene	II	Separable by gravity and sand filtration	22
Cumene	IX	100% reduction @ 100 μg/L	14
	X	100% reduction @ 100 μg/L	14
9,10-dimethyl-anthracene	I	Degradable @ 500 mg/L	55
9,10-dimethyl-1,2-benzanthracene	I	Slowly oxidized @ 500 mg/L	55
Dimethylnaphthalene	IX	80% reduction @ 100 μg/L	14
	X	100% reduction @ 100 μg/L	14
1,1-diphenyl-hydrazine	IX	Adsorbed	25
1,2-diphenyl-hydrazine	I	28% reduction @ 341 μg/L	65
Fluoranthene	IX	80% reduction @ 100 μg/L	14
	X	100% reduction @ 100 μg/L	14

Chemical Treatability Summary (continued)

Chemical		Process–Treatability	Ref.
7-methyl-1,1-benzanthracene	I	Inhibitory @ 500 mg/L	55
20-methylchol-anthrene	I	Toxic or inhibitory; able to undergo slow biological oxidation @ 500 mg/L	55
Naphthalene	I	85–95% reduction; inhibitory @ 500 mg/L	41, 43, 55
	II	Separable by gravity or sand filtration	22
	V	Air strippable by 50:1 volume of air	22
	IX	70% reduction	25, 53
Phenanthrene	IX	80% reduction @ 100 μg/L	14
	X	100% reduction @ 100 μg/L	14
2,3-o-phenylene pyrene	II	Separable by gravity or sand filtration	22
Pyrene	II	Separable by gravity or sand filtration	22
	IX	80% reduction @ 100 μg/L	14
	X	100% reduction @ 100 μg/L	14

Appendix 3-2
References

1. R. K. Anderson, J. M. Nystron, R. P. McDonnell, and B. W. Stevens, 1975, Explosives removal from munitions wastewater, in *Proceedings of the 30th Industrial Waste Conference,* Purdue University, W. Lafayette, Ind. pp. 816–825.
2. E. F. Barth, et al. 1965, Field survey of four municipal wastewater treatment plants receiving metallic wastes, *Water Pollution Control Federation Journal* 37(8):1101–1117.
3. D. A. Bailey, and K. S. Robinson, 1970, The influence of trivalent chromium on the biological treatment of domestic sewage, *Water Pollution Control (G.B.)* 69:100ff.
4. E. F. Barth, et al., 1965, Summary report on the effects of heavy metals on the biological treatment processes, *Water Pollution Control Federation Journal* 37:(1):86.
5. B. A. Beaudet, 1979 (May), *Study of Effectiveness of Activated Carbon Technology for the Removal of Specific Materials from Organic Chemical Processes,* U.S. Environmental Protection Agency Contract No. 68-03-2610. Environmental Science and Engineering Inc., Gainesville, Fla.
6. D. L. Becker, and S. C. Wilson, 1978, *The use of activated carbon for the treatment of pesticides and pesticidal wastes,* in *Carbon Adsorption Handbook,* P. N. Cheremisinoff and F. Ellerbusch, eds., Ann Arbor Science, Ann Arbor, Mich. pp. 167–213.

7. E. E. Berkau, C. E. Frank, and I. A. Jefcoat, 1979 (Aug.), A Scientific Aproach to the Identification and Control of Toxic Chemicals in Industrial Wastewater, paper presented at the A.I.Ch.E. 87th National Meeting, Boston, Mass.
8. F. E. Bernardin, Jr., and E. M. Froelich, 1975, Practical removal of toxicity by adsorption, in *Proceedings of the 30th Industrial Waste Conference*, Purdue University, W. Lafayette, Ind. pp. 548-560.
9. F. D. Bess, and R. A. Conway, 1966, Aerated stabilization of synthetic organic chemical wastes, *Water Pollution Control Federation Journal* **38**(6):939-956.
10. D. Bhattacharyya, K. A. Garrison, and R. B. Grieves, 1976, Membrane Ultrafiltration of nitrotoluenes from industrial wastes, in *Proceedings of the 31st Industrial Waste Conference*, Purdue University, W. Lafayette, Ind., pp. 139-149.
11. D. E. Bruderly, J. D. Crane, and J. D. Riggenbach, 1977, Feasibility of zero aromatic hydrocarbon discharge from a styrene monomer facility, in *Proceedings of the 32nd Industrial Waste Conference*, Purdue University, W. Lafayette, Ind., pp. 726-732.
12. E. S. K. Chian, 1975, Renovation of vehicle washrack wastewater for reuse, *A.I.Ch.E. Symposium Series, Water-1975*, vol. 71, no. 151, pp. 87-92.
13. E. S. K. Chian, and H. H. P. Fang, Removal of Toxic Compounds by Reverse Osmosis, unpublished report available from Abcor, Inc., Wilmington, Mass.
14. C. D. Criswell, R. L. Ericson, G. A. Junk, K. W. Lee, J. S. Fritz, and H. J. Svec, 1977, Comparison of macroreticular resin and activated carbon as sorbents, *American Water Works Association Journal* **69**(12):669-674.
15. J. H. Coco, et al., 1979, *Development of Treatment and Control Technology for Refractory Petrochemical Wastes*, EPA-600/2-79-080, U.S. Environmental Protection Agency, Ada, Okla.
16. J. M. Cohen, 1979 (June), *Briefing for Dr. Gage on Treatability/Removability of Toxics from Wastewater*. U.S. Environmental Protection Agency, Cincinnati, Ohio.
17. R. W. Okey, and R. W. Bogan, 1963, Synthetic organic pesticides, An evaluation of their persistence in natural waters, in *Proceedings of the 11th Pacific Northwest Industrial Waste Conference*, Oregon State University, Corvallis, Ore., pp. 222-251.
18. E. F. Barth, et al., 1965, Field survey of four municipal wastewater treatment plants receiving metallic wastes, *Water Pollution Control Federation Journal* **37**(8):1101-1117.
19. D. A. Bailey, and K. S. Robinson, 1970, The influence of trivalent chromium on the biological treatment of domestic sewage, *Water Pollution Control (G.B.)* **69**:100ff.
20. P. S. Dawson, and S. H. Jenkins, 1950, The oxygen requirements of activated sludge determined by monometric methods, II: Chemical factors affecting oxygen uptake, *Sewage and Industrial Wastes* **22**(4):490.
21. B. W. Dickerson, et al., 1954, Further operating experiences in biological purification of formalydehyde wastes, in *Proceedings of the 9th Industrial Waste Conference*, Purdue University, W. Lafayette, Ind. pp. 331-351.
22. E. S. K. Chian, W. N. Bruce, and H. H. P. Fang, 1975, Removal of pesticides

by reverse osmosis, *Environmental Science and Technology* 9(1):52–59.

23. J. P. Earhart, K. W. Won, H. Y. Wong, J. M. Prausnitz, and C. J. King, 1977, Recovery of organic pollutants via solvent extraction, *Chemical Engineering Progress* 73(5):67–73.

24. H. H. P. Fang, and E. S. K. Chian, 1976, Reverse osmosis separation of polar organic compounds in aqueous solution, *Environmental Science & Technology* 10(4):364–369.

25. E. G. Fochtman, and R. A. Dobbs, *Adsorption of Carcinogenic Compounds by Activated Carbon,* U.S. Environmental Protection Agency, Municipal Environmental Research Laboratory, Cincinnati, Ohio.

26. C. R. Fox, 1978, Plant uses prove phenol recovery with resins, *Hydrocarbon Processing,* November, pp. 269–273.

27. C. R. Fox, 1979, Toxic Organic Removal from Waste Waters with Polymeric Adsorbent Resins, paper presented at the 86th National American Institute of Chemical Engineers Meeting, Houston, Tex.

28. M. M. Ghosh, and P. D. Zugger, 1973, Toxic effects of mercury on the activated sludge process, *Water Pollution Control Federation Journal* 45(3):424–433.

29. D. M. Giusti, R. A. Conway, and C. T. Lawson, 1974, Activated carbon adsorption of petrochemicals, *Water Pollution Control Federation Journal* 46(5):947–965.

30. M. M. Gurvitch, 1979, Description of an Advanced Treatment Plant to Produce Recycled Water at a Chemical R & D Facility, paper presented at the 34th Industrial Waste Conference, Purdue University, W. Lafayette, Ind.

31. D. G. Hager, 1976, Waste water treatment via activated carbon, *Chemical Engineering Progress* 72(10):57–60.

32. S. A. Hannah, M. Jelus, and J. M. Cohen, 1977, Removal of uncommon trace metals by physical and chemical treatment processes, *Water Pollution Control Federation Journal* 23(11):2297–2309.

33. M. W. Hay, et al., 1972 (Sept./Oct.), Factors affecting color development during treatment of TNT wastes, *Industrial Wastes* 18(5).

34. R. P. Heck II, 1978, Munitions plant uses adsorption in wastewater treatment, *Industrial Wastes,* pp. 35–39, March/April.

35. J. A. Heidman, et al., 1967, Metabolic response of activated sludge to sodium pentachlorophenol, in *Proceedings of the 22nd Industrial Waste Conference,* Purdue University, W. Lafayette, Ind. pp. 661–674.

36. J. V. Hunter, and H. Heukelekian, 1964, Determination of biodegradability using Warburg respirometric techniques, in *Proceedings of 19th Industrial Waste Conference,* Purdue University, W. Lafayette, Ind. pp. 616–627.

37. Hydroscience, Inc., 1971 (March), *The Impacts of Oily Materials on Activated Sludge Systems,* Water Pollution Control Research Series, EPA Report No. 12050 DSH 03/71, U.S. Environmental Protection Agency, Cincinnati, Ohio.

38. E. G. Isacoff, and J. A. Bittner, 1979, Resin adsorbent takes on chlororganics from well water, *Water and Sewage Works* 126(8):41–42.

39. The city of Jacksonville, Arkansas, 1971, (June), *The Demonstration of a Facility for the Biological Treatment of a Complex Chlorophenolic Waste,* Water Pollu-

tion Control Research Series, EPA Report No. 12130 EGK 06/71, U.S. Environmental Agency, Cincinnati, Ohio.

40. D. C. Kennedy, 1973, Treatment of effluent from manufacture of chlorinated pesticides with a synthetic, polymeric adsorbent, amberlite XAD-4, *Environmental Science & Technology* 7(2):138-141.

41. G. W. Kimke, J. F. Hall, and R. W. Oeben, 1968, Conversion to activated sludge at Union Carbide's Institute Plant, *Water Pollution Control Federation Journal* 40(8):1408-1422.

42. S. L. Klemetson, and M. D. Scharbow, 1977, Removal of phenolic compounds in coal gasification wastewaters using a dynamic membrane filtration process, in *Proceedings of the 32nd Industrial Waste Conference,* Purdue University, W. Lafayette, Ind., pp. 786-796.

43. G. W. Kimke, et al., 1968, Conversion to activated sludge at Union Carbide's Institute Plant, *Water Pollution Control Federation Journal* 40(8):1408-1422.

44. J. C. Lamb III, et al., 1964, A technique for evaluating the biological treatability of industrial wastes, *Water Pollution Control Federation Journal* 36(10):1263-1284.

45. J. Lawrence, and H. M. Tosine, 1976, Adsorption of polychlorinated biphenyls from aqueous solutions and sewage, *Environmental Science & Technology* 10(4):381-383.

46. R. D. Leary, 1971 (March), *Phosphorus Removal with Pickle Liquor in an Activated Sludge Plant,* Water Pollution Control Research Series, EPA Report No. 11010 FLQ 3/71, U.S. Environmental Protection Agency, Cincinnati, Ohio.

47. N. A. Leipzig, and M. R. Hockenbury, 1979 Powdered activated carbon/activated sludge treatment of chemical production wastewaters, in *Proceedings of the 34th Industrial Waste Conference,* Purdue University, W. Lafayette, Ind.

48. J. T. Ling, 1963, Pilot study of treating chemical wastes with an aerated lagoon, *Water Pollution Control Federation Journal* 35(8):963-972.

49. C. X. Lopez, and R. Johnston, 1977, Industrial wastewater recycling with ultrafiltration and reverse osmosis, in *Proceedings of the 32nd Industrial Waste Conference,* Purdue University, W. Lafayette, Ind., pp. 81-91.

50. J. E. Loveless, and N. A. Painter, 1968, The influence of metal ion concentration and pH value on the growth of a Nitrosomonas strain isolated from activated sludge, *Journal of General Microbiology* 52(3):1ff.

51. H. F. Lund, 1971, *Industrial Pollution Control Handbook,* McGraw-Hill, New York.

52. P. A. Lutin, 1970, Removal of organic nitrites from wastewater systems, *Water Pollution Control Federation Journal* 42(9):1632-1642.

53. P. L. McCarty, M. Reinhard, C. Dolce, H. Nguyen, and D. G. Argo, 1978, *Water Factory 21: Reclaimed Water, Volatile Organics, Virus, and Treatment Performance,* EPA-600/2-78-076, U.S. Environmental Protection Agency, Cincinnati, Ohio.

54. G. W. Malaney, and R. M. Gerhold, 1969, Structural determinants in the oxidation of aliphatic compounds by activated sludge, *Water Pollution Control Federation Journal* 41(2):R18-R33.

55. G. W. Malaney, 1967, Resistance of carcinogenic organic compounds to oxidation by activated sludge, *Water Pollution Control Federation Journal* **39**(12):2029.
56. Manufacturing Chemists Association, 1972 (March), *The Effects of Chlorination on Selected Organic Chemicals,* Water Pollution Control Research Series, EPA Report No. 12020 EXG-03/72, U.S. Environmental Protection Agency, Cincinnati, Ohio.
57. C. V. Marion, and G. W. Malaney 1963, Ability of activated sludge to oxidize aromatic organic compounds, in *Proceedings of the 18th Industrial Waste Conference,* Purdue University, W. Lafayette, Ind. pp. 297–308.
58. T. Maruyama, S. A. Hannah, and J. M. Cohen, 1975, Metal removal by physical and chemical treatment processes, *Water Pollution Control Federation Journal* **47**(5):962–975.
59. J. W. Masselli, et al., 1965, (June), The Effect of Industrial Waste on Sewage Treatment, prepared for the New England Interstate Water Pollution Control Commission by Wesleyan University, Middletown, Connecticut, No. TR-13.
60. Municipal Environmental Research Laboratory, 1977 (March), *Survey of Two Municipal Wastewater Treatment Plants for Toxic Substances,* U.S. Environmental Protection Agency, Cincinnati, Ohio.
61. M. F. Nathan, 1978, Choosing a process for chloride removal, *Chemical Engineering* **85**(3):93.
62. R. W. Okey, and R. W. Bogan, 1963, Synthetic organic pesticides, an evaluation of their persistence in natural waters, in *Proceedings of the 11th Pacific Northwest Industrial Waste Conference,* Oregon State University Corvallis, Ore. pp. 222–251.
63. R. J. Pilie, R. E. Baier, R. C. Ziegler, R. P. Leonard, J. G. Michalovic, S. L. Pek, and D. H. Bock, 1975, *Methods to Treat, Control and Monitor Spilled Hazardous Materials,* EPA-670/2-75-042, U.S. Environmental Protection Agency, Cincinnati, Ohio.
64. O. R. Placak, and C. C. Ruchhoft, 1947, Studies of sewage purification, XVII: The utilization of organic substrates by activated sludge, *Sewage Works Journal* **19**(3):440.
65. SCS Engineers, 1979, *Selected Biodegradation Techniques for Treatment and/or Ultimate Disposal of Organic Material,* EPA-600/2-79-006, U.S. Environmental Protection Agency, Cincinnati, Ohio.
66. T. J. Sorg, M. Csanady, and G. S. Logsdon, 1978, Treatment technology to meet the interim primary drinking water regulations for inorganics, part 3, *Water Pollution Control Federation Journal* **70**:12, 680.
67. T. Stones, 1959, The fate of nickel during the treatment of sewage, *Institute of Sewage Purification (London), Journal and Proceedings,* Part 2, pp. 252ff.
68. R. G. Swedes, Jr., 1977, *Report on Carbon Adsorption Treatment of Contaminated Groundwater at Rocky Mountain Arsenal,* 1977-781-590/139, Department of Army, Rocky Mountain Arsenal, Colo.
69. R. D. Swicher, et al., 1967, Biodegradation of nitrilotriacetate in activated sludge, *Environmental Science & Technology* **1**(10):820–827.

70. Robert A. Taft Sanitary Engineering Center, Chemistry and Physics Center, 1965 (May), *Interaction of Heavy Metals and Biological Sewage Treatment Process,* U.S. Public Health Service, Cincinnati, Ohio.

71. W. Teng-Chung, 1963, Factors Affecting Growth and Respiration in the Activated Sludge Process, Ph.D. dissertation, Case Institute of Technology (now Case Western Reserve University), Cleveland, Ohio.

72. L. Thiem, D. Badorek, and J. T. O'Connor, 1976, Removal of mercury from drinking water using activated carbon, *American Water Works Association Journal* **68**(8):447–451.

73. B. Volesky, N. Czornyj, T. A. Constantine, J. E. Zajic, and K. Ya, 1974, Model treatability study of refinery phenolic wastewater, *A.I.Ch.E. Symposium Series, Water-1974* **70**(144):31–38.

74. R. R. Wilkinson, G. L. Kelso, and F. C. Hopkins, 1978, *State-of-the-Art Report, Pesticide Disposal Research,* EPA-600/2-78-183, U.S. Environmental Protection Agency, Cincinnati, Ohio.

APPENDIX 3-3
EXAMPLES OF TREATMENT APPLICATIONS

Facility Bioplant, Hercules, Inc., Hopewell, Virginia

Type of Waste   "Waste-activated sludge"

Description of Manufacturing Process

The following operations are associated with the corresponding waste:

Operation	Waste	U.S. EPA Hazardous Waste No.
Caustic-chlorine plant	Waste sulfuric acid	D006
Ethyl cellulose plant	Acetone/methanol still bottoms	F003
Ethyl cellulose plant	Ethyl ether still bottoms	F003
Ethyl cellulose plant	Waste Hydrochloric acid	D002

Waste-Activated Sludge

Daily Quantities (16 DWB)	Solids Conc. (by wt.)	Flow Rate to Lagoon (gal/min)
5,300 (avg)	0.6% (avg)	59 (avg)
18,000 (max)	1.0% (max)	200 (max)

Ignitability Flashpoint exceeds 140 °F

Corrosivity pH below 12.4

The hazard of corrosivity is characterized by a material that has a pH of 12.5 or a pH of 2.0. The HCl waste streams are both below a pH of 2, but upon entering the main holding basin these waste acids are buffered to a pH range of 6.5 to 8.5. This pH range is maintained throughout the treatment facility. In the sludge dewatering facility the solid wastes are stabilized with lime, which results in a pH of 12. A pH of 12.5 or higher is impossible because a saturated lime solution has a pH of 12.4.

	Activated Sludge Samples	Effluent Conc.	EP Toxicity Leachate on Activated Sludge
Acetone (mg/L)	10.0		
Dimethyl ether (mg/L)	10.0		
Methanol (mg/L)	10.0		
Toluene (mg/L)	10.0		
Ethyl chloride (μg/L)	5.0		
Dichloroethane (μg/L)		5.0	
Trichloroethylene (μg/L)	5.0		
Trichloroethane (μg/L)	5.0		
Hexachlorobutadiene (μg/L)	5.0		
Solids (%)	3.1		
Hexachlorobenzene (μg/L)	10.0		
Acrolein		**	
Acrylonitrile		**	
Benzene		*	
Bis(chloromethyl)ether		*	
Bromoform		*	
Carbon tetrachloride		*	
Chlorobenzene		*	
Chlorodibromomethane		*	
Chloroethane		*	
2-chloroethylvinyl ether		*	
Chloroform		*	
Dichlorobromomethane		*	
Dichlorodifluoromethane		*	
1,1-dichloroethane		*	
1,2-dichloroethane		*	
1,1-dichloroethylene		*	
1,2-dichloropropane		*	
1,3-dichloropropylene		*	
Ethylbenzene		*	
Methyl bromide		*	
Methyl chloride		*	
Methylene chloride		*	
1,1,2,2-tetrachloroethane		*	
Tetrachloroethylene		*	
Toluene		*	
1,2-trans-dichloroethylene		*	
1,1,1-trichloroethane		*	
1,1,2-trichlorethane		*	
Trichloroethylene		*	
Trichlorofluoromethane		*	
Vinyl chloride		*	

(continued)

	Activated Sludge Samples	Effluent Conc.	EP Toxicity Leachate on Activated Sludge	
Arsenic			0.004	mg/L
Barium			0.8	mg/L
Cadmium			0.005	mg/L
Chromium			0.05	mg/L
Lead			0.05	mg/L
Mercury			0.001	mg/L
Selenium			0.004	mg/L
Silver			0.01	mg/L
Endrin			1.	ppb
Lindane			1.	ppb
Methoxychlor			20.	ppb
Toxaphene			50,	ppb
2,4-D			100.	ppb
2,4,5-TP			100.	ppb

* = below detection limit—10 μg/L.
** = below detection limit—μg/L.

The information presented and any additional information regarding the treatment process may be obtained from the "Hazardous Waste Delisting Petition" submitted by Hercules Inc. to the U.S.EPA, October 8, 1981.

Facility   Chem-Clear Inc., Cleveland, Ohio

Type of Waste   The following specific U.S.EPA hazardous wastes among those transported to the facility:

D002   Corrosive

D004   EP toxicity arsenic

D006   EP toxicity cadmium

D008   EP toxicity lead

F006   Wastewater treatment sludges from electroplating operations; Cd, Cr, Ni, cyanide complexed.

F007   Spent plating bath solutions from electroplating operations; cyanide salts.

F008   Plating bath sludges from the bottom of plating baths from electroplating operations; cyanide salts.

K062   Spent pickle liquor from steel finishing operations; Cr, Pb.

U188   Phenol

Description of Facility

Chem-Clear is a centralized pretreatment facility that treats aqueous wastes generated by industrial concerns. The types of generators transporting wastes to this facility are electroplaters, steel mills, manufacturers of chemicals, agricultural products, plastic products, paints, motor vehicle parts, machinery, surgical supplies, fabricated metals products, rubber products, plastic materials and resins, electronic components, metal forgings, and stampings. Types of wastes generated by these industries and treated at the facility would include spent electroplating baths, electroplating sludge, pickling liquors, reactor washouts, plant wastewaters, and other aqueous wastes.

Treatment of Waste

All wastes received at the location are eventually mixed together in large 400,000-gal holding tanks prior to their treatment in the plant proper.

Before any waste material can be accepted at the facility for treatment, a representative waste sample must be submitted to the facility for analysis to determine its acceptability. Once accepted, each truckload of waste material is sampled by the facility prior to treatment. All incoming waste loads are received in one of four receiving pumps. Two of these pumps are designated to receive wastes destined for pretreatment (i.e., those containing a strong acid, Cr^{+6}, cyanide, phenol wastes) in a specially designed batch rector vessel, and the other two pumps are designated to receive the balance of the waste streams. The waste material is then pumped to a specific part of the facility depending on the type of waste material. Solid (suspended) contaminated material is pumped to one of a pair of primary settling tanks. Acids, phenol-containing, cyanide-containing, and hexavalent chromium-containing wastes are pumped into the plant for prior treatment. All other wastes are pumped into one of two holding tanks.

In many cases the waste must undergo a prior treatment in order to destroy, neutralize, or change the hazardous component of the incoming wastes. These treatments are performed in the batch reactor vessel, which contains three large reactor chambers—two equipped with slow-speed mixers—and a fourth chamber that acts as a reservoir for the treated waste. Continuous monitoring of pH is done. Waste materials treated in this unit include pickling liquors, strong acids, and wastes containing cyanide, hexavalent chromium, or phenol.

Prior to the treatment of any material in this unit, a laboratory determination is made of the amount of reagent required to treat that material.

Thus far, only a single truckload of waste specifically indicated to contain cyanide salt has been treated in the unit. The initial value of cyanide was 75 ppm; after treatment the value was 0.4 ppm.

Hexavalent chromium-containing wastes are initially acidified to a pH of 2. Conversion of the chromium from the hexavalent to the trivalent state is accomplished by adding sodium bisulfate. In the material thus far received, the results are as follows:

Sample	Hexavalent Chromium (ppm)	
	Before Treatment	After Treatment
1	142	0.5
2	230	0.5
3	100	0.5
4	190	0.5
5	150	0.5
6	190	0.5
7	1580	1.0
8	1260	1.0

Phenol-containing wastes are also oxidized with hydrogen peroxide. The initial pH is adjusted to 4.5 by adding acid. Completion of the phenol oxidation is laboratory determined prior to pumping to the holding tank.

Thus far, one truckload of waste material specifically indicated to contain phenol has been received at the facility. The phenol content of this material before oxidation was 20,000 ppm; after oxidation it was 80 ppm.

The final waste treatment is performed in a specially designed chemical treatment unit. This unit is fed at the rate of up to 150 gal/min from material contained in the holding tank. The contents of this tank are composed of materials received and pumped directly to this tank, overflow from the sludge concentrator, overflow from the two primary settling tanks, and the material produced in the reactor vessel. The material in this tank is mixed and blended by air distributed at the bottom through air diffusers. The chemical treatment unit contains five individual compartments. The first four provide three minutes residence time. The final compartment provides a total of 18 minutes for extended flocculation after adding a long-chain polymer.

The unit is operated as follows: Waste pumped from the mixing and blending tank enters the compartment at 150 gal/min and is preferentially directed to the suction side of a high-speed centrifugal pump. Using a positive-displacement variable-capacity chemical feed pump, chemical 1 is also introduced at the suction side of the centrifugal pump. In this manner efficient and uniform mixing is assured. Gravity overflow from compartment 1 to compartment 2 occurs, and the process for introducing chemical 2 is repeated. If required, chemicals 3 and 4 are added and mixed. As the flow passes by gravity into the third-compartment flocculator section, a long-chain polymer is added.

Each flocculator compartment contains an axial flocculator paddle constructed of epoxy-coated expanded metal. A common drive rotates the first paddle at 9 rpm, the second at 6 rpm, and the third at 3 rpm. Use of gently tapered agitation of this type assures propagation of a floc containing all of the heavy metals and suspended solids precipitated. It has been found in the operation of the facility that the addition of only two treatment chemicals plus the polymer are required to accomplish a satisfactory treatment. These two chemicals are sodium hydroxide to precipitate the heavy metals, and either alum or ferric chloride to coagulate the precipitated metals. The treatment conditions in this unit involve raising the pH to 10–10.5 with sodium hydroxide, lowering the pH to 8–8.5 with alum or ferric chloride and then the adding the polymer. Control of pH is accomplished by periodically sampling the reactor compartments and determining the pH. Treatment chemical feed is adjusted to obtain the proper pH in the reactor compartments.

Normally the influent material to the chemical treatment unit is not analyzed for heavy-metal content, because it is the effluent-metal content that is the facility's main concern.

In the single influent–effluent pair that was analyzed, the following results were obtained:

| | Metal Concentration | | |
Metal	Influent	Effluent	Reduction (%)
Antimony	0.25	0.20	—
Arsenic	0.16	0.029	95
Barium	14	0.10	99
Beryllium	0.0025	0.0020	—
Cadmium	1.0	0.006	94
Chromium	100	6.2	94
Copper	260	1.2	99
Iron	460	4.1	99
Lead	9.0	0.07	99
Mercury	0.11	0.043	61
Nickel	27	1.1	96
Selenium	0.096	0.0005	99
Silver	0.095	0.032	66
Thallium	0.025	0.020	—
Zinc	200	4.1	98

These results would indicate a high metal-removal capability of the method and equipment used for this process in the facility.

The flow from the bottom of the flocculator section of the chemical treatment unit is directed to a pair of clarifiers where separation of solids from the liquid occurs. The solids settle to the bottom and are drawn off to sludge concentration while the supernatant liquid is discharged into a holding tank.

Dewatering of the sludge is accomplished on a filter press. This press is fed from material in the sludge concentrator: the material includes that produced in the acid neutralization, that removed from the bottom of the primary settling tanks, and that produced in chemical treatment of waste in the treatment unit. The dewatered sludge is disposed in a landfill.

The information presented and any additional information regarding the treatment process may be obtained from the Hazardous Waste Delisting Petition submitted by Chem-Clear Inc. to the U.S. EPA.

Facility   E. I. DuPont De Nemours, Repauno Plant, Gibbstown, New Jersey

Type of Waste   Combined wastewater streams generated from nitroenzene/aniline production (U.S. EPA Hazardous Waste No. K104).

Description of Manufacturing Process

Nitrobenzene Process.   Nitrobenzene is produced by reacting benzene with nitric acid in a sulfuric acid medium. Nitrobenzene separates as an insoluble organic layer that is decanted and washed with water and alkali, removing all acids and other impurities. The mixed acid is extracted with benzene, then recovered for recycling. The nitrobenzene is dehydrated in a distillation column to produce the product. The alkaline wash water and water taken overhead from the distillation column are extracted with nitrobenzene because of its affinity for residual organic impurities. The resultant nitrobenzene organic layer containing the extracted by-products of reaction is collected and incinerated. The extracted wastewater is combined with the other wash water.

Aniline Process.   Nitrobenzene is reacted with hydrogen to produce aniline. The water of reaction is separated, and the aniline is distilled to purify the product. The distillation bottoms (designated U.S. EPA Hazardous Waste No. K083) are burned in a boiler for fuel value. Process water and water from distillation are combined and other reaction products. The extracted wastewater (designated U.S. EPA Hazardous Waste No. K103) flows through a separator to a collector where it combines with the wastewater from the nitrobenzene process.

Treatment of Wastes

The combined wastewater stream from the nitrobenzene and aniline manufacturing process is adjusted for pH and fed to a steam distillation column. The organics are taken overhead with the steam and condensed. The water phase is recycled to the column and the organics incinerated. The treated wastewater stream, stripped of organics, is discharged from the bottom of the column to the plant ditch system; there it combines with cooling water and effluents from other plant processes and flows to the plant outfall through the ditch system and tidal basin.

Wastewater Stream after Steam Distillation

Average volume	
180,000	gal/day
5,400,000	gal/month
56,000,000	gal/yr[a]
Composition (ppm)[b]	
Benzene	0.0091
Aniline	2.12
Nitrobenzene	0.00028
Diphenylamine	0.0011
Phenylenediamine	0.00066
pH	4.8

[a] Annual average is calculated using 85% utility attained by the manufacturing process.
[b] The values were obtained over a 5-day sampling period, which adequately covered the range of normal operating conditions; process controls would prevent operation at any extreme condition that might generate other materials.

The information presented and any additional information regarding the treatment process may be obtained from the Hazardous Waste Delisting Petition submitted by E. I. DuPont DeNemours to the U.S. EPA, July 13, 1981.

Facility   Eli Lilly and Company, Tippecanoe Laboratories, Lafayette, Indiana

Type of Waste

Liquid effluent from the oxidative treatment of waste streams from the company's chemical manufacturing (Waste 307).

Liquid waste effluent from biological oxidation of Waste 312.

Sludge from neutralization and oxidation treatment of waste from company's chemical manufacturing area (Waste 720).

Sludge from the tertiary waste-treatment plant (Waste 730).

Description of Manufacturing Process

The company is a major manufacturer of human-health-care, agricultural, and cosmetic products. During its manufacturing operations, it generates certain waste streams containing small amounts of listed hazardous wastes.

Treatment of Waste

Wastes from the chemical manufacturing are treated in three functional areas: neutralization, air stripping and air oxidation, and biological reduction.

Waste 307.   All wastes from the chemical manufacturing area flow into aerated neutralization tanks. Some of these wastes contain spent solvents listed as hazardous (U.S. EPA Hazardous Waste Nos. F002, F003, F005). The processes of clarification and solid separation follow neutralization. The liquid effluent then undergoes extensive nonbiological aeration. After this aeration process there is further clarification and solid separation. The liquid effluent from these treatment facilities (Waste 307) flows into two aerated lagoons. Twenty-four hours of aeration time prior to its entering the two aerated lagoons reduces the hazardous content of Waste 307. As Waste 307 enters the aerated lagoons, it contains low levels of methanol (250 ppm). At this stage the waste stream is not ignitible. In addition, the waste stream contains trace levels of methylene chloride (2 ppm), isobutanol (4 ppm), and pyridine (ppm).

Waste 312.   After microbiological treatment in two aerated lagoons, Waste 307 undergoes treatment for clarification and solids separation. As a result of microbiological treatment and additional aeration, the effluent from the aerated lagoon surface impoundment, Waste 312, does not contain any detectable elements of the spent solvents listed as hazardous wastes except for methanol (10 ppm).

Waste 720.   The solids from all neutralization, oxidation, and microbiological treatment described above are combined and centrifuged. Assays show the presence of small quantities of methanol (50 ppm), methylene chloride (20 ppm), pyridine (10 ppm), toluene (20 ppm), and chlorobenzene (15 ppm).

Waste 730. After biological treatment in two aerated lagoons, Waste 312 flows to a tertiary treatment plant where it is mixed with large quantities of nonhazardous waste. Biological processes in the tertiary treatment plant produce solids (Waste 730). Representative samples of these solids have been assayed for the presence of solvents, of which none were present at detectable levels.

Chemical (ppm)	Waste			
	720	730	307	312
Methanol	20	10	250	10
Methyl chloride	20	1	2	1
Isobutyl alcohol	10	10	4	0.1
1,1,1-trichloroethane	1	1	1	1
Pyridine	10	10	1.4	1
Toluene	15	0.1	0.1	0.1
Chlorobenzene	15	0.1	0.1	0.1

Note: Ignitability tests were all negative.

The information presented and any additional information regarding the treatment process may be obtained from the "Hazardous Waste Delisting Petition" submitted by Eli Lilly and Company to the U.S. EPA, November 24, 1980.

Facility Hewlett-Packard Company, Loveland Instrument Division, Loveland, Colorado

Type of Waste Wastewater treatment sludges from electroplating operations (U.S. EPA Hazardous Waste No. F006).

Description of Manufacturing Process

The company primarily manufactures instruments for measuring the characteristics of electricity and electrical signals. The operation includes several electroplating processes: printed circuit board manufacturing, small-parts plating of common and precious metals, chromating aluminum parts.

Feed materials producing the wastewater for treatment are primarily rinse waters following process solutions. The primary process solutions are copper, nickel, and gold. The process solutions not reclaimed are also treated through the waste-treatment facilities by bleeding small volumes continuously into the collected rinse waters for treatment.

The process solutions used are the following:

Printed Circuit Board Electroless Copper Plates.   Concentrated sulfuric acid; alkaline cleaner; black oxide (NaOh, $NaOCl_3$); acid salts (bifluoride); conditioner (soap solution); pre-acid etch (H_2SO_4); predip (acid salts); catalyst (palladium, tin, chlorides); accelerator (fluoboric acid); electroless copper (formaldehyde, NaOh, copper).

Printed Circuit Board Plating.   Acid cleaner (phosphoric); 10% sulfuric acid; acid copper plate ($CuSO_4$, H_2SO_4); watts nickel plate ($NiSO_4$, $NiCl_2$); acid tin plate ($SnSO_4$); Gold plate (potassium gold cyanide); ammonia etchant rinse.

Small-Parts Plating.   Brite dip (H_2SO_4, HCl, HNO_3); silver plate (contains cyanide); alkaline cleaner; 15% sulfuric acid; 15% HCL; copper strike (contains cyanide); watts nickel plate ($NiSO_4$, $NiCl_2$); rhodium plate (sulfate base); gold plate (potassium gold cyanide); acid tin plate ($SnSO_4$); chromate (chromic acid); acid cleaner; zinc (sulfate based).

Chromating Aluminum Parts.   Caustic etch (NaOH); desmut (H_2SO_4); alodine (chromic acid).

Currently the waste flows from the processes are fairly uniform, and the production capacity is at the design levels. Based on the operation, the sludge samples analyzed are representative of the process, and changes in operations and feed materials would only result when additions or modifications to the process have been made.

Description of Waste

The company operates two basic treatment processes. System I receives metal-bearing wastewaters from noncomplexed and nonchelated waste streams and the wastewater from cyanide-bearing streams after destruction of cyanides using chlorination. The system uses a lime/anionic polymer with two-stage lime addition to regulate pH for metal hydroxide formation, metered polymer for proper polyelectrolyte addition, and flocculation followed by clarification. The settled sludge slurry is pumped to a sludge holding tank for processing through a filter press that dewaters the sludge slurry to form the sludge cake of 30-40% solids.

System II serves the complexed or chelated waste streams. The wastewaters are first pH regulated to 2-3 with sulfuric acid, and ferrous sulfate is added for metal reduction. After reduction the wastewaters are pH regulated by sodium hydroxide for metal hydroxide formation followed by metered anionic polymer addition for flocculation prior to clarification. The sludge slurry is maintained in a separate sludge holding tank for processing through the filter press for dewatering into a sludge cake.

Chrome reduction is also conducted on the wastewater from the chromating of aluminum parts. The reduction is done with hydrazine followed by lime addition to form the metal hydroxide slurry. The chrome slurry is combined with the System II sludge slurry for filtering.

Estimated Sludge Cake Quantities (tons)

	System I	System II
Average month	4.4	17.6
Average annual	52.8	211.1
Maximum month	6.2	24.6
Maximum annual	64.0	256.0

EP Extraction for Waste Sludge

	System I	System II
Lead	0.03 + 0.003	0.05 + 0.005
Arsenic	0.01	0.01
Mercury	0.01	0.01
Selenium	0.01	0.01
Zinc	0.20 + 0.02	0.37 + 0.04
Cadmium	0.003 + 0.0003	0.008 + 0.0008
Copper	7.6 + 0.8	18.0 + 0.2
Silver	0.01	0.01
Gold	—	—
Nickel	0.79 + 0.08	0.66 + 0.07
Chromium	0.020 + 0.002	0.020 + 0.0002
Barium	0.050 + 0.005	0.080 + 0.0080
Cyanide total	0.005	0.005
Cyanide amendable to chlorination	0.005	0.005
Chromium hexavalent	0.01	0.01

The information presented and any additional information regarding the treatment process may be obtained from the "Hazardous Waste Delisting Petition" submitted by Hewlett-Packard to the U.S. EPA, July 13, 1981.

Facility   National-Standard Company, Woven Products Division, Corbin, Kentucky

Type of Waste   Sludge from the lime treatment of galvanizing and metal-finishing process wastewaters and spent pickle liquor from steel-finishing operations.

Description of Manufacturing Process

The plant manufactures wire cloth of various sizes and surface finishes; all types are made from steel wire, and some are plated with other metals. Some wire cloth is cleaned after weaving and before painting. Some unplated wire is drawn from rod after pickling and coating.

The plant has five chemical processing lines that generate wastewaters. Two are hot-dip galvanizers, an electrogalvanizer, a paint cleaning line, and a wire cleaning line.

Hot-Dip Galvanizer(s) Process.   Caustic cleaner—a proprietary NaOH-based cleaner; water rinse; acid cleaning—HCl; water rinse; flux—$ZnCl_2$; molten zinc (no discharge).

Electrogalvanizer Process.   Caustic cleaner—a proprietary NaOH-based cleaner; water rinse; acid cleaning—HCl; water rinse; plating tanks, $ZnSO_4$ H_2SO_4 (The zinc sulfate used on this line is manufactured on site by digesting zinc dross in sulfuric acid).

Paint Cleaning Process.   Caustic cleaner—a proprietary NaOH-based cleaner; water rinse; water rinse; water rinse; acid cleaning—HCl; water rinse; water rinse; phosphate coating—$Zn_3(PO_4)_2$ H_3PO_4; water rinse; water rinse; caustic cleaner—same as tank 1; water rinse; water rinse; chromate sealer—a proprietary finish, parcolene 8, H_3PO_4, H_2CrO_4; water rinse; water rinse. (The paint cleaning line is used to clean cloth before it is painted. A variety of products are processed through this line, and not all tanks are used for every product.)

Wire Cleaning Process. The wire cleaning line is used to batch clean coils of wire or rod. The tanks are caustic cleaner NaOH, $KMnO_4$, $K_2S_2O_8$; water rinse; water rinse; acid cleaning—HCl; water rinse; water rinse; water rinse; copper/tin coating—H_2SO_4, $CuSO_4$, $SnSO_4$; liquor finish—H_2SO_4; $CuSO_4$ $5H_2O$; water rinse; phosphate coating—$Zn_3(PO_4)_2$, H_3PO_4; soap dip—a proprietary soap material.

Wastewater Treatment Process

All process rinses flow into a common drain system leading to the wastewater treatment plant. Acid and alkaline solutions flow through separate drain systems to the wastewater treatment plant. All wastes are collected in a large PVC lined pit. The pit contents are agitated by the injection of a low-pressure air through perforated pipes forming a grid on the pit bottom. The pH of the pit contents is sensed, and provision made to add lime slurry or sulfuric acid to bring the wastes to approximately neutral pH. At this point the low-pressure air also oxidizes any ferrous iron present to the ferric state. From the pit the pH is again sensed and automatically raised with lime slurry to approximately 8.5. The wastes then overflow into a pumping well where submersible pumps pick up the wastes and deliver them to one of a pair of pressure filters precoated with Perlite. The filter leaves are stainless steel. The filters are used alternatively to provide time for dumping and precoating. After filtration the clear effluent is monitored again by automatic pH sensors and transferred to a water storage tank for recycle. Water from this tank is pumped into the production areas for use as required.

Some rinses require a better rinse, so fresh water only is used in them. This results in more water in the recycle system than is required. The excess water passes through an overflow and is discharged to surface waters.

**Sludge Composition
(mg/L)**

Sample	Total Chromium	Total Lead
1	0.074	0.120
2	0.030	0.078
3	0.030	0.060
4	0.031	0.175
5	0.035	0.106
6	0.017	0.120
7	0.018	0.071
8	0.018	0.061

Note: All testing performed according to 40 CFR 261 Appendix II.

REFERENCES

1. U.S. Environmental Protection Agency, 1981 (Sept.), *Treatability Manual* (revised), vols. 1–4, EPA-600/2-82-00-e, Industrial Environmental Research Laboratory, Cincinnati, Ohio.
2. Burns & Roe Industrial Services Corp., 1981 (Nov. 15), *Fate of Priority Toxic Pollutants in Publicly Owned Treatment Works. Final Report, vol. 1*, U.S. Environmental Protection Agency, Washington, D.C.
3. E. C. Jordan Co., 1982 (Jan.), *Fate of Priority Toxic Pollutants in Publicly Owned Treatment Works:* 30-day Study Draft Report (T. P. O'Farrell, proj. officer), prepared for Effluent Guidelines, U.S. Environmental Protection Agency, Washington, D.C.
4. Touhill, Shuckrow Associates, Inc., 1981, *Concentration Technologies for Hazardous Aqueous Waste Treatment* (S. C. James, proj. officer), U.S. Environmental Protection Agency, Solid and Hazardous Waste Research Division, Cincinnati, Ohio.
5. Current Treatment Technologies of Hazardous and Industrial Wastes, Maswatduaka, the 5th Japan–U.S. Achievemental Conference on Solid Waste Management, September 28–29, 1982, Tokyo, Japan.
6. Destroying dioxin—A unique approach, *Waste Age*.
7. U.S. Environmental Protection Agency, 1980, *Dioxins*, EPA-600/2-80-197, Industrial Environmental Research Laboratory, Cincinnati, Ohio.
8. R. I. Berry, 1981 (Aug. 10), New ways to destroy PCBs, *Chemical Engineering*.
9. Earhart, J. P., et al., 1977 (May), Recovery of organic pollutants via solvent extraction, *Chemical Engineering Progress*.
10. H. H. Tabak, et al., 1980 (Oct.), Biodegradability studies with organic priority pollutant compounds, *Water Pollution Control Federation Journal*.
11. E. J. Martin, 1985 (Jan.), *Assessment of Innovative Cleanup Technologies for Hazardous Waste Uncontrolled Sites*, Office of Technology Assessment, U. S. Congress, Washington, D.C.
12. U.S. Environmental Protection Agency, 1977 (Aug.), *Assessment of Techniques for Detoxification of Selected Hazardous Materials*, EPA-600/2-77-143, Washington, D.C.
13. U.S. Department of Interior, Bureau of Mines, 1982, *Solvent Extraction of Nickel and Copper from Laterite—Ammoniacal Leach Liquors*, RI-8605, Washington, D.C.
14. A. Mehta, 1980, Recovery of Metal Values from Hydroxide Sludges, in-house EPA Research, Industrial Environmental Research Lab, Cincinnati, Ohio.
15. B. Olexsey, 1981, Supercritical Fluid Processing of Industrial Wastes, paper presented at Annual Meeting of the American Institute of Chemical Engineers.
16. R. E. Speece, 1983 (Sept.), Anaerobic biotechnology for industrial wastewater treatment, *Environmental Science & Technology*.
17. H. Kobayashi and B. E. Rittman, 1982 (March), Microbial removal of hazardous organic compounds, *Environmental Science & Technology*.

4

Incineration of Hazardous Waste

Eugene Crumpler
U.S. Environmental Protection Agency
Washington, D.C.

Edward J. Martin
Howard University, Washington, D.C.

Incineration as a specific waste-management technique has come to the forefront. Combustion of the organic components of wastes is an effective disposal method if the combustion is conducted properly. The U.S. Environmental Protection Agency (EPA) has stated that incineration is one of the best techniques for treating hazardous organic wastes. With such an endorsement, one should expect incineration to be widely used. However, many people think incineration is an expensive process that is not cost competitive with lower-cost land-disposal methods.

The long-term cost of land disposal (really land storage in the case of contained chemical landfills) is likely to be greater than the short-term cost of incineration. If organic wastes are contained in landfills, these wastes ultimately must be retrieved and properly treated. If organic wastes escape from a landfill, expensive treatment of contaminated groundwater or other remedial actions must be practiced in the future. Also liability for damages to human health and property increases the potential long-term cost of land disposal. Proper treatment by methods like incineration eliminates the long-term costs, although the short-term costs may be higher.

Determining if incineration of a waste stream or a mix of wastes is appropriate involves three basic decisions:

1. Is the waste or combination of wastes suitable for incineration?
2. What type of incineration equipment is required? What capacity is needed?
3. What is the incineration cost vis-a-vis other management options?

This chapter provides guidance to make these decisions. Incineration is a versatile process. It can be used to recover energy if the heat generated is converted into steam and power or is put to some other beneficial use, such as preheating combustion air. It can be considered as a volume-reduction process in that many of the component elements of organic materials, including the most common ones (carbon, hydrogen, oxygen, chlorine, and sulfur), are converted wholly or partially to gaseous form, leaving only the noncombustible inorganic volume. Detoxification of organic materials is accomplished by destroying the organic molecular structure through oxidation or thermal degradation.

A PROFILE OF EXISTING INCINERATION TECHNOLOGIES

In 1982 the EPA completed a telephone survey of facilities that had notified EPA (as required by current regulations) that they were incinerating hazardous waste [1]. Privately owned and operated facilities represented 80% of the total of approximately 300 incinerators identified in the survey. The remaining 20% included commercial facilities burning waste from a variety of generators, as well as military incinerators. Just over 80% of the incinerators are less than 10 years old. The data presented here are based on the telephone survey of these incinerators.

Technologies

Three principal types of incinerators are used in the United States: liquid injection, rotary kiln, and hearth. However, other technologies or variations of them identified in the survey were fume incinerators with liquid-waste capability, ammunition burners, drum reclaimers, fluidized beds, and others.

Liquid-injection incinerators account for 52% of the incinerators by number, and all types of incinerators with liquid-injection capability represent 79% of the incinerators. Hearth incinerators represent 34% of the incinerators, rotary kilns 7%.

Capacities

Data on the capacities of 180 liquid-burning incinerators and 44 solids-burning incinerators are presented in Table 4-1. The median capacity for all incinerators with liquid-injection capability is 150 gal (568 L)/h. The median capacity for solids-burning incinerators is 295 k/h—equivalent to approximately 78 gal (295 L) of liquid (aqueous) waste.

Table 4-1. Capacities of Hazardous-Waste Incinerators

Capacity (gal/h)	No. of Incinerators	Cumulative No.
0–50	48	48
51–100	28	76
101–200	22	98
201–300	22	120
301–500	12	132
501–1,000	23	155
1,001–2,000	17	172
2,001–5,000	4	176
5,001–10,000	4	180
Unknown	28	208
0–100	4	4
101–300	5	9
301–500	7	16
501–1,000	12	28
1,001–2,000	6	34
2,001–5,000	7	41
5,001–10,000	1	42
10,001–20,000	2	44
Unknown	17	61

Source: E. Kertz, L. Boberschmidt, D. O'Sullivan, and N. Sanders, 1982 (April 23), Incineration—A Profile of Existing Facilities, unpublished draft report by the Mitre Corporation to the U.S. Environmental Protection Agency, Washington, D.C.

Operating Conditions

Table 4-2 presents the distribution of combustion temperatures for 173 facilities that provide this data. The median temperature is roughly 927–982 °C. The survey indicated that liquid-injection units usually operate at slightly higher temperatures than other incinerator types, with a median of about 982 °C. Fifteen of the 23 units reporting temperatures greater than 1204 °C were liquid-injection incinerators.

Table 4-2 also shows the distribution of gaseous residence times for 104 operating incinerators: 86% have a residence time of 1 s or greater; 48% have a residence time of 2 s or greater. Additional data show that there is a definite tendency for incinerators with higher combustion zone temperatures to have longer residence times. Of incinerators having combustion temperatures greater than 1038 °C, 74% also had residence times of 2 s or greater. In contrast, only 24% of incinerators with combustion temperatures of less than 1038 °C had residence times at least 2 s.

Table 4-2. Distribution of Combustion Temperatures and
Gaseous Residence Times

Range (°C)	No. of Incinerators	Residence Time (s)	No. of Incinerators
871	36	1.0	14
871–1038	64	1.0–1.49	20
1038–1204	50	1.5–1.99	20
1204	23	2.0	50
Sample size	173	Sample size	104

Source: E. Keitz, L. Boberschmidt, D. O'Sullivan, and N. Sanders, 1982 (April 23), Hazardous Waste Incineration—A Profile of Existing Facilities, unpublished draft report by the Mitre Corporation to the U.S. Environmental Protection Agency, Washington, D.C.

Pollution Control and Heat Recovery

Forty-five percent of the respondents in the survey reported the use of one or more air-pollution-control devices (APCD) on their incinerators. Over half (57%) of the APCDs are in the scrubber category. Most of the remainder are also wet devices, such as quenches and packed-spray towers. Only 7% were dry devices, such as electrostatic precipitators or fabric filters.

Approximately 22% of survey respondents reported use of heat recovery at the incinerator. Large liquid-injection incinerators are more likely to have heat recovery than other units. Recent discussions with incinerator manufacturers indicate that a significant portion of new orders are for incinerators with heat recovery.

PERFORMANCE OF HAZARDOUS-WASTE INCINERATORS

Performance of hazardous-waste incinerators can be measured in terms of destruction efficiency (DE) or destruction and removal efficiency (DRE). Destruction and removal efficiency accounts for both the destruction in the combustion chamber(s) and the removal of organics in any air-pollution-control equipment. DRE may be calculated as the percentage mass difference of input (feed) and output (stack emission) waste constituents through the incinerator [2]. DRE has been defined for regulatory purposes on a compound-specific basis and thus must be calculated for each constituent of interest separately.

In the mid-1970s EPA conducted a series of incinerator test burns in which destruction efficiency was measured. More recently, tests have been conducted by EPA or by others to measure destruction and removal efficiency. These tests are presented here.

Tests of Destruction Efficiency

EPA has identified 54 test burns of destruction efficiency carried out on a number of hazardous wastes [3]. Fourteen of these tests were in full-scale commercial incinerators, and 40 were in pilot-scale incinerators. Incinerator types represented included liquid injection, rotary kiln, multiple and single hearth, fluidized bed, molten salt, and pyrolysis.

All but nine of the tests achieved a destruction efficiency of at least 99.99%. Limitations of the sampling and analysis methods presented documentation of higher destruction efficiencies. The test burns that achieved lower destruction efficiencies generally were traced to identifiable and correctable problems. For example, some burns several incinerator operating conditions were tested to determine the effects of lower temperatures, reduced excess air, or changes in turbulence. Thus, lowering of destruction efficiencies was a predictable result.

The results of the 54 test burns are summarized in Table 4-3. These test burns became the basis for development of EPA's incinerator regulations described later in this chapter. Additional test data are now available and are discussed here.

Tests of Destruction and Removal Efficiency

A series of test burns was conducted by EPA at one hazardous-waste incinerator to determine the effects of different conditions of incineration on the DREs for various chlorinated organic compounds [4]. The configuration of this incinerator includes a rotary kiln and a cyclone furnace, both of which lead into a common single secondary combustion chamber. Liquid wastes are injected into the kiln and the cyclone furnace.

Data from the test burns are presented in Table 4-4. The tests were conducted at three temperatures (899 °C, 1093 °C, and 1316 °C) and two ranges of residence time (3.3–3.7 s and 1.5–2.3 s). The waste stream was a mixture of organic compounds, and the feed rate in Table 4-4 indicates the total feed rate. Stack gas samples were analyzed for the listed chlorinated organic compounds.

With each set of conditions, the destruction efficiencies varied somewhat, and there seems to be some correlation to the specific compound. For example, regardless of the particular incinerator conditions, the DREs for hexachlorobenzene and bromodichloromethane were consistently low for the set, and the DREs for hexachlorocyclopentadiene were consistently high for the set.

As would be expected, the DREs were highest with the test burn conducted at the highest temperature and the longest residence time (i.e., at 1316 °C for 3.4 s) for the first set of compounds. This pattern also pertains to the

Table 4-3. Summary of Incinerator Destruction Test Work

Waste	Incineration[a]	Destruction Eff. of Principle Comp. (%)
Shell aldrin (20% granules)	MC	99.99
Shell aldrite	MC	99.99
Atrazine (liquid)	MC	99.99
Atrazine (solid)	MC	99.99
Para-arsanilic acid	MS	99.999
Captan (solid)	MC	99.99
Chlordane 5% dust	LI	99.99
Chlordane, 72% emulsifiable concentrate and no. 2 fuel oil	LI	99.999
Chlorinated hydrocarbon, trichloropropane, trichlorethane, and dichloroethane predominating	HT	99.92 99.98
Chloroform	MS	99.999
DDT 5% oil solution	LI	99.99
DDT (solid)	MM	99.970 to 99.98
DDT 10% dust	MC	99.99
20% DDT oil solution	LI	99.98
DDT 25% emulsifiable concentrate	LI	99.98
DDT 25% emulsifiable concentrate	MC	99.98 to 99.99
DDT oil 20% emulsified DDT waste oil-1.7% PCB	TO	99.9999
DDT powder	MS	99.998
Dieldrin-15% emulsifiable concentrate	LI	99.999
Dieldrin-15% emulsifiable concentrates and 72% chlordane emulsifiable concentrates (mixed 1:3 ratio)	LI	99.98
Diphenylamine-HCl	MS	99.999
Ethylene manufacturing waste	LI	99.999
GB ($C_4H_{10}O_2PP$)	MS	99.99999969
Herbicide orange	RL	99.999 to 99.985
Hexachlorocyclopentadiene	LI	99.999
Acetic acid, solution or kepone	RKP	99.9999
Toledo sludge and kepone coincineration	RKP	99.9999
Lindane 12% emulsifiable concentrate	LI	99.999
Malathion	MS	99.999 to 99.9998
Malthion 25% wet powder	MC	99.99
Malathion 57% emulsifiable concentrate	MC	99.99
Methyl mathacrylate (MMA)	FB	99.999
0.3% Mirex bait	MC	98.21 to 99.98
Mustard	MS	99.999982
Nitrochlorobenzene	LI	99.99 to 99.999
Nitroethane	MS	99.993

Table 4-3. *(continued)*

Waste	Incineration[a]	Destruction Eff. of Principle Comp. (%)
Phenol waste	FB	99.99
Picloram	MC	99.99
Picloram, (tordon 10K pellets)	MC	99.99
PCBs	RK	99.999964
PCB capacitors	RK	99.5 to 99.999
PCB	CK	99.9998
Polyvinyl chloride waste	RK	99.99
Toxaphene 20% dust	MC	99.99
Toxaphene 60% emulsifiable concentrate	MC	99.99
Trichlorethane	MS	99.99
2,4-D low-volatile liquid ester	LI	99.99
2,4,5-T (Weedon™)	MM	99.990 to 99.996
2,4,5-T	SH	99.995
2,4,5-T	SH	99.995
2,4,5-T	SH	92
2,4,5-T	SH	99.995
VX ($C_{11}H_{26}O_2PSN$)	MS	99.999989 to 99.9999945
Zineb	MC	99.99

Source: J. Corini, C. Day, and E. Temrowski, 1980 (Sept. 2). Trial Burn Data (unpublished draft) Office of Solid Waste, U.S. Environmental Protection Agency, Washington, D.C.

[a]MC = multiple chamber; MS = molten salt combustion; LI = liquid injection; HT = 2 high-temp. incinerators; MM = municipal multiple-hearth sewage sludge incinerator; TO = thermal oxidizer waste incinerator; RL = 2 identical refractory-lined furnaces; RKP = rotary kiln pyrolyzer; FB = fluidized bed; CK = cement kiln; SH = single-hearth furnace.

second set of compounds for which the DREs were highest at the highest temperature (1316 °C). There does not, however, seem to be a consistent relation thereafter between temperature and residence time and DRE. For example, when the first waste stream was incinerated for the shortest time, the DREs of four of the six compounds were highest when the temperature was lowest (i.e., 899 °C for 2.2 s). No conclusions should be drawn, however, from this limited set of test burns regarding DRE versus operating conditions, because it is not certain that the differences discussed are statistically significant.

The data in Table 4-5 are from industry test burns of incinerators with various organic compounds as the waste feed [5,6]. Incinerator B consists of a rotary kiln and a secondary combustion chamber. The waste can be fed into the kiln, thereby being exposed to combustion temperatures in both the

Table 4-4. Test Burns of Chlorinated Organic Compounds in a Hazardous-Waste Incinerator

Temp.	899	1093	1316	899	1093	1316	899	1093	1316
Residence time (s)	3.3	3.7	3.4	2.2	1.8	1.7	2.3	2.9	1.5
Waste feed rate (l/min)	37	32–37	41	51–67	41–64	49–77	51	62	67
Destruction and removal efficiencies (%)									
Chloroform	99.998	—	99.9995	99.9997	99.9989	99.998	—	—	—
Carbon tetrachloride	99.995	—	99.9993	99.999	99.96	99.9	—	—	—
Tetrachlorethylene	99.999	—	99.999	99.997	99.99	99.97	—	—	—
Hexachloroethane	99.997	99.996	99.997	99.995	99.994	99.995	99.9997	99.9997	99.9997
Hexachlorobenzene	99.996	99.995	99.999	99.993	99.994	99.995	—	—	—
Hexachloro-cyclopentadiene	99.9997	99.9994	99.99998	99.9994	99.9996	99.9997	—	—	—
Trichloroethane	—	—	—	99.991	—	99.99996	—	—	—
Tetrachloroethane	—	—	—	99.9997	—	99.9998	—	—	—
Bromodichloromethane	—	—	—	99.97	—	99.995	—	—	—
Pentachloroethane	—	—	—	99.9998	99.9997	99.9998	—	—	—
Dichlorobenzene	—	—	—	99.997	00.997	99.998	—	—	—

Notes: Data are presented as the destruction and removal efficiency (in percent) for each compound as achieved under the specified conditions of incineration. The combustion units of the incinerator consist of a rotary kiln and a cyclone furnace with a common single combustion chamber. Waste is fed by liquid injection into both the rotary kiln and the cyclone furnace.

Table 4-5. Industry Test Burns of Organic Compounds in Incinerators

Incinerator	Incineration Type	Organic Destruc. and Removal Eff. (%)	Temperature (°C)	Residence Time (s)	Organics Feed Rate (kg/h)	Waste Feed
A	Fluidized bed	99.995	1117	1.7	212	Nonchlorinated and flammable solvents[a]
A	Fluidized bed	99.997	1114	2.2	163	Chlorinated and flammable solvents[b]
B[c]	Kiln and chamber	99.96	816 + 982	3.3	NR	Acetonitrile
	Chamber	99.67	982	1.2		
B[c]B[c]	Kiln and chamber	99.96	816 + 982	3.3	NR	Methyl ethyl ketone
	Chamber	99.82	982	1.2		
B[c]	Kiln and chamber	99.99	816 + 982	3.3	NR	1,4-diethylene dioxide
	Chamber	99.71	982	1.2		
B[c]	Kiln and chamber	99.98	816 + 982	3.3	NR	Toluene
	Chamber	99.99	982	1.2		
B[c]	Kiln and chamber	99.995	816 + 982	3.3	NR	1,2-dichloroethane
	Chamber	99.995	982	1.2		
B[c]	Kiln and chamber	99.995	816 + 982	3.3	NR	Methylene dichloride
	Chamber	99.995	982	1.2		

Note: NR = feed rate not reported.
[a] Waste feed: 55% methyl ethyl ketone.
[b] Waste feed: 29% methyl ethyl ketone, 19% xylene, 10% toluene.
[c] Waste can be fed into the kiln or into the secondary combustion chamber.

kiln and the secondary combustion chamber; alternatively, the waste can be fed directly into the secondary combustion chamber, thereby bypassing the kiln. In these paired tests the same waste was fed to the incinerator by both routes, and DREs for the compound were calculated. For acetonitrile, methyl ethyl ketone, and 1,4-diethylene dioxide, the increased residence time resulting from feed to the kiln (3.3 vs. 1.2 s) resulted in higher DREs. The difference in residence time had little or no effect on the DREs for toluene, 1,2-dichloroethane, and methylene dichloride.

Eight of the 14 tests achieved 99.99% DRE or greater. Three additional tests achieved at least 99.96% DRE, and the remaining three tests achieved at least 99.6% DRE.

Table 4-6 presents data from test burns of polychlorinated biphenyls (PCBs) conducted on six incinerators [7]. These incinerators represent different incinerator types, with combustion sections ranging from one to three different units. In addition, the operating conditions for the incinerators vary, with the temperatures ranging from 1180 °C to 1347 °C, and residence time from 1.17 s to 6 s.

The two incinerators with residence times less than 2 s operate at combustion temperatures in excess of 1300 °C. The differences in PCB destruction and removal efficiencies are not significant but reflect the differences in sensitivity of the analytical methods used. Nevertheless, the lowest value for DRE reported is 99.9971%, and all but one burn achieved at least 99.9999% DRE.

INCINERATOR DESIGN CONCEPTS

In the combustion of hazardous (chemical) waste one of the major considerations is the complete destruction of the chemical molecules presenting a hazard to human health and the environment if released to the air, water, or land. If incineration of a waste results in air pollution from the exhaust gases formed, release of toxic materials to surface water by contaminated scrubber water, or potential groundwater and land pollution by disposal of toxic ash and residues, then the incineration process has not fulfilled its role as an effective waste-management tool.

Traditional incineration practices address the three "t's": time, temperature, and turbulence. The wastes, as gases, must be exposed to a temperature high enough to allow complete oxidation of the organic materials, the waste gases must have some period of time exposed to these maximum temperatures, and the gases must be mixed well enough so that each molecule is exposed to enough oxygen molecules to allow complete oxidation. The ability to measure the gas temperatures in incinerators is well developed. Residence time can also be estimated by measuring combustion gas velocities in the

Table 4-6. Test Burns of PCBs in Incinerators

Incinerator Type	Date of Test Burn	PCB Destruct. and Removal Eff. (%)	Temp. Avg. (°C)	Residence Time (s)	PCB Feed Rate (kg/h)
Rotary kiln	10/1979	99.99998	760	2	289[a]
primary combustion chamber			1254	2	
secondary combustion chamber			1046	2	
Loddby liquid burner	11/1979	99.99997	1304	2.68	1102[b]
afterburner			1183		
Single combustion chamber	1/1980	99.9971	1347	1.59	1[b]
Single combustion chamber	5/1980	99.99999	1316	1.17	22[b]
Single combustion chamber	7/1981	99.9999	1271	2.14	1.5[b]
Single combustion chamber	7/1981	99.999996	1180	2.96	6[b]

[a] Waste feed: liquids + PCB-containing capacitors.
[b] Waste feed: liquids.

Table 4-7. Typical Operating Conditions for Incinerators

Incinerator Type	Temperature Range (°C)	Solid Residence Time (min)	Liquid and Gas Residence Time (s)
Rotary kiln	820–1,600	20–60	1–3
Kiquid injection	650–1,600	—	0.1–2
Fluidized bed	760–980	10–hours	1–12
Catalytic reactor	320–820	—	1 or less
Multiple hearth	370–520 (drying zone)	30–90	0.25–3
	720–980 (incineration zone)	—	—
Multiple chamber	800–1,000	5–30	1–4
Pyrolysis	480–820	12–15	1–3
Molten salt	800–980	—	0.75

system. Turbulence is not easily measured, and a turbulence "parameter" has not been generally agreed upon by practitioners in the field.

Typical values of temperature and residence-time ranges are shown in Table 4-7 for various types of incinerators (note the differences between gas and solid residence times). No attempt at quantifying a turbulence factor is presented in the table. A description of the various types of incinerators follows in the section on types of incinerators.

Design Considerations

The selection and design of an incinerator for effective destruction of a hazardous waste(s) can be outlined as follows:

1. Consider the physical, chemical, and thermodynamic properties of the waste
2. Determine the basic design parameters: temperature, excess air rate, residence time, and mixing consideration
3. Determine auxiliary fuel needs
4. Select the air-pollution system design
5. Develop the detailed design considerations, such as construction materials
6. Select the process control systems and instrument design

Waste Characterization

To evaluate the feasibility of destroying a hazardous-waste material by incineration, one must know the physical form, elemental composition, heat content, water content, ash, and the organic compounds of the waste. The necessary equipment can be determined and regulatory concerns can better be addressed with such a complete knowledge of the waste(s).

The design engineer must understand that an evaluation of waste disposal is never complete until an actual test is conducted in the type of equipment being considered. Waste-incineration design has not been developed to a rigorous engineering science. The complexity of most organic chemical wastes reduces the ability to deal with their combustion on a purely theoretical basis. The test burning of hazardous organic wastes in pilot facilities before developing the final design is consistent with conventional engineering wisdom. Most engineers will do test work at the pilot scale before committing large sums of money to a full-scale operating facility. Because major incinerator installations can cost several millions of dollars, this approach is essential. Even small incinerators, which may be installed for as little as $50,000, should be subjected to the pilot-test-burn approach. Purchasing the wrong type or the wrong size of equipment can be very expensive. Incinerators, being specifically designed to handle high temperatures, are expensive to modify. More than one incinerator has been abandoned because it could not be made to perform as anticipated.

Increased regulation of hazardous-waste disposal also makes selection of the correct equipment even more critical. Federal, state, and local standards are requiring hazardous wastes to be incinerated with the best state-of-the-art equipment.

The relevant waste characteristics needed to select, design, and evaluate an effective incineration system are summarized as follows [2]:

1. General waste characteristics
   a. Composition weight fractions (paper, metals, glass, other specific waste)
   b. Weight fractions of combustibles, noncombustibles, recoverable materials
2. Physical characteristics
   a. For solid wastes: solubility, combustibles, volatiles, ash, size distribution, shape, melting point
   b. For liquid wastes: temperature, viscosity, pour point, percent solids, vapor pressure, flash point, specific gravity
   c. For gaseous wastes; temperature, pressure, suspended liquids and solids, concentration, density
3. Chemical characteristics
   a. General: pH, explosivity, reactivity
   b. Thermochemical: heating value, enthalphy, stoichiometic air requirements, adiabatic flame temperature

   c. Elemental composition: carbon, hydrogen, oxygen, sulfur, phosphorus, heavy metals, alkali metals, toxic materials (asbestos, chromium, arsenic, etc.)

   d. Organics: organic analysis for regulated compounds, volatiles, fixed carbon.

In matching wastes with incinerator types, the physical form (solid, liquid, percent solids, viscosity, etc.) of the wastes is more important than chemical properties. The most important chemical properties are the halogen (Cl, F, Br, I) content and alkali metals (Na, K, etc.). The most important thermo-chemical properties are the stoichiometric air requirements and net heating value.

Wastes containing a large concentration of halogens require gas-cleaning equipment to control acid gas emissions. A high halogen content also requires sufficient hydrogen to be available to maximize HCl formation and to minimize Cl_2 formation in the combustion process. This additional hydrogen can be added as natural gas, steam, or water.

Wastes containing substantial amounts of sodium or other alkali metals can cause fouling of waste heat boilers and damage to refractories. Defluidization of fluidized beds is also caused by forming low-melting eutectic mixtures such as $NaCl-Na_2CO_3$ or $NaCl-Na_2SO_4$. If the particles of the fluidized bed are silicasand, Na_2SO_4 will react with the silica for form a viscous sodium-silicate glass, which will cause rapid defluidization [5].

The following list contains criteria used for matching different wastes to the various incineration facilities:

1. Physical form: gas, liquid, slurry, sludge solids, solids content
2. Temperature range required for destruction: 1100 °C (2000 °F); 760–1100 °C (1400–2000 °F); 370–760 °C (700–1400 °F); 370 °C (700 °F);
3. Off-gases generated: oxides of carbon and nitrogen; water; halogens, sulfur, and phosphorous; volatile metals
4. Ash generated: nonfusible, fusible, or metallic; particles
5. Heating value: 23 MJ/kg (10,000 Btu/1b); 12–23 MJ/kg (5,000–10,000 Btu/1b; 12 MJ/kg (5,000 Btu/1b)

TYPES OF INCINERATORS

Several types of incinerators can provide adequate destruction of hazardous wastes. Some types, such as flares and open-pit methods, are considered inadequate for hazardous-waste disposal due to uncontrolled air addition, uncontrolled disposal of combustion products, and formation of incomplete combustion products.

Air-emission control usually is required for hazardous-waste combustion. Emission-control devices are normally applicable to fuel gas cleanup regardless of incinerator type. Wet-scrubbing devices are the most common emission-control equipment encountered. These devices are useful in reducing sulfur oxides, hydrogen, halogens (HCl), nitrogen oxides, and particulates. Packed towers and venturi scrubbers are the most widely encountered types of wet scrubbers, although many other types of emission-control devices are applicable. Other types used for particulate control are electrostatic precipitators, demisters, bag houses, and other mechanical devices.

Liquid injection and rotary kiln are the two common designs of combustion systems currently used for hazardous-waste combustion. These designs are effective and represent the state-of-the-art. Hence our discussion of general incinerator designs and operations principles will focus on these two types of combustion systems. Many of the thermochemical principals presented are equally applicable to other types of combustion systems, such as fluidized beds, pyrolysis units, molten salts, and others. The details of waste handling and waste movement through the system can be quite different, however.

Liquid-Injection Systems

Liquid-waste combustors (liquid-injection combustors) are flexible units that can dispose of virtually any combustible liquid waste. The heart of the liquid-injection system is the waste-atomization device or nozzle (burner). Because a liquid-combustion device is essentially a suspension burner, efficient and complete combustion is obtained only if the waste is adequately divided or atomized and mixed with the oxygen source. Atomization is usually achieved by mechanical means, such as rotary cup or pressure atomization systems, or by gas fluid nozzles using high-pressure air or steam. A forced draft must be supplied to the combustion chamber to provide the necessary mixing and turbulence.

A number of liquid-injection configurations are available. The most common type is the horizontally fired unit illustrated in Figure 4-1. Other types are down-fired and up-fired. Each configuration has unique advantages. Liquid-injection units are also often part of rotary kiln incinerators, the solid wastes being first vaporized in the kiln and the bases then fed to a liquid-fired unit to assure complete combustion. Generally, only low-ash liquid wastes are fired into liquid-injection units because there is usually no provision for ash removal. The exception is the down-fired design, which can handle salt- and particulate-laden wastes. The down-fired salt unit collects the salts in molten form in the bottom of the injection chamber; these salts can be removed as molten material and quenched outside the incinerator. Down-

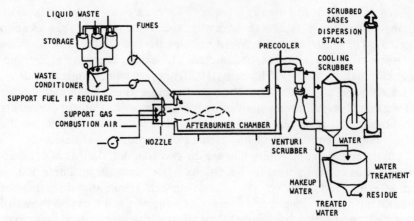

Figure 4-1. Horizontally fired liquid injection incineration.

fired units can effectively handle solids-laden liquids by having a water quench at the bottom of the combustion chamber. Combustion gases then pass to air-pollution equipment.

Advantages

1. Capable of incinerating a wide range of liquid wastes
2. No continuous ash removal system is required other than for air-pollution control
3. Capable of a fairly high turndown ratio
4. Fast temperature response to changes in the waste-fuel flow rate
5. Virtually no moving parts
6. Low maintenance costs
7. Proven technology

Disadvantages

1. Must be able to atomize liquids through a burner nozzle except for certain limited applications
2. Must provide for complete combustion and prevent flame impingement on the refractory
3. Burners are susceptible to plugging. High percent solids can cause problems
4. No bulk solids capability

Rotary Kiln

The rotary kiln is a cylindrical, horizontal refractory-lined shell mounted at a slight incline. Rotating the shell mixes the waste with the combustion air. This rotation provides turbulence and agitation to maximize burnout. Rotary kilns can handle solid wastes with heating values from 550 to 8250 kcal/kg (100 to 15,000 Btu/1b). Typical throughput of solids range from 45 kg (100 lb) to 1.8 metric tons (2 tons)/h. Kilns are usually cofired with fuel or liquid wastes to sustain the combustion of solids since solids are usually charged intermittently; the cofiring maintains temperatures and aids combustion of solids with low heat content.

The rotary kiln incinerator is a cylindrical shell lined with fire brick or other refractory and mounted with its axis at a slight angle from the horizontal. It is a highly efficient unit when applied to solids, liquids, sludges, and tars because of its ability to attain excellent mixing of unburned waste and oxygen as it revolves.

Rotary kiln incinerators generally have a length-to-diameter ratio (L/D) between 2 and 10. Smaller L/D ratios result in less particulate carryover. Rotational speeds of the kiln are usually much slower than those for kilns used as calciners or dryers and are on the order of 5 to 25 mm/s 1–5 ft/min measured at the kiln periphery. Both the L/D ratio and the rotational speed strongly depend on the type of waste being combusted. In general, larger L/D ratios along with slower rotational speeds are used when the waste material requires longer residence times in the kiln for complete combustion.

The residence time and combustion temperature required for proper incineration are totally dependent on the combustion characteristics of the waste materials. Combustion temperatures usually range from 820 to 1600 °C (1500 °C to 3000 °F). For instance, a finely divided propellant may require 0.5 s whereas wooden boxes, municipal refuse, and railroad ties may require 5, 15, and 60 min, respectively [8].

The rotary kiln incinerator is generally applicable to the ultimate disposal of any form of combustible waste material and represents proven technology. It can incinerate combustible solids (including explosives), liquids (including chemical warfare agents), gases, sludges, and tars.

Slag and metallic objects are discharged at the lower end of the kilns and are usually water quenched. Gases enter a secondary combustion chamber that permits sufficient dwell time to assure complete combustion. The secondary chamber may or may not be fired with either fuel or liquid wastes. The gases then pass through an air-pollution-control system. Figure 4-2 is a schematic of a rotary kiln incinerator.

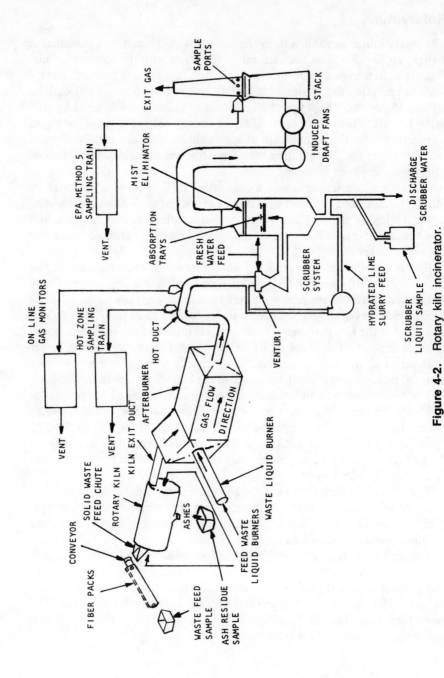

Figure 4-2. Rotary kiln incinerator.

Advantages

1. Can incinerate a wide variety of liquid and solid wastes
2. Capable of receiving liquids and solids independently or in combination
3. Not hampered by materials passing through a melt phase
4. Feed capability for drums and bulk containers
5. Wide flexibility in feed mechanism design
6. Provides high turbulence and air exposure of solid wastes
7. Continuous ash removal does not interfere with waste burning
8. No moving parts in the kiln
9. Adaptable for use with a wet-gas-scrubbing system
10. Retention or residence time of the waste can be controlled by adjusting rotational speed of the kiln
11. Many wastes can be fed directly into the incinerator without any preparation, such as preheating, mixing, and so forth
12. Can be operated at temperatures in excess of 1400 °C (2500 °F), making them well suited for destroying toxic compounds that are difficult to thermally degrade
13. Proven technology

Disadvantages

1. High capital cost for installation, especially for low feed rates
2. Operating care necessary to prevent refractory damage from bulk solids
3. Airborne particles may be carried out of kiln before complete combustion
4. Spherical or cylindrical items may roll through kiln before complete combustion is achieved
5. Kiln incinerators frequently require large excess air intakes to operate due to air leakage into the kiln by the kiln end seals and feed chute. This large excess air intake lowers supplementary fuel efficiency
6. High particulate loadings into the air-pollution-control system
7. Relatively low thermal efficiency

GENERAL DESIGN PRINCIPLES

Physical, Chemical, and Thermodynamic Waste Property Considerations

General Waste Characteristics.   Before a liquid or solid waste can be combusted, it must be converted to the gaseous state. This change from a liquid to a gas occurs inside the combustion chamber and requires heat transfer from the hot combustion gases to the injected liquid. For rapid vaporization (i.e., increase heat transfer), the exposed liquid surface area must be increased. Most commonly the amount of surface exposed to heat is increased by finely atomizing the liquid to small droplets, usually to a 40μ size or smaller. Good atomization is particularly important when high aqueous wastes or other low-heating-value wastes are being burned. It is usually achieved in the liquid burner directly at the point of air–fuel mixing.

The degree of atomization achieved in any burner depends on the kinematic viscosity of the liquid and on the amount of solid impurities present. Liquids should generally have a kinematic viscosity of 10,000 ssu or less to be satisfactorily pumped and handled in pipes. For atomization they should have a maximum kinematic viscosity of about 750 ssu. If the kinematic viscosity exceeds this value, the atomization may not be fine enough. This may cause smoke or other unburned particles to leave the unit. However, this is only a rule of thumb. Some burners can handle more-viscous fluids, while others cannot handle liquids approaching this kinematic viscosity.

Viscosity can be reduced by heating with tank coils or in-line heaters. However, 400–500 °F (200–260 °C) is normally the limit for heating to reduce viscosity, because pumping a hot tar or similar material becomes difficult above these temperatures. If gases are evolved in any quantity before the desired viscosity is reached, they may cause unstable fuel feed and burning. If this occurs, the gases should be trapped and vented safely, either to the incinerator or elsewhere. Before heating a liquid-waste stream, one should check that undesirable preliminary reactions, such as polymerization, nitration, oxidation, and so forth, will not occur. If preheating is not feasible, based on these considerations, a lower-viscosity miscible liquid may be added to reduce the viscosity of the mixture (e.g., fuel oil).

Solid impurities in the waste can interfere with burner operation by plugging, erosion, and ash buildup. Both the concentration and size of the solids, relative to the diameter of the nozzle, need to be considered. Filtration may be employed to remove solids from the waste prior to injection through the burner.

Liquid waste atomization can be achieved by any of the following means:

Rotary cup atomization

Single-fluid-pressure air atomization

Two-fluid, low-pressure air atomization

Two-fluid, high-pressure air atomization

Two-fluid, high-pressure steam atomization

In air or steam atomizing burners, atomization can be accomplished internally by impinging the gas and liquid stream inside the nozzle before spraying, and externally by impinging jets of gas and liquid outside the nozzle; sonic means can also be used (see Figs. 4-3 through 4-5). Sonic atomizers use compressed gas to create high-frequency sound waves that are directed on the liquid stream. The liquid nozzle diameter is relatively large, and little waste pressurization is required. Some slurries and liquids with relatively large particles can be handled without plugging problems.

The rotary cup consists of an open cup mounted on a hollow shaft. The cup is spun rapidly, and liquid is admitted through the hollow shaft. A thin film of the liquid to be atomized is centrifugally torn from the lip of the cup, and surface tension reforms it into droplets. Conical-shaped flames are achieved by directing an annular high-velocity jet of air (primary air) axially around the cup. If too little primary air is admitted, the fuel will impinge on the sides of the incinerator. If too much primary air is admitted, the flame will be unstable and blown off the cup. For fixed firing rates the proper adjustment can be found and the unit operated for long periods of time without cleaning. This requires little liquid pressurization and is ideal for atomizing

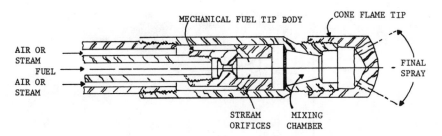

Figure 4-3. Internal mix nozzle.

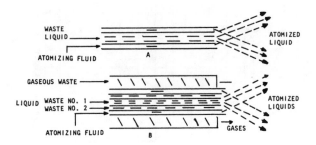

Figure 4-4. External mix nozzle.

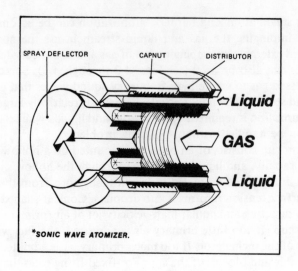

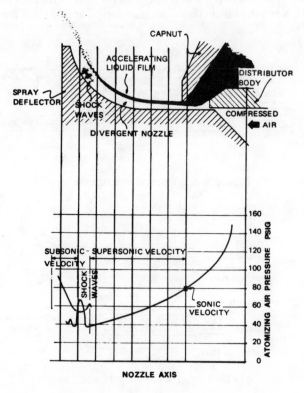

Figure 4-5.   Sonic atomizing nozzle.

liquids with relatively high solids content. Burner turndown is about 5:1, and capacities from 1 gal/h to 265 gal/h (1–280 cm³/s) are available.

In single-fluid pressure atomizing nozzle burners, the liquid is given a swirl as it passes through an orifice with internal tangential guide slots. Moderate liquid pressures of 100–150 psi provide good atomization with low-to-moderate liquid viscosity. In the simplest form the waste is fed directly to the nozzle, but turndown is limited to 2.5 to 3:1, because the degree of atomization drops rapidly with decrease in pressure. In a modified form, involving a return flow of liquid, turndown up to 10:1 can be achieved.

When this type of atomization is used, secondary combustion air is generally introduced around the conical spray of droplets. Flames tend to be short, bushy, and low velocity. Combustion tends to be slower because only secondary air is supplied, and a larger combustion chamber is usually required.

Typical burner capacities are in the range of 10 to 105 gal/h. Disadvantages of single-fluid pressure atomization are erosion of the burner orifice and a tendency to be plugged with solids or liquid pyrolysis products, particularly in smaller sizes.

Two-fluid atomizing nozzles may be of the low-pressure or high-pressure variety, the latter being more common with high-viscosity materials. In low-pressure atomizers air from blowers at pressures from 0.5 to 5 psig is used to aid atomization of the liquid. A viscous tar, heated to a viscosity of 15–18 cs, requires air at a pressure of somewhat more than 1.5 psig, whereas a low-viscosity or aqueous waste can be atomized with 0.5 psig air. The waste liquid is supplied at a pressure of 4.5–17.5 psig. Burner turndown ranges from 3:1 up to 6:1. Atomization air required varies from 370 ft³/gal to 1000 ft³/gal of waste liquid. Less air is required as atomizing pressure is increased. The flame is relatively short, because up to 40% of the stoichiometric air may be mixed with the liquid in atomization.

High-pressure two-fluid burners require compressed air or steam at pressures from 30 to 150 psig. Air consumption is from 80 ft³/gal to 210 ft³/gal of waste, and steam requirements may be 2.1–4.2 lb/gal with careful control of the operation. Turndown is relatively poor (3:1 or 4:1), and considerable energy is employed for atomization. Because only a small fraction of stoichiometric air is used for atomization, flames tend to be relatively long. The major advantage of such burners is the ability to burn barely pumpable liquids without further viscosity reduction. Steam atomization also tends to reduce soot formation with wastes that would normally burn with a smoky flame.

Table 4-8 identifies typical kinematic viscosity and solids-handling limitations for the various atomization techniques. In evaluating a specific incinerator design, however, the viscosity and solids content of the wastes should be compared with manufacturer specifications for the particular burner employed.

**Table 4-8. Kinematic Viscosity and Solids-Handling Limitations
of Various Atomization Techniques**

Atomization type	Maximum Kinematic Viscosity (ssu)	Maximum Solids	
		Mesh Size	Concentration (%)
Rotary cup	175–300	35–100	20
Single-fluid pressure	150	—	Essentially zero
Internal low-pressure air (30 psi)	100	—	Essentially zero
External low-pressure air	200–1,500	200 (depends on nozzle ID)	30 (depends on nozzle ID)
External high-pressure air	150–5,000	100–200 (depends on nozzle ID)	70
External high-pressure air	150–5,000	100–200 (depends on nozzle ID)	70

Stoichiometry.   Chemical and thermodynamic properties of the waste that need to be considered when evaluating incinerator design are the elemental composition, the net heating value, and any special properties of the waste (e.g., explosive properties) that may interfere with incinerator operation or require special design considerations. The percentages of carbon, hydrogen, oxygen, nitrogen, sulfur, halogens, and phosphorous in the waste, as well as its moisture content, need to be known to calculate stoichiometric combustion air requirements and to predict combustion gas flow and composition. In these calculations the following reactions are assumed:

$$C + O_2 \longrightarrow CO_2$$
$$H_2 + \tfrac{1}{2}O_2 \longrightarrow H_2O$$
$$H_2O \longrightarrow H_2O$$
$$N_2 \longrightarrow N_2$$
$$Cl_2 + H_2O \longrightarrow 2HCl + \tfrac{1}{2}O_2$$
$$F_2 + H_2O \longrightarrow 2HF + \tfrac{1}{2}O_2$$
$$Br_2 \longrightarrow Br_2$$
$$I_2 \longrightarrow I_2$$
$$S + O_2 \longrightarrow SO_2$$
$$2P + 2.5O_2 \longrightarrow P_2O_5$$

Table 4-9 shows the stoichiometric or theoretical oxygen requirements and combustion product yields for each of these reactions. Once the weight fraction of each element in the waste has been determined, the stoichiometric

Table 4-9. Stoichiometric Oxygen Requirements and Combustion Product Yields

Elemental Waste Component	Stoichiometric Oxygen Requirement	Combustion Product Yield
C	2.67 lb/lb C	3.67 lb CO_2/lb C
H_2	8.0 lb/lb H_2	9.0 lb H_2O/lb H_2
O_2	-1.0 lb/lb O_2	—
Cl_2	-0.23 lb/lb Cl_2	1.03 lb HCl/lb Cl_2
		-0.25 lb H_2O/lb Cl_2
F_2	-0.42 lb/lb F_2	1.05 lb/HF/lb F_2
		-0.47 lb H_2O/lb F_2
Br_2	—	1.0 lb Br_2/lb Br_2
I_2	—	1.0 lb I_2/lb I_2
S	1.0 lb/lb S	2.0 lb SO_2/lb S
P	1.29 lb/lb P	2.29 lb P_2O_5/lb P
Air N_2	—	3.31 lb N_2/lb $(O_2)_{stoich}$
Stoichiometric air requirement	—	4.31 lb Air/lb $O_{2(stoich)}$

oxygen requirements and combustion product yields can be calculated in a 1b/1b waste basis. The stoichiometric air requirement is determined directly from the stoichiometric oxygen requirement by using the weight fraction of oxygen in air.

Of course, the listed reactions are not the only ones occurring in combustion processes. Carbon, carbon monoxide, free hydrogen, nitrogen oxides, free chlorine and fluorine, hydrogen bromide and iodide, sulfur trioxide, and hydrogen sulfide, among other compounds, are also formed to some extent when the corresponding elements are present in the waste or fuel being burned. However, these combustion product yields are usually small compared to the yields of the primary combustion products listed and need not be considered in gas-flow scoping calculations. (They do, however, need to be considered to determine the potential products of incomplete combustion). For most organic wastes and fuels, nitrogen, carbon dioxide, and water vapor are the major combustion products. When excess air is factored into the combustion gas flow, oxygen also becomes a significant component of the gas.

Exceptions to the aforementioned combustion stoichiometry can occur when highly chlorinated or fluorinated wastes are being burned and insufficient hydrogen is present for equilibrium conversion to the halide form. Because hydrogen halides are more readily scrubbed from combustion gases than are halogens themselves, sufficient hydrogen should be provided for this equilibrium conversion to take place. If the waste contains insufficient hydrogen, auxiliary fuel or steam injection is needed to supply the necessary hydrogen equivalents. The stoichiometric (absolute minimum) requirements are 1 lb H_2/35.5 lb Cl_2 and 1 lb H_2/10 lb F_2 in the waste.

Equilibrium between halogens and hydrogen halides in incinerator gases is given by

$$X_2 + H_2O = 2HX + 1/2\ O_2 \tag{4-1}$$

where X_2 represents any free halogen. For chlorine, this expression becomes

$$Cl_2 + H_2O = 2HCl + 1/2\ O_2 \tag{4-2}$$

At equilibrium the concentration of Cl_2, H_2O, HCl, and O_2 in the combustion gas (at essentially atmospheric pressure) is given by

$$k = \frac{(P_{HCl})^2\ (P_{O_2})^{1/2}}{(P_{Cl_2})\ (P_{H_2O})} \tag{4-3}$$

where K is the equilibrium constant and P_i is the partial pressure of ith component in atm.

Figure 4-6 presents a plot of the equilibrium constant, K, versus temperature for converting Cl_2 to HCl. If the combustion temperature is known, it can be identified from Figure 4-6 and the following equation can be used to predict the extent of conversion of Cl_2 to HCl:

$$K_p = \frac{4x^2{}^P Cl_2 i \left({}^P O_2 i = {}^{1/2} \times {}^P Cl_2 i\right)^{1/2}}{{}^P H_2 O I^{(1-x)^2}} \tag{4-4}$$

where

$$x = \text{fractional conversion of } Cl_2$$
$${}^P Cl_2 i,\ {}^P O_2 i,\ {}^P H_2 O i = \text{calculated partial pressures of } Cl_2,\ O_2,\text{ and}$$
$$H_2O,\text{ assuming that all organic chlorine is}$$
$$\text{converted to } Cl_2 \text{ before the reaction to form}$$
$$HCl \text{ occurs}$$

In addition to the aforementioned waste constituents, metallic elements present in the waste influence the assessment of air-pollution-control re-

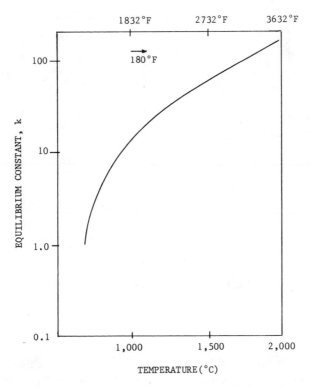

Figure 4-6. HCl equilibrium constant vs. temperature.

quirements and materials of construction (e.g., refractory type). However, the metals content of a waste will not significantly affect the stoichiometric air requirements or combustion gas flow rate.

Heating Value Considerations. The heating value of a waste corresponds to the quantity of heat released when the waste is burned, commonly expressed as Btu/1b. Because combustion reactions are exothermic, all organic wastes have some finite heating value. However, the magnitude of this heating value must be considered in establishing an energy balance for the combustion chamber and in assessing the need for auxiliary fuel firing. For combustion to be maintained, the amount of heat released by the burning waste must be sufficient to heat incoming waste up to its ignition temperature and to provide the necessary activation energy for the combustion reactions to occur. Activation energy, expressed as Btu/1b or the equivalent, is the quan-

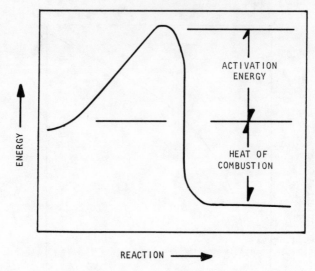

Figure 4-7.   Relationship between activation energy and heat of combustion.

tity of heat needed to destabilize molecular bonds and to create reactive intermediates so that the exothermic reaction with oxygen will proceed. Figure
4-7 shows the general relationship between activation energy and heating
value.

Waste heating values needed to sustain combustion without auxiliary fuel
firing depend on the following criteria:

Physical form of waste (i.e., gaseous vs. liquid vs. solid)

Temperature required for refractory waste component destruction

Excess air rate

Heat transfer characteristics of the incinerator

In general, higher heating values are required for solids versus liquids versus
gases, for higher operating temperatures, and for higher excess air rates if
combustion is to be sustained without auxiliary fuel consumption. Gases can
sustain combustion at heating values as low as 3000 Btu/1b, whereas 4500
to 5500 Btu/1b may be considered minimum heating value requirements for
combustion of liquid wastes in high-efficiency burners. Higher heating values

are needed for solid wastes, but the requirements depend on particle size and, thus, the area available for heat and mass transfer. In the hazardous-waste incineration industry, it is common practice to blend wastes (and fuel oil, if necessary) to an overall heating value of 8000 Btu/1b.

When an organic waste exhibits a low heating value, it is usually due to high concentrations of moisture of halogenated compounds. Because water is an ultimate oxidation product, it has no heating value. In fact, a portion of the heat generated by combustion of the organic waste fraction is consumed in vaporizing and heating the moisture up to incinerator temperature. Therefore, an increase in the moisture content of an organic waste proportionately decreases the overall heating value on a Btu/1b waste basis.

The heating value of a waste also decreases as the chlorine (or other halogen) content increases, although there is no simple mathematical relationship. At chlorine contents of 70% or greater, auxiliary fuel is needed to maintain combustion. Auxiliary fuel may also be required for less highly chlorinated waste unless high-efficiency burners are used. In hazardous-waste incineration it is common practice to blend waste so that the chlorine content does not exceed 30%. This is done to maintain sufficient heating value for sustained combustion and to limit free chlorine concentration in the combustion gas.

When heating-value data are reported for a given waste, it is desirable to know whether they are "higher heating values," "lower heating values," or "net heating values." The difference between the higher heating value and lower heating value of a material is that the higher value includes the heat of condensation of water formed in the combustion reaction. In the combustion of methane, for example, the higher heating value is based on the following stoichiometry:

$$CH_{4(g)} + 2O_{2(g)} \rightarrow CO_2 + 2H_2O_{(l)} \qquad \textbf{(4-5)}$$

where the subscripts g and l represent gaseous and liquid states, respectively. The lower heating value is based on

$$EH_{4(g)} + 2O_{2(g)} \rightarrow CO_{2(g)} + 2H_2O_{(g)} \qquad \textbf{(4-6)}$$

The net heating value of a waste is determined by subtracting from its lower heating value the energy necessary to vaporize any moisture initially present in the waste. Thus, high aqueous wastes may exhibit a negative net heating value. Since this quantity represents the true energy input to the combustion process, only net heating values should be used in developing energy balances for incinerators.

The heating value of a complex waste mixture is difficult to predict a priori. Therefore, these values should be measured experimentally. Because heating values measured using oxygen bomb calorimeters are higher heating values, conversion to the net heating value is required for energy balance calculations. This conversion from the higher heating value obtained from bomb calorimeters to the net heating value is done as follows:

$$NHV = HHV - 1050 \left[H_2O + 9(H - Cl/35.5 - F/19) \right] \quad \text{(4-7)}$$

where

NHV = net heating value in Btu/lb
HHV = higher heating value in Btu/lb
H_2O = wt. fraction of water in earth
Cl = wt. fraction of chlorine in waste
F = fraction of fluorine in waste

Approximate net heating values for common auxiliary fuels are

17,500 Btu/lb for residual fuel oil (e.g., No. 6)

18,300 Btu/lb for distillate fuel oil (e.g., No. 2)

19,700 Btu/lb (1000 Btu/scf) for natural gas

Other Considerations.   Special characteristics of a waste, such as extreme toxicity, mutagenicity or carcinogenicity, corrosiveness, fuming, odor, pyrophoric properties, thermal instability, shock sensitivity, and chemical instability should also be considered when designing an incinerator facility. Thermal or shock instability is of particular concern from a combustion standpoint because wastes with these properties pose an explosion hazard. Other special properties relate more directly to the selection of waste-handling procedures and air-pollution-control requirements. If potentially explosive wastes are encountered, technical assistance is advised.

Although liquid wastes are frequently incinerated in rotary kilns, kilns are primarily designed for combustion of solid wastes. They are exceedingly versatile in this regard, capable of handling slurries, sludges, bulk solids of varying size, and containerized wastes. The only wastes that create problems in rotary kilns are (1) aqueous organic sludges that become sticky on drying and form a ring around the kiln's inner periphery, and (2) solids (e.g., drums) that tend to roll down the kiln and are not retained as long as the bulk of solids. To reduce this problem, drums and other cylindrical containers are usually not introduced to the kiln when it is empty. Other solids in the kiln help to impede the rolling action.

Chemical and thermodynamic properties of the waste that need to be considered in evaluating rotary kiln design are its elemental composition, its net heating value, and any special properties (e.g., explosive properties, extreme toxicity) that may interfere with incinerator operation or require special design considerations. These are essentially the same properties that must be considered in evaluating liquid-injection incinerator design.

Unlike liquid-injection incinerators, which have no moving parts, rotary kiln designs incorporate high-temperature seals between the stationary end plates and rotating section. These seals are inherently difficult to maintain airtight, thus creating a potential for release of unburned wastes. Rotary kilns burning hazardous wastes are almost always operated at negative pressure to circumvent this problem; however, difficulties can still arise when batches of wastes are fed semicontinuously. When drums containing relatively volatile wastes are fed to the kiln, for example, extremely rapid gas expansion occurs. This results in a positive pressure surge at the feed end of the kiln (even though the discharge end may still be under negative pressure), which forces unburned waste out through the end plate seals. This phenomenon is known as "puffing;" it can pose a major problem when extremely toxic or otherwise hazardous materials are being burned.

Selecting Operating Conditions

Temperature, residence time, oxygen concentration, and the degree of air-waste mixing achieved are the primary variables affecting combustion efficiency in any incinerator design. In general, two major factors are involved in evaluating these variables as they relate to incinerator design. The first factor is whether or not the temperature, residence time, and excess air level, along with the degree of mixing achieved in the incinerator, are *adequate* for waste destruction. The second factor is whether or not the operating conditions are *achievable,* because temperature, excess air, residence time, and mixing are all interrelated.

At the current state-of-the-art, the *adequacy* of incinerator operating conditions can only be determined by past experience with the waste or by actual testing. Therefore, this factor is not addressed per se in this chapter. The major focus of the following evaluation procedures is whether a proposed set(s) of operating conditions is *achievable.* Basically, this involves a series of internal consistency checks.

Air Requirements. The most basic requirement of any combustion system is a sufficient supply of air to completely oxidize the feed material. The stoichiometric, or theoretical, air requirement is calculated from the chemical composition of the feed material. If perfect mixing could be achieved and liquid-waste burnout occurred instantaneously, then only the stoichiometric requirement of air would be needed. Neither of these phenomena occur in real-world applications, however, so some excess air is always required to ensure adequate waste–air contact. Excess air is usually expressed as a percentage of the stoichiometric air requirement. For example, 50% excess air implies that the total air supply to the incinerator is 50% greater than the stoichiometric requirement.

The amount of excess air used or needed in a given application depends on the degree of air–waste mixing achieved in the primary combustion zone, process-dependent secondary combustion requirements, and the desired degree of combustion gas cooling. Because excess air acts as a diluent in the combustion process, it reduces the temperature in the incinerator (e.g., maximum theoretical temperature is achieved at 0% excess air). This temperature reduction is desirable when readily combustible, high-heating-value wastes are being burned to limit refractory degradation. When high-aqueous or other low-heating-value waste is being burned, however, excess air should be minimized to keep the system temperature as high as possible. Even with highly combustible waste, it is desirable to limit excess air to some extent so that combustion chamber volume and downstream air-pollution-control-system capacities can be limited.

In liquid-injection incinerators two excess air rates must be considered: (1) the excess air present in the primary combustion air introduced through the burner, and (2) the total excess air, which includes secondary combustion air. Normally, 10–20% excess air (i.e., 1.1 to 1.2 times the stoichiometric requirement) is supplied to the burner to prevent smoke formation in the flame zone. When relatively homogeneous wastes are burned in high-efficiency burners, 5% excess air may be adequate. Too much excess air through the burner is also undesirable because this can blow the flame away from its retention cone. Burner manufacturer specifications are the best source of information for analysis.

In general, the total excess air rate should exceed 20–25% to insure adequate waste–air contact in the secondary combustion zone. However, the minimum requirement for a given incinerator depends on the degree of mixing achieved and waste-specific factors.

Temperature.   Four basic questions should be considered when determining if a proposed operating temperature is sufficient for waste destruction:

1. Is the temperature high enough to heat all waste components (and combustion intermediates) above their respective ignition temperatures and to maintain combustion?
2. Is the temperature high enough for complete reaction to occur at the proposed residence time?

3. Is this temperature within normal limits for the generic design and/or attainable under the other proposed operating conditions?
4. At what point in the combustion chamber is the proposed temperature to be measured?

Complete waste combustion requires a temperature and heat release rate in the incinerator high enough to raise the temperature of the incoming waste constituents above their respective ignition temperatures (i.e., to provide energy input in excess of their respective activation energies). In cases where combustion intermediates are more stable than the original waste constituents, higher temperatures are required for complete combustion of the intermediates than are required for parent compound destruction.

Since heat transfer, mass transfer, and oxidation all required a finite length of time, temperature requirements must also be evaluated in relation to the proposed residence time in the combustion chamber. Heat transfer, mass transfer, and kinetic reaction rates all increase with increasing temperature, lowering the residence time requirements. For extremely short residence times, however, temperatures higher than those needed for ignition may be required to complete the combustion process.

The current state-of-the-art in combustion modeling does not allow a purely theoretical determination of temperature and residence time requirements for waste and combustion intermediate destruction. Therefore, the only reasonable alternative is an examination of temperature–residence time combinations used to destroy the same or similar waste in a similar or identical incinerator.

After addressing the temperature requirements for waste destruction, it is reasonable to determine whether the proposed temperature is within normal limits for the generic incinerator design and whether this temperature can be attained under the proposed firing conditions. Generally, liquid-injection incinerator temperatures range from 1400 °F to 3000 °F, depending on the generic design, type of waste burned, and location within the combustion chamber. Usually 1400 °F is the minimum temperature needed to avoid smoke formation. A more typical hazardous-waste incineration temperature is 1800 °F, although temperatures of 2000 °F–2200 °F or higher are usually employed for halogenated aromatic wastes.

The question whether the proposed temperature and excess air rate are attainable can be resolved by approximate calculations based on a heat balance around the combustion chamber. Figure 4-8 shows the heat inputs and outputs for the combustion chamber.

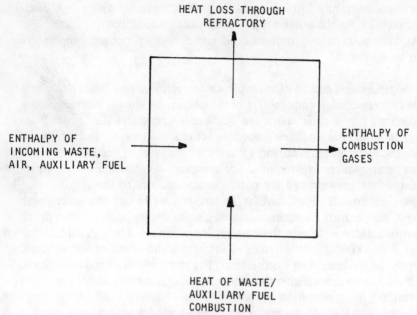

Figure 4-8. Energy balance for combustion chamber.

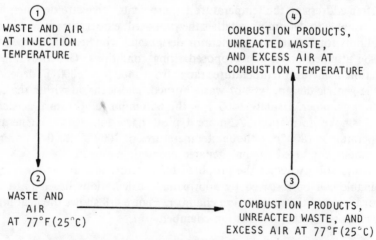

Figure 4-9. Enthalpy balance for combustion processes.

Because liquid-waste incineration is generally a steady-state process, the enthalpy of the waste–fuel–combustion air feed plus the heat released by combustion must equal the enthalpy of the combustion gases leaving the unit plus the heat loss through the refractory walls. This yields the general relationship:

Heat loss through refractory = Enthalpy of incoming feed + Heat released by combustion − Enthalpy of combustion gases

$$Q = \Delta H \qquad\qquad (4\text{-}8)$$

where Q is heat loss through refractory in Btu/lb waste and ΔH is overall enthalpy change in the combustion chamber in Btu/lb waste. In mathematical terms these incremental enthalpy changes are expressed as (refer to Fig. 4-9)

$$\Delta H_{1\text{-}2} = \underset{\substack{i=1 \\ \text{waste} \\ \text{components}}}{\overset{k}{\sum}} n_i C_{pi}(77 - T_{\text{in}})$$

$$+ \; 4.31 \, C_{p \, \text{air}}(77 - T_{\text{air}}) \, (O_2)_{\text{stoich}}(1 + EA)$$

$$\Delta H_{2\text{-}3} = \underset{\substack{i=1 \\ \text{reactive} \\ \text{waste} \\ \text{components}}}{\overset{k}{\sum}} n_i X_i (H_c)_i \; 77\,^{\circ}\text{F}$$

$$\Delta H_{3\text{-}4} = \left\{ \underset{\substack{i=1 \\ \text{reaction} \\ \text{products}}}{\overset{k}{\sum}} n_i C_{pi} X_i + \underset{\substack{i=1 \\ \text{reactive} \\ \text{waste} \\ \text{components} \\ \text{remaining}}}{\overset{k}{\sum}} n_i C_{pi}(1 - X_i) \right.$$

$$+ \; 4.31 \, C_{p \, \text{air}}(O_2)_{\text{stoich}}(EA) \left[1 + \underset{\substack{i=1 \\ \text{reactive} \\ \text{components}}}{\overset{k}{\sum}} n_i \, (1 - X_i) \right]$$

$$\left. + \; \underset{\substack{i=1 \\ \text{inert} \\ \text{waste} \\ \text{components}}}{\overset{k}{\sum}} n_i C_{pi} \right\} (T_{\text{out}} - 77)$$

$$(4\text{-}9)$$

where

n_i =   lb ith component/lb waste

C_{pi} =   mean heat capacity of ith component over the temperature range involved in Btu/lb °F

T_{in} =   waste injection temperature in °F

T_{air} =   air inlet temperature in °F

X_i =   fractional conversion of ith component (X_i − 1.0 at 100% combustion of ith component)

$(\Delta H_c)_i$ =   heat of combustion of ith component at 77°F (25°C) in Btu/lb 77°F

T_{out} =   temperature at the combustion chamber outlet in °F

(O_2) =   stoichiometric oxygen requirement in lb O_2/lb waste, stoich

EA   =   excess air in %/100

To determine if the proposed temperature–excess air combination is achievable, one must specify the desired temperature and calculate the corresponding excess air rate for comparison with the proposed value. However, there are far too many unknowns in these equations to solve for EA.

These equations can be simplified considerably by assuming that the combustion reactions go to essentially 100% completion. Acceptance of the proposed temperature/residence combination should ensure combustion efficiencies close enough to 100% for this value to be used in heat balance calculations. With this assumption the overall energy balance reduces to

$$Q = C_{p \text{ waste}}(77 - T_{in}) + 4.31 C_{p \text{ air}}(77 - T_{air})(O_2)_{\text{stoich}}(1 + EA)$$

$$+ 1.0(-NHV)_{77°F} + (T_{out} - 77)\left[\sum_{\substack{i=1 \\ \text{combustion} \\ \text{products}}}^{k} n_i C_{pi} + 4.31 C_{p \text{ air}}(O_2)_{\text{stoich}}EA\right]$$

$$(4\text{-}10)$$

where NHV = net heating value of the waste in Btu/lb.

From the empirical wastes composition (carbon content, hydrogen content, etc.), proposed excess air rate, and combustion stoichiometry, all variables in this equation are fixed, except the outlet temperature or excess air rate, mean heat capacities of the combustion gases, and the heat loss through the walls of the combustion chamber. To avoid rigorous heat transfer calculations, we can assume that this heat loss is about 5% of the heat released in the combustion chamber, based on operating experience with hazardous-waste incinerators. With this assumption the energy balance reduces to

$$C_{p\ \text{waste}}\ (77 - T_{\text{in}}) + 4.31\ C_{p\ \text{air}}\ (77 - T_{\text{air}})\ (O_2)_{\text{stoich}}(1 + EA)$$

$$+ \ 0.95(-\text{NHV})$$

$$+ \ \left[\sum_{\substack{i=1 \\ \text{combustion} \\ \text{products}}}^{k} n_i C_{pi} + 4.31\ C_{p\ \text{air}}(O_2)_{\text{stoich}}EA \right] (T_{\text{out}} - 77) = 0$$

$$\text{(4-11)}$$

from which the first two terms can be deleted if neither waste nor air preheating is employed. (The waste enthalpy term can almost always be deleted.) This yields

$$0.95(-\text{NHV}) \ + \ \left[\sum_{\substack{i=1 \\ \text{combustion} \\ \text{products}}}^{k} n_i C_{pi} + 4.31 C_{p\ \text{air}}(O_2)_{\text{stoich}}EA \right] (T_{\text{out}} - 77) = 0$$

$$\text{(4-12)}$$

The mean heat capacities of the combustion gases will vary slightly, depending on the incinerator outlet temperature. For the purposes of approximate calculations, however, the following values can be assumed:

Gas component C_p	Btu/lb°F
Excess air	0.26
N_2	0.26
CO_2	0.26
H_2O	0.49
HCl	0.20
SO_2	0.18

Thus for wastes containing only C, H, O, and N, we get

$$0.95(-\text{NHV}) \ + \ \left[0.26(n_{N_2} + n_{CO_2}) + 0.49N_{H_2O} \right.$$

$$\left. + \ 1.12(O_2)_{\text{stoich}}EA \right] (T_{\text{out}} - 77) = 0 \qquad \text{(4-13)}$$

where N_2 refers to the nitrogen present in the combustion gases under stoichiometric conditions. It does not include excess air nitrogen. If other gas components constitute more than a few percent of the total flow, additional heat capacity terms must be added.

If auxiliary fuel is burned in conjunction with the waste, the following modification of the previous equation is needed:

$$0.95 \left[(-\text{NHV})_w + n_f(-\text{NHV})_f \right]$$

$$+ \left[\underbrace{\sum_{i=1}^{k} n_i C_{pi}}_{\substack{\text{waste} \\ \text{combustion} \\ \text{products}}} + n_f \underbrace{\sum_{i=1}^{k} (n_{if} C_{pi})}_{\substack{\text{fuel} \\ \text{combustion} \\ \text{products}}} + 4.31 C_{p\,\text{air}}(O_2)_{\text{stoich}}\text{EA} \right] (T_{\text{out}} - 77) = 0$$

(4-14)

where

n_{if} = lb ith combustion gas component/lb fuel

C_{pi} = mean heat capacity of fuel over applicable temperature range in Btu/lb °F

$(\text{NHV})_f$ = heating value of fuel in Btu/lb

n_f = lb fuel/lb waste

$(\text{NHV})_w$ = heating value of waste in Btu/lb

If only carbon, hydrogen, oxygen, and nitrogen are present, the equation can be simplified to

$$0.95 \left[(-\text{NHV})_w + n_{f\text{N}_2} + (-\text{NHV})_f \right]$$

$$+ \left[0.26(n_{\text{N}_2} + n_{\text{CO}_2}) + 0.49\, n_{\text{H}_2\text{O}} + 1.12(O_2)_{\text{stoich}}\text{EA} \right](T_{\text{out}} - 77) = 0 \quad \text{(4-15)}$$

In this equation, n_{N_2}, n_{CO_2}, $n_{\text{H}_2\text{O}}$, $(O_2)_{\text{stoich}}$, and EA apply to the combined waste–auxiliary fuel mix, and n_{N_2} refers to the nitrogen present in the combustion gases under stoichiometric conditions.

By fixing the outlet temperature at the proposed value, the equations can be used to estimate the maximum achievable excess air rate for comparison with the proposed rate. Thus, the equations provide an internal consistency check for proposed temperatures–excess air combinations.

When identifying a minimum temperature acceptable for waste destruction, it is also important to identify the location in the combustion chamber at which this temperature should be measured. Temperature varies from one point to another in the combustion chamber, being highest in the flame and lowest at the refractory wall or at a point of significant air infiltration (e.g., in the vicinity of secondary air ports). Ideally, temperature should be measured in the bulk gas flow at a point after which the gas has traversed the combustion chamber volume that provides the specified residence time for the unit. It should not be measured at a point of flame impingement or at a point directly in sight of radiation from the flame.

Residence Time. In addition to temperature and excess air, residence time is a key factor affecting the extent of combustion. This variable, also referred to as retention time or dwell time, is the mean length of time that the waste is exposed to the high incinerator temperatures. It is important in designing and evaluating incinerators, because a finite amount of time is required for each step in the heat transfer–mass transfer reaction pathway to occur.

In liquid-waste combustion, discrete (although short) time intervals are required for heat transfer from the gas to the surface of the atomized droplets, liquid evaporation, oxygen mixing in the gas stream, and reaction, which itself involves a series of individual steps depending on the complexity of the waste's molecular structure. The total time required for these processes depends on the temperature of the combustion zone, the degree of mixing achieved, and the size of the liquid droplets. Residence time requirements increase as combustion temperature is decreased, as mixing is reduced, and/or as the size of the discrete waste particles is increased. Typical residence times in liquid-injection incinerators range from 0.5 s to 2.0 s.

Gas residence times are defined by the following formula:

$$\Theta = \int_0^V \frac{dV}{q}$$

(4-16)

where

Θ = mean residence time in s
V = combustion chamber volume in ft^3
q = gas flow rate in ft^3/s within the differential volume dV

Gas flow rate is given by

$$q = \left(\frac{0.79}{Y_{N_2}}\right)\left(\frac{T + 460}{528}\right) 431(O_2)_{\text{stoich } 68\,°F}(1 + EA)$$

(4-17)

where

Y_{N_2} = mole fraction N_2 in the gas within the differential volume

T = gas temperature in °F within the differential volume

$(O_2)_{stoich}$ = stoichiometric oxygen requirement in scf/S

EA = excess oxygen fraction in %/100 within the differential volume

As indicated in this equation, residence time is not an independent variable. For an incinerator of fixed volume and relatively constant feed, residence time is influenced by the temperature and excess air rate employed.

Gas flow at any point along the length of the combustion chamber is a function of the temperature at that point, the amount of excess air added up to that point, and the extent to which the combustion reactions are completed at that point, and the extent to which the combustion reactions are completed at that point. Therefore, solving Eq. (4-16) requires a knowledge of the temperature profile, excess air profile, and waste-conversion profile along the combustion chamber. These factors must be expressed as functions of combustion chamber length (i.e., volume) in order for the integration to be performed.

Because this detailed information can rarely, if ever, be determined with a reasonable degree of accuracy, an alternative approach is normally adopted. In this approach the flow rate, q, is specified at the desired operating temperature (measured at the incinerator outlet) and total excess air rate. The equation is then simplified to

$$\Theta = \frac{V}{q_{out}} \tag{4-18}$$

The chamber volume used in this calculation is an estimated value, corresponding to the volume through which the combustion gases flow after they have been heated to the desired operating temperature. Thus, the chamber volume used in residence time calculations should be at least somewhat less than the total volume of the chamber. However, an upper bound residence time can be estimated by

$$\Theta_{max} = \frac{V_T}{q_{out}} \tag{4-19}$$

where V_T = total volume of the chamber.

Any residence times calculated by this equation should only be used for general comparison purposes.

Mixing. Temperature, oxygen, and residence time requirements for waste destruction all depend to some extent on the degree of mixing achieved in the combustion chamber. This parameter is difficult to express in absolute terms, however. Many of the problems involved in interpreting burn data relate to the difficulty involved in quantifying the degree of mixing achieved in the incinerator, as opposed to the degree of mixing achieved in another incinerator of different design.

In liquid-waste incinerators the degree of mixing is determined by the specific burner design (i.e., how the primary air and waste—fuel are mixed), combustion product gas and secondary air flow patterns in the combustion chamber, and turbulence. Turbulence is related to the Reynolds number for the combustion gases, expressed as

$$\text{Re} = \frac{Dvp}{\mu} \tag{4-20}$$

where

D = combustion chamber diameter in ft
v = gas velocity in ft/s
p = gas density in lb/ft^3
μ = gas viscosity in lb/ft-s

Turbulent flow conditions exist at Reynold's numbers of approximately 2300 and greater. Below this Reynold's number laminar or transition flow prevails, and mixing occurs only by diffusion.

In conventional liquid-injection incinerators or afterburners, it is possible to simplify the Reynold's number to consideration of superficial gas velocity only. Adequate turbulence is usually achieved at superficial gas velocities of 10–15 ft/s. Superficial gas velocities are determined by

$$v = \frac{q}{A} \tag{4-21}$$

where q is gas flow rate at operating temperature in ft^3/s and A is cross-sectional area of the incinerator chamber in ft^2.

When primary combustion air is introduced tangentially to the burner (e.g., vortex burners), secondary air introduced tangentially, or burner alignment is such that cyclonic flow prevails in the incinerator, actual gas velocities exceed the superficial velocity. Thus, adequate turbulence may be achieved at superficial velocities less than 10 ft/2 in cyclonic flow systems. However, the tradeoff is difficult to quantify. Turbulence can also be increased by installing baffles in the secondary combustion zone of the incinerator that abruptly change the direction of gas flow. However, this also increases pressure drop across the system and is not a common practice in liquid-injection incinerator design. Steam jets can also be used to promote turbulence.

Excess Air.   In a rotary kiln–afterburner incineration systems, three excess air rates must be considered:

1. Excess air present in the primary combustion air introduced through liquid-waste burners in the kiln or afterburner section
2. Total excess air fed to the kiln
3. The excess air percentage maintained in the afterburner

A practical minimum for liquid-injection incinerators to achieve adequate air–waste contact is 20–25% total excess air. Higher excess air rates are needed in rotary kilns, however, because the efficiency of air–solids contact is less than that for air and atomized liquid droplets. Typical excess air rates range from 100–200% or greater, depending on the desired operating temperature and the heating value of the waste. When high aqueous wastes are being burned, lower excess air rates may be needed to maintain adequate temperature. However, less than 100% excess air in the kiln may not provide adequate air–solids contact nor provide for peaks caused by sudden rapid combustion.

Because it is usually desirable to maintain the afterburner at a higher temperature than the kiln, and because only liquid wastes or auxiliary fuel is fired in the afterburner, the excess air rate in the afterburner is usually less than that in the kiln. In a typical system operating at 1500 °F in the kiln and 1800 °F in the afterburner, approximately 60–70% excess air would be maintained in the afterburner compared to 110% in the kiln. Considering 100% excess air in the kiln as a practical minimum, approximately 80% excess air or more should be maintained in the afterburner. This includes air contained in the kiln exit gases as well as air introduced in the afterburner itself and is based on the total stoichiometric oxygen requirement for all wastes and fuel burned in the system.

Heat Balance.   In evaluating temperature requirements for a rotary kiln-afterburner system, seven basic questions should be considered:

1. Is the kiln temperature high enough to volatize, partially oxidize, or otherwise convert all organic components of the waste to a gaseous state?
2. Is this temperature high enough for the aforementioned process to occur within the proposed solids retention time?

3. Is the afterburner temperature high enough to heat all volatilized wastes (and combustion intermediates) above their respective ignition temperatures and maintain combustion?
4. Is the temperature high enough for complete reaction to occur within the proposed afterburner residence time?
5. Is the kiln operating temperature within normal limits and/or attainable under the other proposed operating conditions?
6. Is the afterburner temperature within normal limits and/or attainable under the other proposed operating conditions?
7. At what points in the system are the temperatures to be measured?

The current state-of-the-art is combustion modeling does not allow a purely theoretical determination of time and temperature requirements for solid-waste burnout or combustion in the gas phase. Therefore, the only reasonable alternative is an examination of temperature–time combinations used in practice to destroy similar wastes. When examining such data, the comparability of the incinerator used must also be considered. This information is needed to address questions 1–4. The last three questions are addressed in the following paragraphs.

Temperatures in rotary kiln incinerators usually range from about 1400 to 3000 °F, depending on the types of waste burned and the location in the kiln. Common operating temperatures, measured outside of the flame zone, are 1500–1600 °F. Whether these or other proposed temperatures are attainable at the proposed excess air rate can be resolved by approximate calculations based on a heat balance around the kiln.

The difficulty that arises in this calculation is that the extent of combustion, or actual heat release to the maximum attainable heat release, is unknown. However, the maximum achievable excess air rate in the kiln at the specified operating temperature can still be estimated by assuming complete combustion. This corresponds to a worst-case analysis. The maximum calculated excess air rate must exceed the proposed excess air rate, or the proposed operating temperature will not be attainable.

The applicable heat balance equation for the kiln, assuming complete combustion is shown in Eq. (4-22), which is also based on the assumptions that (a) heat loss through the kiln walls is about 5% of the heat released on combustion, and (b) waste preheating, if employed, results in negligible heat input compared to the heat released on combustion (which is almost always the case). Equation (4-22) can be solved directly for EA, the maximum attainable excess air rate in the kiln, once the desired operating temperature is specified.

$$4.31\ C_{p\ \text{air}}(77 - T_{\text{air}})\ (O_2)_{\text{stoich (k)}}(1 + EA_k)$$

$$- 0.95 \left(\frac{n_1 NHV_1 + n_2 NHV_2 + n_{\text{fk}} NHV_{\text{fk}}}{1 + n_{\text{fk}}} \right)$$

$$+ \left[\sum_{\substack{i=1 \\ \text{combustion} \\ \text{products} \\ \text{from kiln}}}^{k} (n_i C_{pi}) + 4.31\ C_{p\ \text{air}}\ (O_x)_{\text{stoich}} EA \right] (T_{\text{out}} - 77) = 0$$

$$\text{(4-22)}$$

where

C_{pi} = mean heat capacity of ith component over the temperature range involved in Btu/lb °F

T_{air} = total stoichiometric oxygen requirement for wastes and auxiliary fuel fed to the kiln in lb O_2/lb feed

EA_k = percent excess air/100 (in kiln)

n_1 = lb liquid waste/lb waste

n_2 = lb solid waste/lb waste

n_{fk} = lb fuel/lb waste

NHV_1 = net heating value of liquid waste in Btu/lb

NHV_2 = net heating value of solid waste in Btu/lb

NHV_{fk} = net heating value of fuel in Btu/lb ·

n_i = lb ith combustion product/lb feed

T_{out} = desired temperature at the kiln outlet in °F

When no combustion air preheating is employed, this equation simplifies to

$$- 0.95 \left(\frac{n_1 NHV_1 + n_2 NHV_2 + n_{\text{fk}} NHV_{\text{fk}}}{1 + n_{\text{fk}}} \right)$$

$$+ \left[\sum_{\substack{i=1 \\ \text{combustion} \\ \text{products} \\ \text{from kiln}}}^{k} (n_i C_{pi}) + 4.31\ C_{p\ \text{air}}\ (O_x)_{\text{stoich}} EA \right] (T_{\text{out}} - 77) = 0$$

$$\text{(4-23)}$$

Using the heat capacities presented and assuming that CO_2, H_2O, N_2, and O_2 are the only significant components of the combustion gas, we can further simplify the equation to

$$- 0.95 \left(\frac{n_1 NHV_1 + n_2 NHV_2 + n_{fk} NHV_{fk}}{1 + n_{fk}} \right)$$

$$+ \left[0.26 \, (n_{CO_2} + n_{N_2}) + 0.49 \, n_{H_2O} \right.$$

$$\left. + 1.12 \, (O_2)_{stoich} \right] EA_k (T_{out} - 77) = 0 \qquad \text{(4-24)}$$

The term n_{N_2} in this equation relates to the nitrogen present in the combustion gas under stoichiometric conditions. It does not include excess air nitrogen.

Excess air in the afterburner can be estimated in similar fashion after the desired operating temperature is specified. In this calculation heat inputs to and from the entire system (kiln and afterburner) are considered. The resulting heat balance equation is shown in Eq. (4-25). This equation is also based on assumptions of 5% heat loss from the system and negligible energy input due to waste–auxiliary fuel or air preheating.

$$- 0.95 \left(\frac{\dfrac{n_1 NHV_1 + n_2 NHV_2 + n_{fk} NHV_{fk}}{1 + n_{fk}} + n_{Ak} \dfrac{NHV_3 + n_{fA} HV_{fA}}{1 + n_{fA}}}{1 + n_{Ak}} \right)$$

$$+ \left[\frac{\overset{k}{\underset{i=1}{\sum}} (n_i C_{pi}) + n_{Ak} \overset{k}{\underset{i=1}{\sum}} n_{iA} C_{pi}}{1 + n_{Ak}} \right] (T_{out} - 77)$$

combustion products from kiln   combustion products afterburner feed

$$+ 4.31 \, C_{p \, air} \left[(O_2)_{stoich(k)} + (O_2)_{stoich(A)} \right] EA (T_{out} - 77) = 0 \qquad \text{(4-25)}$$

where

$(O_2)_{stoich(A)}$ = stoichiometric oxygen requirement for waste and auxiliary fuel fed to the afterburner in lb/O_2/lb feed

EA = percent excess air/100 (in afterburner)

n_{Ak} = lb afterburner feed/lb kiln feed

NHV_3 = net heating value of liquid waste fed to the afterburner in Btu/lb

n_{fA} = lb fuel/lb waste in afterburner

HV_{fA} = heating value of auxiliary fuel burned in the afterburner in Btu/lb

n_{ik} = lb ith combustion product from kiln/lb kiln feed

n_{iA} = lb ith combustion product from afterburner feed/lb afterburner feed

T_{out} = desired afterburner outlet temperature in °F

Once the major components of the combustion gas have been identified (CO_2, H_2O, N_2, and O_2 in most cases), the latter two terms in this equation can be simplified by substituting in the heat capacities of the gases. A similar substitution is shown for the rotary kiln heat balance equation (4-24).

In rotary kiln incineration systems the solids retention time in the kiln and the gas residence time in the afterburner must be considered. Afterburner residence time considerations are essentially the same as those for liquid-injection incinerators. Therefore, the following discussion focuses primarily on solids retention time estimates.

Solids retention times in rotary kilns are a function of the length-to-diameter ratio of the kiln, the slope of the kiln, and its rotational velocity. The functional relationship between these variables is [4].

$$O = \frac{0.19(L/D)}{SN} \qquad \text{(4-26)}$$

where

O = retention time in min

L = kiln length in ft

D = kiln diameter in ft

S = kiln slope in ft/ft

N = rotational velocity in rpm

Typical ranges for the parameters are L/D = 2–10, 0.03–0.09 ft/ft slope, and 1–5 ft/min rotational speed measured at the kiln periphery (which can be converted to rpm by dividing by the kiln circumference measured in ft).

Some examples of retention time requirements are 0.5 s for fine propellants, 5 min for wooden boxes, 15 min for refuse, and 60 min for railroad ties [9]. However, the retention time requirements for burnout of any particular solid waste should be determined experimentally or extrapolated from operating experience with similar wastes.

Air–solids mixing in the kiln is primarily a function of the kiln's rotational velocity, assuming a relatively constant gas flow rate. As rotational velocity is increased, the solids are carried up higher along the kiln wall and showered down through the air–combustion gas mixture. Typical rotational velocities are in the range of 1–5 ft/min, measured at the kiln periphery.

Since solids retention time is also affected by rotational velocity, there is a tradeoff between retention time and air–solids mixing. Mixing is improved to a point by increased rotational velocity, but the solids retention time is reduced. Mixing is also improved by increasing the excess air rate, but this reduces the kiln operating temperature. Thus, there is a distinct interplay between all four operating variables.

Auxiliary-Fuel Requirements

Liquid-Injection Incinerators. As discussed earlier, liquid-injection incinerators should be equipped with an auxiliary-fuel firing system to heat the unit to operating temperatures before waste is introduced. Although not essential from an engineering standpoint, it is desirable for the auxiliary-fuel system to have sufficient capacity to attain this temperature at the design air flow rate for waste combustion. This capacity requirement can be approximated by the following heat balance Eq. (4-28):

$$0.95 \ m_f \ \text{NHV}_f = m_f \sum_{i=1}^{k} N_{if} C_{pi} (T_{out} - 77)$$

$$+ \ 4.31 \ m_w (O_2)_{\text{stoich(w)}} \ (1 \ + \ \text{EA}) \ C_{p \ \text{air}} \ (T_{out} - 77)$$

$$- \ 4.31 \ m_f \ (O_2)_{\text{stoich(f)}} \ C_{p \ \text{air}} (T_{out} - 77) \qquad \textbf{(4-27)}$$

where

m_f	= required auxiliary capacity in lb/hr
NHV_f	= net heating value of auxiliary fuel in Btu/lb
N_{if}	= lb combustion ith product/lb fuel
C_{pi}	= heat capacity of ith component in Btu/lb °F

T_{out}	=	proposed operating temperature measured at the incinerator outlet in °F
$4.31(O_2)_{stoich(w)}$	=	stoichiometric air requirement for waste combustion in lb air/lb waste
m_w	=	proposed waste feed rate (average) in lb/hr
EA	=	proposed excess air rate in %/100
$4.31(O_2)_{stoich(f)}$	=	stoichiometric air requirement for fuel combustion in lb air/lb fuel

This equation is b ased on these assumptions: (a) air is not preheated, (b) there is a 5% heat loss through the refractor walls, and (c) the air flow rate for normal waste-burning operation exceeds the air requirements for fuel combustion during startup.

Since CO_2, H_2O, and N_2 are the only major components of fuel combustion gases at stoichiometric firing conditions, this equation can be further simplified by using the heat capacities:

$$0.95 \ m_f NHV_f = m_f \left[0.26(n_{CO_2} + n_{N_2}) + 0.49 \ n_{H_2O} \right] (T_{out} - 77)$$

$$+ 1.12 \ m_w(O_2)_{stoich(w)}(1 + EA) \ (T_{out} - 77)$$

$$- 1.12 \ m_f(O_2)_{stoich(f)} \ (T_{out} - 77) \tag{4-28}$$

where n_{CO_2}, n_{N_2}, n_{H_2O} are based on the stoichiometric air–fuel ratio.

Rotary Kiln Incinerators.   In rotary kiln incinerators the kiln and afterburner need to be heated to operating temperature before waste is introduced. Because afterburner temperature is usually higher than kiln temperature and more critical in terms of emissions, it should be sufficient to limit the auxiliary-fuel-capacity evaluation to the afterburner section. The evaluation procedure described for liquid-injection incinerators can be modified for this purpose in the following manner:

The proposed average waste-feed rate (m_w) and stoichiometric air requirement for waste combustion should be based on the combined kiln and afterburner waste feed.

Temperature (T_{out}) should be specified at the afterburner outlet.

The excess air rate (EA) used in the calculation should be the proposed excess air level for the afterburner section.

With these modifications, the auxiliary-fuel start-up requirements for rotary kiln incinerators can be evaluated.

Although liquid-injection and rotary kiln incinerators are the most widely used types of incinerators for industrial hazardous waste, several other types of incinerators are being used. These types range from fluidized bed incinerators, which have been successfully used at full scale, to the molten salt type, which has only been demonstrated on a pilot scale.

Also included in this section are single- and multiple-hearth incinerators, which have limited application to hazardous wastes. These hearth-type units are designed to handle sludges and solid waste, but experience has shown they can handle certain types of hazardous materials.

Fluidized Bed. The fluidized bed incinerator is a simple device consisting of a refractory-lined vessel containing inert granular material such as sand. Gases are blown through this material at a rate sufficiently high to cause the bed to expand and act as a theoretical fluid. The gases are injected through nozzles that permit flow up into the bed but restrict downflow of the material. Waste feed enters the bed through nozzles located either above or within the bed. Preheating of the bed to start-up temperatures is accomplished by a burner located above and impinging down on the bed.

In the combustion process, heat transfer occurs between the bed materials and the injected waste materials. Typical bed temperatures are in the range of 760–870 °C (1400–1600 °F). Heat from combustion is transferred back to the bed material [12]. Solid materials remain in the bed until they have become small and light enough to be carried off with the flue gases as particulates. Collected ash is land disposed.

Fluidized bed incinerators are versatile devices that can dispose of solid, liquid, and gaseous combustible wastes. The technique is a relatively new method for ultimate disposal of waste materials. It was first used commercially in the United States in 1962 and has found limited use in the petroleum and paper industries and in processing nuclear wastes and sewage sludges [7].

A typical fluidized bed incinerator is shown schematically in Figure 4-10. Air driven by a blower enters a plenum at the bottom of the combustor and rises vertically through a distributor plate into a vessel containing a bed of inert granular particles, typically sand. The upward flow of air through the sand bed results in a dense turbulent mass that behaves similarly to air flowing through a liquid. Waste material to be incinerated is injected into the bed, and combustion occurs within the bed. Air passage through the bed strongly agitates bed particles. This promotes rapid and relatively uniform mixing of the injected waste material within the fluidized bed.

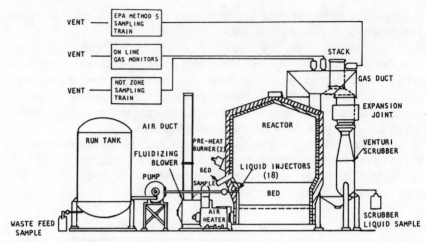

Figure 4-10.   Fluidized bed incinerator schematic.

The mass of the fluidized bed is largely relative to the mass of the injected material. Bed temperatures are quite uniform and typically in the 750–870° C (1400 to 1600° F) range. At these temperatures heat content of the fluidized bed is approximately 600 MJ/m³ (16,000 Btu/ft³), thus providing a large heat reservoir. By comparison, the heat capacity of flue gases at similar temperatures is three orders of magnitude less than that of a fluidized sand bed.

Heat is transferred from the bed into the injected waste materials to be incinerated. Upon reaching ignition temperature (which takes place rapidly), the material combusts and transfers heat back into the bed. Continued bed agitation by the fluidizing air allows larger waste particles to remain suspended until combustion is completed. Fines are carried off the bed by the exhausting flue gases at the top of the combustor. These gases are subsequently processed and/or scrubbed before atmospheric discharge.

Primary factors to be considered in specifying or designing a fluidized bed combustor are gas velocity, bed diameter, bed temperature, and waste composition and type.

Gas velocities are typically low, in the order of 1.5–2.4 m/s (5–8 ft/s). Maximum gas velocity is constrained by the terminal velocity of the bed particles and is therefore a function of particle size. Higher velocities result in bed attrition and an increased particulate load on downstream particulate collection equipment. Relatively low velocity reduces pressure drop and therefore lowers power requirements, but it increases equipment size.

The largest fluidized beds are on the order of 15 m (50 ft) in diameter. At nominal values of gas velocity and temperature, the maximum volumetric flow would be approximately 1200 m^3/s (2.5 × 10^6 acfm).

Bed depths range from about 0.4 m (16 in.) to several feet. Variations in bed depth affect waste-particle residence time and system pressure drop. One therefore desires to minimize bed depth consistent with complete combustion and minimum excess air.

The type and composition of the waste is a significant design parameter in that it will impact storage, processing, and transport operations (prior to incineration) as well as the combustion. If the waste is a heterogeneous mixture such as municipal refuse and has a relatively low [19 MJ/kg (8000 Btu/1b)] heating value, processing (shredding, sorting, drying, etc.) operations will be more complex and auxiliary fuel must be added to the combustor.

Advantages

1. Can burn solid, liquid, and gaseous wastes
2. Simple design concept with no moving parts
3. Compact design due to high heating rate per unit volume [900,000–1,800,000 kg-cal/h-m^3 (100,000–200,000 Btu/h-ft^3)]
4. Low gas temperatures and excess air requirements minimize nitrogen oxide formation
5. Long incinerator life and low maintenance costs
6. Large active surface area enhances combustion efficiency
7. Fluctuation in feed rate and composition are easily tolerated due to the large heat capacity of the bed
8. Currently widely used for combustion of fuels

Disadvantages

1. Difficult to remove residual materials from the bed
2. Requires fluid bed preparation and maintenance
3. Feed selection must avoid bed damage
4. May require special operating procedures to avoid bed damage
5. Incineration temperatures limited to a maximum of about 1500 °F to avoid fusing of the bed material
6. Operating costs are relatively high, particularly power costs
7. Little experience on hazardous-waste combustion

Multiple Hearth. The multiple-hearth incinerator (commonly called a Herreshoff furnace) is a versatile unit used to dispose of sewage, sludges, tars, solids, gases, and liquid combustible wastes. This type of unit was initially designed to incinerate sewage plant sludges in 1934. In 1968, there were over 125 installations in operation with a total capacity of 180 kg/s (17,000 tons/day) (wet basis) for this application alone. Currently, numerous industrial installations are primarily used for chemical sludge and tar incineration as well as activated carbon regeneration [9]. A typical multiple-hearth furnace (shown in Fig. 4-11) includes a refractory-lined steel sheet, a central rotating shaft, a series of solid flat hearths, a series of rabble arms with teeth for each hearth, an air blower, fuel burners mounted on the walls, an ash removal system, and a waste feeding system. Side parts for tar injection, liquid-waste burners, and an afterburner may also be included.

Sludge and/or granulated solid combustible waste is fed through the furnace roof by a screw feeder or belt and flapgate. The rotating air-cooled central shaft with air-cooled rabble arms and teeth plows the waste material across the top hearth to drop holes. The waste falls to the next hearth and then the next until discharged as ash at the bottom. The waste is agitated as it moves across the hearths to make sure fresh surface is exposed to hot gases.

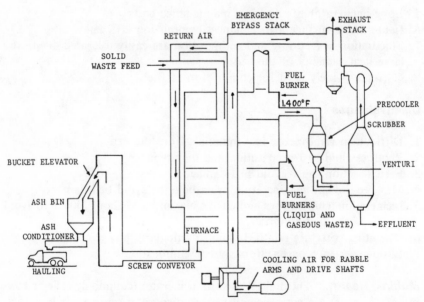

Figure 4-11. Multiple hearth incinerator.

The rabble arms and teeth located on the central shaft all rotate in the same direction; additional agitation of the waste (back rabbling) is accomplished by reversing the angles of the rabble tooth pattern and the rotational speed of the central shaft [2].

Liquid gaseous combustible wastes may be injected into the unit through auxiliary burner nozzles. This utilization of liquid and gaseous waste represents an economic advantage because it reduces secondary fuel requirements, thus lowering operating costs [2].

Multiple-hearth incinerators have three operating zones: (1) the top hearths, where feed is dried to about 48% moisture; (2) the incineration–deodorization zone, which has a temperature of 750 to 1000 °C (1400 to 1800 °F); and (3) the cooling zone, where the hot ash gives up heat to incoming combustion air. Exhaust gases exit at 250 to 600 °C (500 to 1100 °F).

Incineration ash is sterile and inert. Volume discharged from the bottom hearth is about 10% of the furnace feed, based on sludge cake with 75% moisture and 70% volatile content in the solids. The ash usually has less than 1% combustible matter, which is normally fixed carbon.

Current systems include gas-cleaning devices on exhaust air. A number of multiple-hearth incinerators are operating without difficulty in areas with strict air-pollution codes. Although the exhaust does not violate opacity codes, existence of steam plumes has occasionally caused adverse public reaction.

Advantages

1. The retention of residence time in multiple-hearth incinerators is usually higher than that in other incinerator configurations for hazardous materials having low volatility.
2. Large quantities of water can be evaporated.
3. A wide variety of wastes with different chemical and physical properties can be handled.
4. They are able to use many fuels, including natural gas, reformer gas, propane, butane, oil, coal dust, waste oils, and solvents.
5. High fuel efficiency is allowed by the multizone configuration.

Disadvantages

1. Due to the longer residence times of the waste materials, temperature response is very slow.
2. It is difficult to control the firing of supplementary fuels because of this slow response.
3. Maintenance costs are high because of the moving parts (rabble arms, main shaft, etc.) subjected to combustion conditions.

4. They are susceptible to thermal shock resulting from frequent feed inter-ruptions and excessive amounts of water in the feed. These conditions can lead to early refractor and hearth failures.
5. If used to dispose of hazardous wastes, a secondary combustion chamber probably will be necessary, and different operating temperatures might be necessary.
6. They are not well suited for wastes containing fusible ash, wastes requiring extremely high temperature for destruction, or irregular bulky solids.

Multiple-Chamber Incinerators. Multiple-chamber incinerators (see Fig. 4-12) are divided into three separate zones: (1) an ignition or primary com-bustion chamber; (2) a downdraft mixing chamber; and (3) an up-pass second-ary combustion chamber. Solid wastes are either manually or automatically fed into the incinerator through charging doors to grates at the bottom of the ignition chamber. Here, the wastes are dried, ignited, volatilized, and partially oxidized into gases and particulates. As more waste is charged to the system, the pile of burning waste is pushed farther along the hearth toward the ash pit. The amount of ash entrained as particulate versus the amount leaving the system as bottom ash is primarily a function of the underfire-

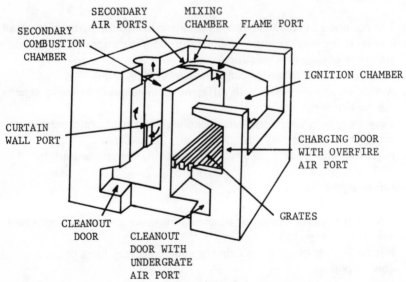

Figure 4-12. Multiple chamber incinerator.

overfire air ratio. Multiple-chamber incinerators are generally classified as the retort type or the in-line type.

The retort multiple-chamber incinerator is identified by an arrangement of the chambers that forces the combustion gases to make a 90° change in the direction in both horizontal and vertical axes. The primary and secondary reaction chambers are separated by a common wall.

The in-line multiple chamber incinerator is distinguished by an intermediate secondary burner–mixing zone, followed by a third chamber, which is a secondary combustion chamber. The combustion gases only change direction in the vertical plane. Multiple-chamber incinerators are generally more labor intensive than other incineration equipment because of variations in the form of feed waste and the special handling that is therefore required.

Retort-type incinerators are considered to have an upper capacity limit of approximately 450 kg/h (1000 1l/h), above which the effective turbulence in the mixing chamber is reduced, resulting in incomplete combustion of the vaporized waste components. In-line multiple chamber incinerators are not well suited to applications with a waste capacity of less than 340 kg/h (750 lb/h) due to flame propagation restrictions in these small-sized units. Although the upper limit for using in-line units has not been formally established, rates in the range of 910 kg/h (2000 1b/h) are common. Specifically designed units with considerably larger capacities are possible.

The requirement for supplementary gas burners must be determined. In general, when the moisture content of the waste is less than 10% by weight, burners are usually not required. Moisture contents from 10% to 20% normally necessitate installing mixing-chamber burners, and moisture contents over 20% usually necessitate inclusing ignition-chamber burners.

The average temperature of the combustion products, 540°C (1000°F), is not high enough to ensure complete destruction of many hazardous materials. High temperatures could be maintained by additional auxiliary burners and proper construction materials.

Fly ash and other solid particulate matter are collected in the combustion chamber by wall impingement and simple settling. The gases finally discharge through a stack directly or may be quenched beforehand.

Advantage   Represents a proven technology

Disadvantages

1. Requires a high degree of labor
2. Inability to process liquids, gases, sludges, and tars
3. Temperature not high enough to ensure acceptable destruction of hazardous waste

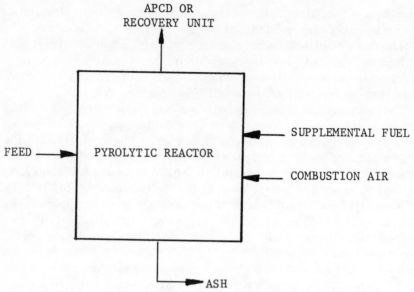

Figure 4-13.   Starved air combustion/pyrolysis schematic.

Pyrolysis.   Pyrolysis is generally defined as the thermal decomposition of a compound in the absence of oxygen. The heart of the pyrolytic waste-conversion process is the pyrolytic converter. The unit consists of a sealed, airtight retort cylinder inside a heavy insulating jacket. The gas-dried rotary revolves slowly on a slight decline from infeed to outfeed. Wastes are injected through a seal area that intermittently opens. A flapper valve seal minimizes oxygen entry. A pyrolysis unit is schematically shown in Figure 4-13.

The major attraction of pyrolysis is the potential for recovery of economic value from waste products. Pyrolysis of hazardous waste is relatively new. Designs of pyrolysis have been developed by several organizations, including the Bureau of Mines, West Virginia University, the Union Carbide Corp., the Garret Research and Development Corp., Monsanto Inc., the Battelle Northwest Laboratories, and Midland Ross Inc.

Pyrolysis is expected to be limited to those wastes for which the pyrolyzed product, gas or char, does not contain toxic or hazardous substances such as chlorine, fluorine, nitrogeneous substance, or heavy metals because there is too great a potential for their introduction into the environment. Using pyrolysis gases as a low-Btu gas for process heating is being investigated as a method of recovering heat value from solid and semisolid (tars, etc.) wastes.

Advantages

1. Potential for by-product recovery
2. Reduction of sludge volume without large amount of supplementary fuel
3. Thermal efficiency higher than for normal incineration due to the lower quantity of air required for this process
4. Reduced air emissions sometimes possible
5. Converts carbonaceous solids into a more easily combustible gas
6. Allows suppression of particulate emissions
7. Allows treatment of the hot fuel gas stream prior to combustion to suppress the formation of acid gases

Disadvantages

1. Potential source of carcinogenic decomposition product formation
2. Not capable of functioning very well on sludge-like or caking material alone unless cake-breaking capabilities are included in the design

Starved Air Combustion. Starved air combustion utilizes equipment and process flows similar to those for incineration, but in this process less than the theoretical amount of air for complete combustion is supplied. Because the process is neither purely pyrolytic nor purely oxidative, it is called starved air combustion or thermal gasification rather than pyrolysis or incineration. An auxiliary fuel may be required, depending on the proportion of volatiles in the solids. High temperatures decompose and vaporize the waste. The gas-phase reactions are pyrolytic or oxidative, depending on the concentration of oxygen remaining in the stream. Under proper control the gas leaving the vessel is a low-Btu fuel gas (up to 130 Btu/ft^3) that can be burned in an afterburner, a boiler, or another furnace. Some processes utilize pure oxygen instead of air and thus produce a higher-Btu fuel gas. The solid residue is a char with more or less residual carbon, depending on how much combustion air has to be supplied to reach the proper operating temperatures.

Furnaces may be operated in one of three modes, resulting in substantially different heat generation and residue characteristics. The low-temperature char (LTC) mode only pyrolyzes the volatile material, thereby producing a charcoal-like material converted to fixed carbon and ash. The char-burned (CB) mode reacts away all carbon and produces ash as a residue. Heat recovered is maximum for the CB mode, less for the HTC mode, and substantially less for the LTC mode.

Pollution equipment normally associated with the starved air combustion process includes exhaust gas scrubbers and afterburners.

Advantages

1. Potential for by-product recovery
2. Reduction of sludge volume without large amount of supplementary fuel
3. Thermal efficiency higher than incineration due to the lower quantity of air required for this process

Disadvantage   Not proven with hazardous waste

Molten Salt.   Molten salt incinerators (see Fig. 4-14) have recently been developed to pilot and demonstration scales for incineration of organic waste compounds. In the basic molten salt concept for waste disposal, the waste is injected below the surface of a molten salt bath, which usually contains approximately 90% sodium carbonate and 10% sodium sulfate and is designed to operate in the range of 820 to 980 °C (1500 to 1800 °F). Substituting

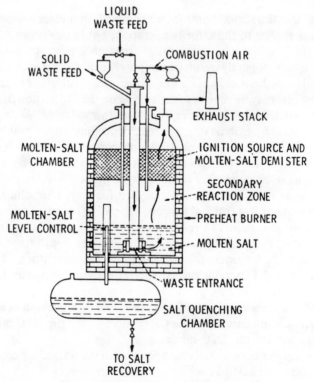

Figure 4-14.   Molten salt incinerator.

other salts, such as potassium carbonate, allows even lower incineration temperatures. Reactive salts, such as the eutectic mixtures NaOH–KOH and Li_2CO_3–$NA_2CO_3K_2CO_3$, produce the additional benefit of entrapping potentially toxic or objectionable off-gas constituents such as heavy metals (mercury, lead, cadmium, arsenic, selenium). This reduces or eliminates the need for pollution abatement equipment. The spent salt often can be regenerated, or it may be land disposed.

Wastes such as free-flowing powders and shredded materials may be directly fed to molten salt incinerators. Residence times average three-fourths of a second. Waste liquids may be sprayed into the combustion air and fed to the unit.

Project developers feel that for most materials no subsequent pollution abatement equipment is necessary. However, hazardous wastes containing metallic compounds with high-vapor-pressure metallic-oxide potential produce metallic-oxide fumes that will condense as difficult-to-remove submicrometer particulate.

Advantages

1. Salt reacts with pollutant off-gas such that scrubbing may not be needed
2. Salt media entrap particulates
3. Rapid and complete destruction of carbonaceous materials in the salt media at lower than normal temperatures
4. Compact and fuel efficient

Disadvantages

1. Demonstrations have been on pilot-scale incinerators and do not represent a proven technology
2. Estimated to have high operating costs

Open-Pit Incinerators. Open-pit incineration is the burning of waste materials on open land without the use of combustion equipment. This form of incineration is utilized mainly for the disposal of waste explosives. It is generally unacceptable for the disposal of other forms of waste because of the associated lack of combustion product effluent control.

Open-pit incinerators vary from the pedestal-mounted oil burner used by Union Carbide to the DuPont pit. The Union Carbide installation burns without visible smoke organic liquids containing up to 25% water. The installation consists of burners firing into a pit surrounded by earthworks that provide personnel protection. Heat is dissipated by direct convection and radiation. By the very nature of the design, there is always excess air.

The open-pit incinerator was originally developed at DuPont for the safe destruction of nitrocellulose, which presents an explosion hazard in a conventional closed incinerator. The incinerator has an open top and an array of closely spaced nozzles that create a rolling action of high-velocity air over the burning zone. Very high burning rates, long residence times leading to complete combustion, and high flame temperatures are achieved. Visible smoke is readily eliminated, and smuts are contained by proper screening. Oversized wastes and plastics that create problems in conventional incinerators are easily desroyed in the open-pit incinerator. Note that the concentration of particulates is slightly higher than for conventional incinerators and that there is no way to clean the exit gases. Although these pit incinerators are used for liquid wastes, they are more efficient for solid wastes, especially rubber and plastic. Current federal regulations do not allow open burning of hazardous wastes except for explosive military-type wastes.

Advantages

1. Oversized wastes are easier to destroy by this method than by a conventional incinerator
2. Usually considered safer for personnel than enclosed incinerators

Disadvantages

1. Concentration of particulates is higher
2. Lack of confinement of combustion product effluents
3. No way to clean exit gas

AIR-POLLUTION-CONTROL SYSTEMS

With the current regulatory emphasis on hazardous-waste incineration, few incinerators will be built and operated without some type of air-pollution-control system. Air pollutants from hazardous-waste incineration may arise from incomplete combustion and from the products of combustion of constituents in the wastes and combustion air. The products of incomplete combustion include carbon monoxide, carbon, hydrocarbons, aldehydes, amines, organic acids, polycyclic organic matter (POM), and any other waste constituents or their partially degraded products that escape thermal destruction in the incinerator. In well-designed and well-operated incinerators these incomplete combustion products are emitted in insignificant amounts. The primary overall end products of combustion are in most cases carbon dioxide (CO_2) and water vapor (H_2O). Hydrogen chloride (HCl) and small amounts of chlorine (Cl_2), for example, are formed from the incineration of chlorinated hydrocarbons. Hydrogen fluoride (HF) is formed from the incineration of organic fluorides, and both hydrogen bromide (HBr) and bromine (Br_2) are formed from the incineration of organic bromides. Sulfur oxides, mostly as sulfur dioxide (SO_2), but also including 1–5% sulfur trioxide (SO_3), are formed from the sulfur present in the waste and the auxiliary fuel. Phosphorus pentoxide (P_2O_5) is formed by incinerating organophosphorus compounds. In addition, nitric oxide (NO) is formed by thermal fixation of nitrogen from the combustion air and from nitrogen compounds in the waste material. Particulate emissions include particles of mineral oxides and salts from the mineral constituents in the waste material as well as fragments of incompletely burned combustible matter.

Particulates and acid gases (SO, HCl, HF, etc.) are usually controlled with scrubbers. These scrubbers operate on two mechanisms: (1) physical removal of particulates, and (2) chemical removal by absorption and neutralization of the acid gases. These physical and chemical processes are conducted in several different types of gas-handling equipment.

Afterburners

Afterburners are simple combustors employed to destroy (by oxidation) gaseous hydrocarbons not destroyed in the primary incinerator chamber. Three types of afterburners are described here: direct flame, thermal, and catalytic. Direct flame and thermal afterburners are similar, but they destroy organic vapors by different methods. In a direct flame afterburner a high percentage of the vapors pass directly through the flame in a direct flame unit. In a thermal unit the vapors remain in a high-temperature oxidizing atmosphere long enough for oxidation reactions to take place. Catalytic devices incorporate a catalytic surface to accelerate the oxidation reactions.

Thermal afterburners are usually an integral part of rotary kilns used in hazardous-waste incinerators. Thermal afterburners are also used with liquid-injection incinerators in a few instances, pyrolysis units when chemicals are not being recycled, and co-incineration units where the incinerator used normally requires an afterburner. Catalytic afterburners are a proven technology not yet extensively on hazardous-waste incinerators.

Catalytic afterburners are used to destroy combustible materials in low concentrations but are readily poisoned by chlorinated hydrocarbons due to formation of corrosive HCl.

Thermal afterburners, used commonly with rotary kilns, provide exposure of the organic vapors to a high temperature oxidizing atmosphere to ensure vapor destruction. Temperatures ranging from 650 to 1300 °C are generally required to successfully operate these devices. Hydrocarbon levels can usually be satisfactorily reduced at temperatures of 760 °C, but higher temperatures may be required to oxidize CO [10]. The following temperatures are often used as guidelines:

To oxidize hydrocarbons   500–650 °C

To oxidize carbon monoxide 650–800 °C

Depending on the type of pollutant in the gas stream, residence times ranging from 0.2 to 6.0 s are required for complete combustion. The residence time in most practical afterburner systems is dictated primarily by chemical kinetic considerations. To ensure good mixing, afterburners are operated at high-velocity gas flows. Superficial gas velocities in afterburners are in the order of 50 ft/s [10].

Both gaseous and liquid fuels are used to fire afterburners. Gaseous fuels have the advantage of permitting firing in multiple-jet (or distributed) burners. Oil firing has the disadvantage of producing sulfur oxides (from sulfur in the oil) and normally produces higher nitrogen oxides emissions. Many applications use high-heat-value waste liquids to fuel afterburners. The use of waste can produce significant savings in operating costs.

Catalytic afterburners are applied to gaseous wastes containing low concentrations of combustible materials and air. Usually noble metals such as platinum and palladium are the catalytic agents. A catalyst is defined as a material that promotes a chemical reaction without taking part in it. The catalyst does not change nor is it used up. However, it is subject to contamination and loss of effectiveness.

The catalyst must be supported in the hot waste gas stream in a manner that will expose the greatest surface area to the waste gas so that the combustion reaction can occur on the surface, producing nontoxic effluent gases of carbon dioxide, nitrogen, and water vapor. Most of the combustion occurs during flow through the catalyst bed, which operates at maximum temperatures of 810–870 °C. The ability to carry out combustion at relatively low temperatures while achieving high destruction efficiencies is a major advantage of the catalytic incinerator for gaseous wastes [9]. Eliminating the need for fuel is also a major advantage. Residence time for catalytic oxidation is typically about 1 s.

Generally, catalytic afterburners are considered for operation with waste combustion gases containing hydrocarbon levels that are less than 25% of the lower explosive limit. When the waste gas contains sufficient heating value to cause concern about catalyst burnout, the gas may be diluted by atmospheric air to ensure operating temperature within the temperature limits of the catalyst.

Advantages: Thermal or Direct Flame

1. Destroys organic pollutants not destroyed in primary incineration
2. Allows more flexibility in incinerator operation

Advantages: Catalytic

1. Carries out combustion at relatively low temperatures and is more economical to operate than other afterburners
2. Clean heated gas produced is well suited for waste heat recovery units

Disadvantage: Thermal or direct flame.   Auxiliary-fuel requirements increase fuel costs

Disadvantages: Catalytic

1. Catalyst burnout occurs at temperatures exceeding 1500 °F
2. Catalyst systems susceptible to poisoning agents, activity suppressants, and fouling agents
3. Catalyst needs occasional cleaning and eventual replacing
4. High maintenance costs

Venturi and Orifice Scrubbers

Gas-atomized spray scrubbers utilize the kinetic energy of a moving gas stream to atomize the scrubbing liquid into droplets. Typical scrubbers are the venturi and the orifice. In the venturi scrubber, liquid is injected into the high-velocity gas stream either at the inlet to the converging section or at the venturi throat. In the process the liquid is atomized by the formation and subsequent shattering of attenuated, twisted filaments and thin cup-like films. These initial filaments and films have extremely large surface areas available for mass transfer. It is the gas–liquid contact that permits removal of gaseous contaminants. The venturi scrubber is one of the most predominant air-pollution-control devices for hazardous-waste incinerators. It is commonly used with rotary kiln and liquid-inspection incinerators.

Prior to passage of the incinerator exhaust gas into the venturi, the gas is quenched to reduce the temperature. Although it is recognized that the quench systems will effect some degree of particle removal, the primary function of these units is to reduce flue-gas volume and downstream materials and operating problems through gas cooling. As a result of quenching, inlet temperatures for venturi scrubbers range from 10 to 300° F.

Some hazardous-waste incineration facilities employ sequential venturi and plate-type or packed-bed scrubbers. For these systems a gas quench is optional because the venturi may be used to effect gas cooling by adiabatic expansion of the gases. Such systems are capable of handling a variety of incineration gas compositions and dust loadings. Plate towers or packed beds, when used in conjunction with gas-atomized spray scrubbers, serve the dual function of eliminating the entrainment of liquid droplets from upstream and further reducing the emission levels of gaseous contaminants.

Incinerating hazardous-waste may produce effluent gases with corrosive contaminants, such as CHl. It is possible to neutralize the acid with a caustic solution. Erosion is also a problem in venturi scrubbers due to the high gas velocities and particulate loading encountered during normal duty. Throat and elbow areas are generally subject to the most wear. Acid-resistant tile liners, polymeric liners, and Inconel 625 are often used for scrubber construction.

Venturi scrubbers have been used to control emissions of SO_2, HF, and HCl. Several of the primary operating parameters that affect the removal of these gaseous contaminants are pressure drop, liquid-to-gas ratio, contact time, and gas flow rate. Pressure drops in venturi scrubbers for controlling gaseous emissions from hazardous-waste incineration are typically in the 30–50-in.-water-gage (WG) (7.5–12.5-kPa) range [11]. It is necessary to use the correct pressure drop to ensure efficient removal. A pressure drop that is too high results in wasted energy; one that is too low results in a lower removal efficiency. As a prime operating parameter, the pressure drop should be as low as possible yet yield the needed removal efficiency.

The liquid-to-gas ratio is an important design and operating parameter. It is needed to determine the scrubber diameter and has an effect on the unit dimensions. Normal liquid-to-gas ratios for venturi scrubbers are 5–20 gal/1000 acf (0.7–2.7 L/m³)[11].

Higher efficiencies are attained by allowing the gas and liquid phases to be in contact for a longer time. The contact time required for gas absorption is a function of the rate of mass transfer. The mass transfer rate, in general, depends on four separate resistances: gas-phase resistance, liquid-phase resistance, chemical-reaction resistance, and a solids-dissolution resistance for scrubbing liquids containing solid reactants. For absorpting gaseous contaminants that are highly soluble or chemically reactive with the scrubbing liquid, such as the absorption of HCl by caustic solution, the contact time required for 99% removal is extremely short (about 0.4–0.6 s). The less reactive and less soluble pollutants require a longer contact time [11].

The rate at which a flue gas from waste incineration must be processed by a particle control device depends primarily on the waste composition, the quantity of excess combustion air used, the initial gas temperature, and the method(s) by which the gas has been cooled, if cooling is used. Hence, these parameters, in conjunction with control device size or geometry, dictate the velocity at which the gas passes the particle collection elements [12].

It has been shown that the pressure drop across a venturi is proportional to the square of gas velocity and directly proportional to the liquid-to-gas ratio. Therefore, within limits, increasing gas velocity causes increasing pressure drop, other parameters being equal. Typical gas velocities employed commercially are 100 to 390 ft/s (30 to 120 m/s). The low end of this range, 100 to 150 ft/s (30 to 45 m/s), is typical of power plant applications, and the upper end of the range has been applied to lime kilns and blast furnaces [12].

Particle cut diameter (diameter of particles in which there is a 50% collection) is a frequently used parameter for describing the particle collection performance of venturi scrubbers. One reason is that plots of collection efficiency versus particle diameter tend to be rather steep in the region where inertial impaction is the predominant collection mechanism. High-energy venturi scrubbers provide the highest wet-scrubber efficiency with cut diameters in the (1.17×10^5)–$1.95 \times 10^5)$-in. (0.3–0.5-m) range [12].

Orifice scrubbers are similar to venturi scrubbers, but the orifice creates more turbulence than the venturi type provides and also typically introduces higher gas pressure drops. Thus they are less commonly used.

Wet Spray Towers

Spray towers remove contaminants by a gas-absorption process. The scrubbing liquid is atomized by high-pressure spray nozzles into small droplets, then directed into a chamber through which gases pass in either countercurrent, cocurrent, or crossflow directions. In this case the scrubbing liquid is the dispersed phase, and gas is the continuous phase (see Fig. 4-15). Because mass transfer occurs at the liquid droplet surface, gas absorption is enhanced by finer droplets, that is by the increased droplet surface area.

Several of the primary operating parameters that affect the removal of gaseous contaminants in spray towers are discussed here, including pressure drop, liquid-to-gas ratio, contact time, and gas flow rate. A normal gas pressure drop for a spray tower is 0.5–4 in. WG (0.125–0.996 kPa)[12].

Liquid-to-gas ratios strongly depend on the control device and the specific application. Under normal operating conditions, preformed spray towers employ liquid-to-gas ratios in the order of 30 gal/1000 acf (4–14 L/m³)[8].

In gas-absorption devices, higher efficiencies are attained by allowing the gas and liquid phases to be in contact for a longer time. The contact required

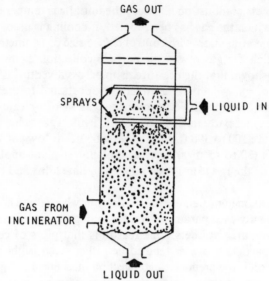

Figure 4-15.   Wet spray tower.

for gas absorption is a function of the rate of mass transfer. The mass transfer rate, in general, depends on four separate resistances: gas-phase resistance, liquid-phase resistance, chemical-reaction resistance, and a solids-dissolution resistance for scrubbing liquids containing solid reactants such as lime. For absorpting gaseous contaminants that are highly soluble or chemically reactive with the scrubbing liquid, such as the absorption of HCl by caustic solution, the contact time required for 99% removal is extremely short (about 0.4–0.6 s). Because inertial impaction is the principal particle collection mechanism, it is beneficial to operate with a high relative velocity between the gas and the collection element. Practical relative velocity limitations occur as a result of the increased operating costs associated with high-pressure drops, flooding, or other considerations. The most common gas velocities in spray towers range from 7 ft/s to 10 ft/s (2.1–3.0 m/s)[12].

Advantages

1. Simultaneous gas absorption and dust removal
2. Suitable for high-temperature, high-moisture, and high-dust loading applications
3. Simple design
4. Scaling problems are rare

Disadvantages

1. High efficiency may require pump-discharge pressures
2. Dust is collected wet
3. Spray nozzles are susceptible to plugging
4. Requires a downstream mist eliminator
5. Has a lower particulate collection efficiency than a venturi scrubber
6. Lower absorption efficiency than a packed tower
7. Structure is large and bulky

Dry Spray Towers

A very recent variation of the spray tower is the dry spray tower. Here the evaporation of the water from the scrubbing solution is carefully controlled, so by the time the materials reach the bottom of the tower, they are a dry powder. This eliminates scrubber-water treatment and allows the effluent from the spray tower to be directly land filled.

As of this writing, only one such system is known to be operational [13]. The technology is currently proprietary, so little information is available on design details, effectiveness, or potential cost savings. However, if it is successful, significant savings in handling scrubber effluents should be realized.

Packed-Bed Scrubber

Contaminants are removed in packed-bed scrubbers by a gas-adsorption process that depends on intimate gas–liquid contact. These scrubbers are vessels filled with randomly oriented packing material, such as saddles and rings. The scrubbing liquid is fed to the top of the vessel, with the gas flowing in either concurrent, countercurrent, or crossflow modes. As the liquid flows through the bed, it wets the packing material and thus provides interfacial surface area for mass transfer with the gas phase (see Fig. 4-16).

Packed-bed scrubbers are a major air-pollution-control device for hazardous-waste incinerators because of their high removal efficiency for gaseous emissions. Designed properly, a packed-bed scrubber will remove 99% of the halogens from incinerator exhaust gases. The inherent nature of the design does not, however, allow for removal of particulates from exhaust gases with high particulate loadings. Unless prior treatment is used, this type of waste stream causes clogging in the packed-bed scrubber [11]. Predominant use of the APCD has been with liquid-injection incinerators, due to their typically low-effluent particulate loadings. When packed beds are used to control gaseous emissions from rotary kilns and fluidized bed incinerators, venturi scrubbers are usually incorporated upstream as the primary APCD to remove particulates [11].

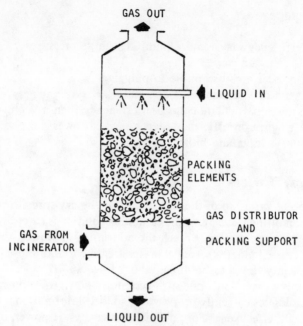

Figure 4-16. Packed wet scrubber.

Differences between packed-bed scrubbers include the flow mode, the packing material, and the depth of packing. The choice of flow mode depends on the particular application. Crossflow scrubbing is generally applied to situations where the bed depth is less than 6 ft, and countercurrent design is applied at bed depths of 6 ft or more. Packing materials are usually made of ceramic or plastic materials that will withstand acid corrosion [11].

Packed beds used for gaseous emission control in hazardous-waste incineration facilities usually have a pressure drop range from 2.0 in. to 7.2 in. WG (0.5–1.9 kPa). The total pressure drop across the packed bed is directly proportional to the depth of packing and affects the gaseous removal efficiency in the packed-bed scrubber. Normal liquid-to-gas ratios in packed beds vary from 6 gal/1000 acf to 75 gal/1000 acf (0.8–10 L/m³), with most units operating between 22 gal/1000 acf and 52 gal/1000 acf (3 L/m³ and 7 L/m³)[12].

In gas-absorption devices higher efficiencies are attained by allowing the gas and liquid phases to be in contact for a longer time. Removal efficiencies for gaseous contaminants in packed beds are directly related to the depth of packing, which in turn determines the contact time [11].

The contact time required for gas absorption is a function of the rate of mass transfer. The mass transfer rate, in general, depends on four separate resistances: gas-phase resistance, liquid-phase resistance, chemical-reaction resistance, and a solids-dissolution resistance for scrubbing liquids containing solid reactants.

In the design of gas-absorption devices, the cross-sectional area for gas–liquid contact is determined by the superficial gas velocity selected. The greater the gas velocity selected, the smaller will be the scrubber diameter but the larger will be the pressure drop.

There are two additional factors to consider when selecting gas velocity. First, the gas velocity through the scrubber should allow sufficient resistance time for gas–liquid contact. Second, in a countercurrent packed bed, the gas velocity should not exceed the flooding velocity. At the flooding point the pressure-drop-versus-gas-rate curve becomes almost vertical, and a liquid layer starts to build up on top of the packing. The flooding point represents the upper limiting conditions of pressure drop and fluid rates for practical tower operation (Fig. 4-17). A margin of 30 to 40% of the flooding velocity should be allowed in designing these scrubber types. The most common gas velocities in packed beds range from 7 ft/s to 10 ft/s (2.1–3.0 m/s) [11].

As with other wet scrubbers, mist eliminators are often used downstream of the packed-bed scrubber for proper pollution control. When a wet scrubber follows a packed-bed scrubber, mist eliminators are often not used. A packed-bed scrubber often follows a venturi scrubber in hazardous-waste incineration facilities.

Advantages

1. High removal efficiency for gaseous and aerosol pollutants
2. Low to moderate pressure drops
3. Engineering principles controlling the performance of packed-bed scrubbers are well developed and understood
4. Corrosion-resistant packings to withstand corrosive materials are available

Disadvantages

1. Low efficiency for fine particles
2. Not suitable for high-temperature or high-dust loading applications
3. Requires downstream mist eliminator

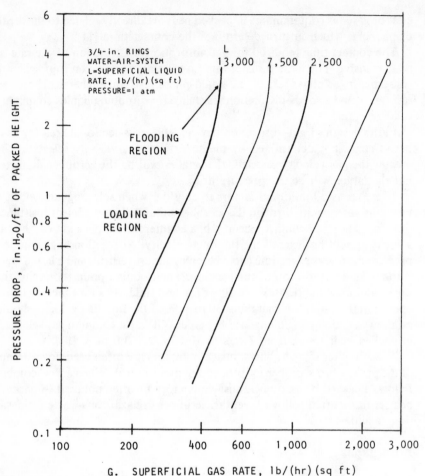

Figure 4-17. Packed tower pressure drop as function of gas rate and liquid rate.

4. Potential scaling and fouling problems and subsequent plugging
5. Possible damage to the scrubber if scrubber solution pumps fail

Plate Scrubbers

A plate scrubber is a type of wet scrubber that relies on gas absorption to remove contaminants. The basic design is a vertical cylindrical column with plates or trays inside. The scrubbing liquid is introduced at the top plate and flows successively across each plate as it moves downward to the liquid outlet

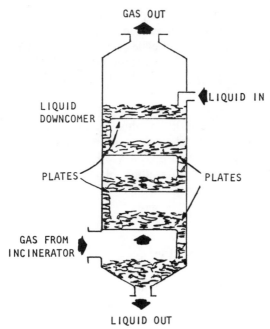

GAS OUT

LIQUID IN

LIQUID
DOWNCOMER

PLATES

PLATES

GAS FROM
INCINERATOR

LIQUID OUT

Figure 4-18.   Plate scrubber.

at the tower bottom. Gas comes in at the bottom of the tower and passes through openings in each plate before leaving through the top. Gas absorption is promoted by the breaking up of the gas phase into small bubbles formed by the plates, which pass through the volume of liquid in each plate (see Fig. 4-18).

Plate towers are not as common as packed-bed towers or venturi scrubbers for the control of air pollution from hazardous-waste incineration. They are capable of controlling gaseous emissions from liquid-injection incinerators.

At hazardous-waste incineration facilities, plate towers with two sieve trays are typically used as an absorber–mist eliminator in conjunction with a high-energy venturi scrubber. The primary operating parameters that affect the removal of gaseous contaminants such as SO_2 are the pressure drop, liquid-to-gas ratio, contact time, and gas flow rate.

Total pressure drop across plate towers is similar to that of packed beds—in the 2.0–7.2 in. WG (0.5–1.8 kPa) range. In plate towers pressure drop is not used as an operating parameter to estimate removal efficiency. Rather, the number of plates is the primary parameter that determines removal efficiency [12].

Plate towers are appropriate when particle size is not less than 1 μm. Unlike absorption efficiency, particle collection efficiency will not necessarily improve with an increased number of plates, but decreased perforation diameter does increase particle collection efficiency [11].

Advantages

1. Simultaneous gas absorption and dust removal
2. High removal efficiency for gaseous and aerosol pollutants
3. Low to moderate pressure drop
4. Mass transfer increases with multiple plates
5. Handles high liquid rates

Disadvantages

1. Low efficiency for fine particles
2. Unsuitable for high-temperature or high-dust loading applications
3. Requires downstream mist eliminator
4. Limestone scrubbing solution causes scaling
5. Unsuitable for foamy scrubbing liquid

Electrostatic Precipitators (ESP)

Electrostatic precipitation is a process by which particles suspended in a gas are electrically charged and separated from the gas stream under the action of an electric field. In this process, negatively charged gas ions are formed between emitting and collecting electrodes by applying a sufficiently high voltage to the emitting electrodes to produce a corona discharge. Suspended particulate matter is charged as a result of bombardment by the gaseous ions and migrates toward the grounded collecting plates due to electrostatic forces. Particle charge is neutralized at the collecting electrode where subsequent removal is effected by periodically rapping or rinsing. A majority of industrial ESPs used today are the single-stage, wire, and plate types; charging and collection take place in the same section of the ESP. Two-stage ESPs, often called electrostatic filters, utilize separate sections for particle charging and collecting and are not generally employed for controlling particulate emissions from combustion sources (see Fig. 4-19).

Electrostatic precipitators have been widely used in conjunction with utility boilers and with municipal and industrial incinerators. Dry ESPs are not capable of removing acid gases; therefore, facilities burning halogenated wastes must employ wet scrubbing of acid halides if ESPs are used for particulate emission control. Electrostatic precipitators effectively collect fine particles [less than 3.9×10^5 in. (1 μm) in diameter] but are unable to cap-

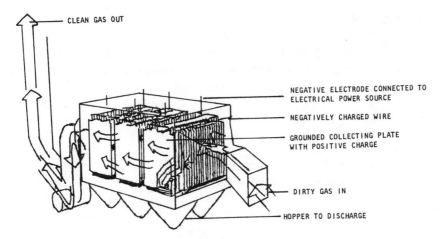

Figure 4-19. Electrostatic precipitator schematic.

ture noxious gases. They perform poorly on particles with high electrical resistivity.

Electrostatic precipitators have been employed by European facilities where hazardous wastes are incinerated, although the wastes generally do not contain very high levels of chlorine. When halogenated wastes are incinerated, careful waste blending is employed to protect ESPs from corrosion, so HCl gas concentrations do not exceed 1000 ppm and usually average 300 ppm [14].

Components of ESPs in direct contact with the process gas stream include the shell, electrodes, high-voltage frames, rapper rods, and gas distribution plates. On the basis of mild-steel construction, such components constitute approximately 68% of the total unit cost [15]. Hence, the applications requiring exposure to corrosive gas streams have substantial impact on ESP design and ultimate cost. Lead linings, used in acid-mist ESPs, are not generally suitable for use in incinerator gas treatment due to poor resistance to attack by gaseous halogens. Fiberglass reinforced plastic (FRP) has been successfully used for inlet and outlet plenums as well as collecting electrodes; however, the latter application must have adequate conductivity to permit current flow to ground.

An ESP is carefully designed and constructed for maximum electrical safety; however, normal high-voltage precautions must be observed. Design features such as interlocks between access doors and electrical elements should be employed. Also, access after deenergizing should be delayed to allow for static charge drainage.

Pressure and temperature drops across ESPs are very small compared to those of wet scrubbers: for example, typically below 1.00 in. WG (0.25 kPa) as compared with up to 60.2 in. WG (15 kPa). Additionally, ESPs provide generally higher removal efficiencies for particles smaller than 3.9×10^5 in. (1 m) in diameter than do wet scrubbers. A standard gas temperature range is up to 370 °C, and the voltage normally applied ranges from 30 to 75 kV. An ESP is often placed downstream of waste heat boilers where gas temperatures of 250–370 °C are encountered.

Advantages

1. Dry collection of dust
2. Low pressure drop and operating cost
3. Efficient removal of fine particles

Disadvantages

1. Relatively high capital cost
2. Sensitive to changes in flow rate
3. Particle resistivity affects removal and economics
4. Not capable of removing gaseous pollutants
5. Fouling potential with tacky particles

Wet Electrostatic Precipitator

The wet electrostatic precipitator (WEP) is a variation of the dry ESP design. The two major added features are (1) a preconditioning step, where inlet sprays in the entry section are provided for cooling, gas absorption, and removal of coarse particles, and (2) a wetted collection surface, where liquid is used to continuously flush away collected materials. Particle collection is achieved by introducing evenly distributed liquid droplets to the gas stream through sprays located above the electrostatic field sections and by migration of the charged particles to the charged surfaces. To control the carryover of liquid droplets and mists, the last section of the WEP is often operated without penetrate, and mists can be collected on baffles.

The WEP overcomes some of the limitations of the dry ESP. The operation of the WEP is not influenced by the resistivity of the particles. Further, since the internal components are continuously being washed with liquid, buildup of tacky particles is controlled and there is some capacity for removal of gaseous pollutants. In general, applications of the WEP fall into two areas: removal of fine particles, and removal of condensed organic fumes. Outlet particulate concentrations are typically in the 2–24-mg/m³ range [15].

Presently there are few WEP installation at hazardous-waste incineration facilities. At these facilities WEP is the last step of the air-pollution train. It is used as a "polishing" process to remove particulates prior to exhausting

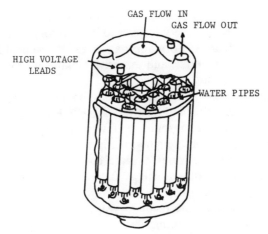

Figure 4-20. Wet electrostatic precipitator.

to the stack (see Fig. 4-20). Increasing stringent particulate emission standards are requiring the installation of WEP at solids-handling kiln incinerators, which typically have high particulate loadings. Due to their very recent introduction to incinerators, little information on their design and operation parameters is available.

Advantages

1. Simultaneous gas absorption and dust removal
2. Low energy consumption
3. No dust resistivity problems
4. Efficient removal of fine particles

Disadvantages

1. Low gas absorption efficiency
2. Sensitive to changes in flow rate
3. Dust collection is wet

Incinerator Energy Recovery Systems

The recent dramatic rise in energy costs has increased interest in recovering heat generated by waste incineration. The recovery of heat from municipal waste incineration has been actively pursued for many years. Heat recovery from incineration of industrial wastes has only recently become of interest.

Two basic approaches have developed in industrial-waste-heat recovery: (1) adding waste-heat recovery systems to incinerators, and (2) firing wastes into existing boiler units. Both approaches can cause problems with emissions of hazardous materials and excessive maintenance with heat transfer surfaces due to plugging and corrosion.

The technique of changing wastes to existing boilers has been encouraged by Resource Conservation and Recovery Act (RCRA) regulatory exemptions for beneficial recovery of resources. However, the design of many boilers may not provide sufficiently high gas-temperature residence times to effectively destroy some hazardous organic compounds. Instead, boilers are designed to extract maximum energy from combustion of fuels with high heat value. The combustion of hazardous waste in boilers is being carefully studied by federal and state regulatory agencies.

Incinerators are designed primarily to insure complete destruction of wastes and secondarily to recover energy. As such, an incinerator with waste-heat recovery will not usually achieve as high a heat-recovery efficiency as a boiler. This fact should not deter the potential incinerator owner or operator from very seriously considering heat recovery as a part of an incineration installation. Recovered energy credits can often offset or significantly impact the cost of installing and operating a hazardous-waste incinerator. The majority of incinerators being ordered, designed, and installed as of this writing have energy recovery systems, mostly stream-producing waste-heat boilers.

Incinerator heat recovery can be accomplished by radiant transfer in the primary combustion chamber (water-wall furnace) and/or convective transfer of the sensible heat in the flue gas (waste-heat boiler or air heater). The water-wall furnace design generally consists of vertically arranged metal tubes in the walls of the primary chamber of the incinerator. Water is circulated and heated in the tubes. Energy is transferred to the tubes by radiant exchange with the burning waste and by gas radiation. Flame temperatures above 1100 °C (2000 °F) and low excess air rates enhance the efficiency of radiant heat transfer [9].

Convective heat exchangers are located in the flue-gas system downstream of the primary combustion chamber. The sensible heat contained in the flue gases is recovered by convection. Steam, hot water, heated air, or a heated process stream can be used as the recovery medium.

Current thinking on incinerator design favors the convective flue gas approach to hazardous-waste incinerators. The water-wall design raises concern about the cooling effect of combustion gases at the colder heat transfer walls. This cooling effect is suspected of causing incomplete destruction of organics and possible production of unacceptable combustion by-products.

The ideal conditions for any heat-recovery application would be a continuous, steady-state waste feed of waste materials and a constant fuel-to-air ratio that would produce steady flue-gas volumes and temperatures. This condition can be approached by selective blending of liquid wastes and careful charging of solid wastes. The larger the waste combustion capacity of the unit, the better variations can be dampened out. Also an adequately sized and controlled auxiliary-fuel system can minimize swings of temperature and flow.

Waste-heat boilers are basically two designs: fire tube and water tube. In the fire-tube design, hot gases pass through the tubes while water is in the shell side. Water-tube designs circulate the water through the tubes and the hot gases surround the tubes.

Summarized below are the major characteristics of both types of boilers which affect their selection for a hazardous-waste disposal application.

Fire-tube Boilers

1. Lower first-time cost than the water-tube design
2. Less suitable for high particulate loads (0.25 g/acf is considered the maximum loading)
3. Limited to steam production rates of 15,000 lb/h and 200 psig steam pressures
4. More difficult to clean than water-tube boilers

Water-tube Boilers

1. Higher first-time cost than the fire-tube design
2. Can handle higher particulate loadings than the fire-tube design can
3. Higher steam output and pressures than the fire-tube design
4. Higher variation in flue-gas temperatures can be tolerated
5. More easily cleaned than the fire-tube design

In selecting a waste-heat boiler, consider the following:

1. The variability and maximum temperature of the flue gases must be considered. A gas temperature of 1100 °C (2000 °F) is the maximum tolerated by standard boiler designs. Variability of gas temperatures can be minimized by careful consideration of the solids-changing rate and the volatility of the solids and/or sludge being charged. A properly designed and operated auxiliary-fuel system can contribute to consistent gas temperatures.

2. Gas flow rates must be as consistent as possible. Severe cycling of the flow rate must be minimized. Combustion of air flow rate control is the primary method of minimizing surges.

3. Particulate loading can cause plugging and physical erosion. Fire-tube designs are limited to particle concentrations at about 0.25 gr/acf; 0.15 gr/acf

is usually recommended. Water-tube designs can tolerate higher particulate loadings. Many wastes, such as paint wastes, inorganic pigments, clays, and other materials, have inherently high particulate emissions.

4. Chemical corrosion potential of the combustion gases must be considered. Chemical corrosion of boiler surfaces is very complex and not fully understood. Major sources of tube corrosion are

a. High-temperature liquid-phase corrosion probably due to molten alkali metal sulfates occurring at metal surface temperatures above 480 °C (900 °F).

b. Hydrogen chloride acid corrosion. Operating above the HCl dew point with metal temperatures below 290 °C (550 °F) can effectively minimize corrosion from HCl.

c. Localized corrosion due to reducing atmospheric conditions. This is avoided by assuring good turbulence with devices such as turbulizers in fire-tube boilers.

The generation of steam is the best-developed method for recovering and using the energy value of wastes. However, steam use is limited to short distances from the generating source and is not storable. Steam can be used in many ways—for example, space heating, process heating, driving turbines for electricity or refrigeration, drying sludge, or for commercial uses such as sterilizers and laundering.

Preheating of incinerator combustion air is also a method of recovering some of the waste heat. Preheating the air allows both high combustion temperatures and reductions in auxiliary fuel to maintain combustion temperatures.

CEMENT KILNS

The combustion of liquid hazardous waste in cement manufacturing kilns has attracted considerable attention. Several firms are currently practicing this form of resource recovery. Burning wastes in cement kilns has a number of advantages:

1. Very high temperatures (3000 °F) and long gas-residence time (10 + s) provide potentially excellent destruction conditions.

2. The heat available in wastes can reduce the fuel required for the process and thus the cost of cement production.

3. The ash content of the waste and the acid gases generated can be effectively absorbed in the cement product if properly maintained within limits.

4. The U.S. EPA has exempted from regulation the disposal of hazardous wastes in cement kilns, avoiding the need to obtain a RCRA permit [2].

There are two basic cement manufacturing processes, wet and dry (Figs. 4-21 and 4-22). The wet process involves preparing raw materials with 30-40% water. It was the original manufacturing process because of the ease of handling and blending of raw materials. However, the requirement to remove this moisture in the kiln creates an energy penalty.

The dry process is not preferred because of advances in handling and blending of dry materials [16]. Eliminating the water from the process has obvious energy-saving benefits.

Portland cement clinker is produced by the controlled high-temperature curing of calcareous material (e.g., limestone, oyster shells), argillaceous material (e.g., clay, shale), and siliceous materials (e.g., sand). These materials provide the basic elements required in cement: calcium, silicon, aluminum, and iron. All elements are usually present in their fully oxidized form in cement clinker. The blended and ground raw materials are fed into a kiln, which is an inclined elongated steel cylinder lined with a refractory brick. Rotary kilns are used exclusively in both the wet and dry manufacturing process in the United States. Kiln lengths can be 60-760 ft with diameters of 6-25 ft.

Raw materials are fed into the raised end of the kiln and travel down the incline to the other end where coal, oil, or gas is burned as a fuel. Chains are employed in most wet-process kilns to air in the transfer of heat from the hot combustion gases to the wet feed. The charge retention time in the kiln is roughly 1-4 h, and the gases experience a retention time of approximately 10 s. Gas temperatures typically reach a maximum of 3000 °F.

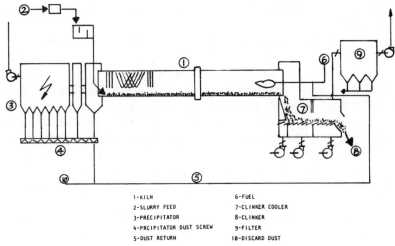

1-KILN	6-FUEL
2-SLURRY FEED	7-CLINKER COOLER
3-PRECIPITATOR	8-CLINKER
4-PRCIPITATOR DUST SCREW	9-FILTER
5-DUST RETURN	10-DISCARD DUST

Figure 4-21. Typical wet process kiln.

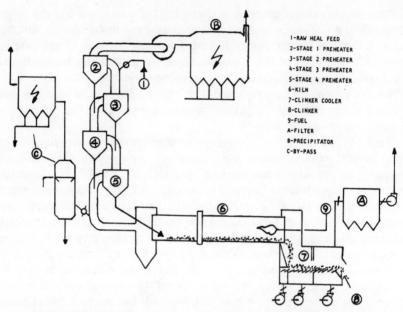

1-RAW HEAL FEED
2-STAGE 1 PREHEATER
3-STAGE 2 PREHEATER
4-STAGE 3 PREHEATER
5-STAGE 4 PREHEATER
6-KILN
7-CLINKER COOLER
8-CLINKER
9-FUEL
A-FILTER
B-PRECIPITATOR
C-BY-PASS

Figure 4-22.   Typical dry process kiln with suspension preheat.

Quality control requires that the clinker be heated to a minimum temperature of 2600 °F. This requirement makes cement kilns attractive as waste combustion devices.

Reactions occurring during conversion of the charge into clinker are the evaporation of free water, evolution of combined water, and evolution of carbon dioxide from carbonates to form the desired clinker compounds.

Cement kilns provide a suitable environment for the complete destruction and assimilation of even the most stable and toxic wastes. The high temperatures, long residence times, and strong turbulence encountered in cement kilns have been shown to destroy waste materials, such as PCBs, to beyond analytical detection limits. In addition, the highly caustic environment of cement kilns acts as an effective scrubbing system for removing acidic residues, such as hydrochloric acid, that are formed during the combustion of chlorinated wastes. The inability of other industries to use halogenated and highly toxic wastes as fuels combined with the demonstrated destructive capacity of cement kilns implies that a tremendous potential exists for using chlorinated wastes as supplemental kiln fuels.

A number of plants have used wastes of relatively low toxicity to supplement their fuel needs. In addition, researchers in Canada, Sweden, and the United States have successfully demonstrated extremely high destruction efficiencies in cement kilns when burning highly toxic organic wastes. These studies indicate that a significant potential exists for the expanded use of cement kilns to safely dispose of many types of hazardous wastes [17-20].

OCEAN INCINERATION

Major interest in the incineration of hazardous waste at sea has developed in the last few years. The United States has significant experience with ocean incineration, beginning in 1974. A number of burns have been successfully completed under the auspices of the Marine Protection Research and Sanctuaries Act (MPRSA) of 1972 (Public Law 92-532), as amended (Public Law 93-254) (results are presented in Table 4-10). This act requires the U.S. EPA to issue permits for ocean disposal of material originating in the United States.

The Convention of the Prevention of Marine Pollution by Dumping of Wastes and Other Matter (1972 London Dumping Convention) was negotiated in 1972 and became effective in August 1975. All nations agreed to regulate ocean dumping through issuing of permits. The Intergovernmental Maritime Consultative Organization (IMCO) is charged with administration of the Act. Guidelines and mandatory requirements for ocean incineration have been added to the 1972 London Dumping Convention. These amendments require, among other things, a minimum combustion efficiency of 99.9% and a minimum combustion temperature of 1200 °C. Ocean incineration may be a viable option for disposing of hazardous materials. A summary of the significant factors that will impact on a decision to burn a waste(s) at sea is as follows: [17]

The only incinerator ship currently operational is the *M/T Vulcanus,* owned by Waste Management Incorporated of the United States. Originally a cargo ship, the *M/T Vulcanus* was converted in 1972 to a chemical tanker fitted with two large incinerators. Two diesel engines drive a single propeller to give cruising speeds of 10–13 knots. The crew numbers 18:12 to operate the vessel and 6 to operate the incinerators. The *Vulcanus* tank capacity is 3500 m³. Waste to be incinerated must be liquid and pumpable, may contain solid substances in pieces up to five centimeters in size, and must not attack mild steel.

Waste is burned in two identical refractory-lined furnaces in the stern. Each incinerator consists of two main sections: a combustion chamber configuration, which uses the first chamber for internal mixing, and a second chamber to provide adequate residence time. Each incinerator can combust a maximum of 12.5 metric tons per hour at temperatures over 1200 °C. The

Table 4-10. U.S. Permitted Burns on the M/T Vulcanus

Date	Waste	Combustion Quant. (metric tons)	Destruct. Eff. (%)	Eff. (%)
1974	Shell Chemical by-product organochlorinated wastes[a]	8,400	NR	99.92–99.98
1977	Shell Chemical by-product organochlorinated wastes[b]	4,100	99.96–99.98	99.92–99.999
1977	Herbicide orange (2,4-D and 2,4,5-T)[c]	10,400	99.983–99.992	99.999

NR = not reported.
[a] T. A. Wastler, C. K. Offut, C. K. Fitzsommons, and P. E. DesRosiers, 1975 (July), *Disposal of Organochlorine Wastes by Incineration at Sea,* U.S. Environmental Protection Agency, Washington, D.C.
[b] J. R. Clausen, et al., 1977 (Sept.), *At-Sea Incineration of Organochlorine Wastes Onboard the M/T VULCANUS,* EPA-600/2-77-196, U.S. Environmental Protection Agency, Research Triangle Park, N.C.
[c] D. C. Ackerman, et al., 1978 (April), *At-Sea Incineration of Herbicide Orange Onboard the M/T VULCANUS,* EPA-600/2-78-086, U.S. Environmental Protection Agency, Research Triangle Park, N.C.

gas residence time is approximately 1 s at 1500 °C. Combustion air is supplied by large fixed-speed blowers with a rates maximum capacity of 90,000 m^3/h for each incinerator. Adjustable vanes are incorporated in the combustion air-supply system that control air flow.

Liquid wastes are fed to the combustion system by electrically driven pumps. Upstream of each burner supply pump is a device (Gorator) for reducing the solids in the waste to a pumpable slurry. The Gorator also acts as a mixing pump by recirculating the waste through the waste tank.

Three burners are located at the same level on the periphery of each furnace near its base. The burners are a rotating cup type, with a concentric design nozzle, in which waste mixes with high-velocity air delivered through an annulus. The ship is equipped to automatically measure the concentrations of CO, CO_2, and O_2 in effluent incinerator gas for each incinerator.

Advantages

1. Ocean-disposal costs can be significantly lower than land-based incineration because scrubbing of the combustion gases is not practiced on the ocean.
2. Ocean incineration can handle large quantities of relatively low-solids-content liquid wastes.

Disadvantages

1. The potential exists for large spills during ship loading operations.
2. A major environmental disaster could occur if a ship loaded with waste should wreck.
3. Current technology cannot handle sludges, solids, and containerized wastes.
4. Solids contents of liquid waste cannot exceed about 1%.
5. Resource recovery cannot be practiced on the high seas.

DESIGN PROBLEM

To illustrate some of the concepts discussed in this chapter, we give a sample design problem.

Problem Statement

The waste stream is generated at a rate of 200 gal/h, 24 h/day. The waste is a water–organic mixture. An analysis of the waste yielded the following data:

Gross heating value	= 3500 Btu/1b
Specific gravity	= 0.93
Water	= 65%
Organics	= 34.6%
carbon	= 69.4%
hydrogen	= 7.6%
oxygen	= 7.5%
chloride	= 15.0%
Solids	= 0.4%
Ash	= 0.2%

Based on this analysis, a significant part of this waste is chlorinated organic compound. As an initial design basis a temperature of 1000 °C (1830 °F) and a gas residence of time of 1.5 s are chosen. These conditions should insure good destruction of the chlorinated wastes.

The high percentage of water in the waste raises the question of potential phase separation in the liquid-feed system. This potential problem dictates a fairly high portion of excess air for a liquid-injection system of 50%. However, the low 3500 Btu/1b gross heating value requires auxiliary fuel (natural gas) to maintain the 1830 °F temperature.

All Calculations Are Based on 1-h Operation

Waste feed rate

200 gal/h × 0.93 × 8.34 lb/gal = 1551.2 lb/h

Compute combustion products (assuming complete combustion)

H_2O from waste

1551.2 lb × 0.65 wt. fraction = 1008.3 lb H_2O

H_2O from waste combustion

1551.2 × 0.346 wt. fraction organic
× 0.076 H_2 in organic phase = 40.8 lb H_2
40.8 lb H_2 × 9.0 lb H_2O/lb H_2 = 367.2 lb H_2O

CO_2 from waste combustion

1551.2 lb × 0.346 wt. fraction
× 0.694 fraction carbon = 372.5 lb C
372.5 lb C × 3.67 lb CO_2/lb C = 1367.1 lb CO_2

HCl from waste

1551.2 lb × 0.346 wt. fraction × 0.15 fraction Cl = 80.5 lb Cl
80.5 lb Cl × 1.03 lb HCl/lb Cl_2 = 82.9 lb HCl

O_2 from waste

1551.2 lb. × 0.346 wt. fraction
× 0.075 fraction O_2 = 40.3 lb O_2

This 40.3 lb of O_2 is subtracted from the O_2 required in the combustion stoichiometric air.

Stoichiometric air required

O_2 required for carbon

372.5 lb C × 2.67 lb O_2/lb C = 994.6 lb O_2

O_2 required for hydrogen

40.3 lb H × 8.0 lb O_2/lb H = 322.4 lb O_2
Subtract O_2 from waste = 40.3 lb O_2
Total stoichiometric O_2 = 1276.7 lb O_2

N_2 for stoichiometric combustion

1276.7 lb O_2 × 3.31 lb N_2/lb O_2 = 4225.9 lb N_2

Summary of Waste Combustion Products

Component	lb
N_2	4225.9
H_2O	1375.5
CO_2	1367.1
HCl	80.5
Ash	3.1

Determine Combustion Products from Natural Gas

Let X = pounds of natural gas (CH_4). The combustion reactions for natural gas are as follows:

$$CH_4 + 2O_2 \rightarrow CO_2 + 2H_2O$$

where CH_4 = 16.0, O_2 = 32.0, CO_2 = 44.0 and H_2O = 18.0 in mole wt.
  CO_2 formed
  $X \times$ 44.0 lb/mole CO_2 + 16.0 lb/mole CH_4 = 2.75 X lb CO_2
  H_2 formed
  $X \times 2 \times$ 18.0 lb/mole H_2O + 16.0 lb/mole CH_4 = 2.25X lb H_2O
  O_2 used by natural gas
  $X \times$ 32.0 lb/mole $O_2 \times 2 \div$ 16.0 lb/mole CH_4 = 4X lb O_2 used
  N_2 from air
          4 lb $O_2 \times$ 3.31 lb N_2/lb O_2 = 13.24 lb N_2
Summary of combustion products from auxiliary fuel
          CO_2 = 2.75X H_2O = 2.25X N_2 = 13.24X
The value of X can be computed by means of a heat balance.

Heat balance
  Heat in
          5.43 $\times 10^6$ (waste heat) + 19,700X (natural gas)
          (1552 lb/h $\times$ 3500 Btu/lb + 19,700X)
  Heat out
    CO_2
  (2.75X + 1367.1) lb $CO_2 \times$ 0.26 Btu/lb CO_2-°F $\times$ (1830°F − 60°F)
          = 462.9X + 0.628 $\times 10^6$ Btu

    N_2
  (4225.9 + 13.24X) lb $N_2 \times$ 0.26 Btu/lb H_2-°F $\times$ (1830°F − 60°F)
          = 1.945 $\times 10^6$ + 6093.0X Btu

    N_2 from 50% excess air
  0.5(1.9435 $\times 10^6$ + 6093.0X) = 0.973 $\times 10^6$ + 3046.5X Btu
    O_2 from 50% excess air
  0.5(1276.7 + 4.0X) $\times$ 0.26 Btu/lb O_2-°F $\times$ (1830°F − 60°F
          = 0.294 $\times 10^6$ + 690.3X Btu

    HCl
    80.5 lb HCl $\times$ 0.20 Btu/lb HCl-°F $\times$ (1830°F − 60°F)
          = 0.028 $\times 10^6$ Btu

  Assume 5% heat loss
    0.05(5.43 $\times 10^6$ + 19,700X) = 0.272 $\times 10^6$ + 985X

H_2O sensible heat
$(1375.3 + 2.25X) \times 0.49$ Btu/lb H 0 °F $\times (1830\,°F - 60\,°F)$
$$= 1.192 \times 10^6 + 1951X \text{ Btu}$$
Latent heat of vaporization of H_2O
$(1375.3 + 2.25X) \times 971$ Btu latent heat/lb
$$= 1.335 \times 10^6 + 2182.5X \text{ Btu}$$

Total Heat Out

Component	Btu
CO_2	$0.628 \times 10^6 + 462.9X$
N_2 (waste and gas)	$1.945 \times 10^6 + 6093.0X$
N_2 (excess air)	$0.973 \times 10^6 + 3046.5X$
O_2 (excess air)	$0.294 \times 10^6 + 690.3X$
HCl	0.028×10^6
H_2O (sensible)	$1.192 \times 10^6 + 1951X$
H_2O (latent)	$1.335 \times 10^6 + 2182.5X$
Heat loss	$0.272 \times 10^6 + 985X$
Total heat out	$6.667 \times 10^6 + 15,411.2X$

Because heat in = heat out, we have
$$5.43 \times 10^6 + 19,700X = 6.667 \times 10^6 + 15,411.2X$$
$$4288.8X = 1.237 \times 10^6$$
$$X = 288.4 \text{ lb}$$
natural gas needed to maintain 1830 °F temperature at 50% excess air

Calculate the Combustion Chamber Volume for 1.5-s Residence Time

	Total Gas Flow (lb)
H_2O (waste)	1375.5
H_2O (natural gas)	648.9
CO_2 (waste combustion)	1367.1
CO_2 (natural gas)	793.1
HCl (waste)	80.5
N_2 (stoichiometric air)	4225.9
N_2 (natural gas)	3819.7
O_2 (excess air)	4022.2
O_2 (waste)	3.1
Total combustion products	17551.2

Assume that the density of air for combustion products is 0.0808 lb/ft³ at standard temperature and pressure (STP).

17551.2 lb ÷ 0.0808 lb/ft³ @ STP = 217,218 ft³ @ STP

At 1830°F we have

217,218 ft³ × (1830 + 460)/460 = 1,011,034 ft³

To compute the chamber volume for a 1.5-s residence time, we calculate as follows:

1,011,034 ft³/h ÷ 3600s/h × 1.5 s = 421.2 ft³ chamber volume

Assuming a 3 : 1 length-to-diameter ratio, we get

$$421.2 = \pi \left(\frac{d}{2}\right)^2 \text{ x } 3d$$

$$d^3 = 178.7$$
$$d = 5.6 \text{ ft}^2$$

The combustion chamber should be approximately 17 ft long by 5½ ft in diameter to give a 1.5-s residence time.

Air-Pollution-Control System

The particulate loading is

3.1 lb × 7,000 gr/lb + 217,218 ft³ @ STP = 0.099 gr/scf

This value is slightly higher than the federal standard and many state standards. However, this level does not warrant a particulate collection device such as a venturi scrubber or an ESP device.

For HCl loading, 80.5 lb of HCl are expected. The emission level needs to be below 4 lb/h, which translates into a minimum 95% removal efficiency for a scrubbing device.

Based on the above, the following design is selected; a water quench followed by a packed water scrubber. This was selected because the particulate level is relatively low, and only a 95% removal efficiency for HCl is needed. The quench system will absorb much of the HCl, and the packed tower will absorb the remainder.

Calculate the Quench Water Needed to Cool Gases to 200°F (93°C)

Heat Balance

Let y = pounds of water at 60°F required.

y lb H_2O × 1.0 Btu/lb H_2O × (212 − 60)°F = 152y Btu required
to raise water to 212°F

Heat in (Btus removed from gases)

	lb		C_p		T		Btu/h
CO_2	2160.2	×	0.26	×	1630	=	0.915 × 10^6
HCl	80.5	×	0.20	×	1630	=	0.0262 × 10^6
N_2	12067.8	×	0.29	×	1630	=	5.704 × 10^6
O_2	1215.2	×	0.26	×	1630	=	0.515 × 10^6
H_2O (sensible)	2024.4	×	0.49	×	1618	=	1.605 × 10^6
H_2O (latent)	2024.4	×	1.0	×	12	=	0.0243 × 10^6
			Total Btu in gas into quench			=	8.79 × 10^6

$$y \times 0.49 \times (212 - 200)\,°F = 5.88y \text{ Btu}$$
given up to get water vapor down to 200 °F
$$y \times 971 \text{ Btu/lb } H_2O = 971y \text{ Btu required to vaporize water}$$

Because heat removed from gases = heat absorbed by quench water, we get
$$8.8 \times 10^6 = 152y + 971y - 5.88y$$
$$y = 8.8 \times 10^6 \div 1117.12$$
$$y = 7877 \text{ lb } H_2O \text{ or } 944 \text{ gal } H_2O$$

The total gas rate out of quench is
$$17,551 \text{ lb gas} + 7877 \text{ lb } H_2O = 25,428 \text{ lb gases (wet)}$$

Assuming density of air = 0.0808 lb/ft^3 at STP gives
$$25,428 \text{ lb} \div 0.0808 \text{ lb/ft}^3 = 314,702 \text{ ft}^3 \text{ @ STP}$$

Correcting to 200 °F gives
$$314,702 \text{ ft}^3 \times (200 + 460)/460 = 451,150 \text{ ft}^3 \text{ @200 °F}$$

For a 0.6-s residence in tower, we get
$$125.3 \text{ ft}^3/\text{s} \times 0.6 \text{ s} = 75.2 \text{ ft}^3$$

We now calculate the diameter of the tower that will have a 7-ft/s superficial gas velocity: 125.3 ft^3/s ÷ 7 ft/s = 17.9 ft^2 area

$$d = 2\sqrt{\frac{A}{\pi}} = 4.77 \text{ ft diameter}$$

The height of the contact section is
$$75.2 \text{ ft}^3 = 17.9 \text{ ft}^3 \times h$$
$$h = 4.2 \text{ ft}$$

The liquid-to-gas rate typically is 0.0299 gas/acf of gas. Hence,
$$0.0299 \text{ gas/acf} \times 451,150 \text{ ft}^3 = 13,489 \text{ gal/h}$$

The scrubber system has a total flow rate of 13,489 gal/h through the tower. Most of this water would be recirculated and a blowdown stream would be controlled to a pH that would be acceptable to a water treatment system (3-4 pH). A stack height for a system of $10-12 \times 10^6$ Btu/h would be typically about 20-25 ft above grade.

REFERENCES

1. E. Keitz, L. Boberschmidt, D. O'Sullivan, and N. Sanders, 1982 (April 23), Hazardous Waste Incineration—A Profile of Existing Facilities, unpublished draft report by the Mitre Corporation to the U.S. Environmental Protection Agency, Washington, D.C.
2. Anonymous, 1981 (Jan 23), Incinerator standards for owners and operators of hazardous waste management facilities; interim final rule. *Code of Federal Regulations* Title 40, Pt. 264, Subpt. O.
3. J. Corini, C. Day, and E. Temrowski, 1980 (Sept. 2), Trial Burn Data (unpublished draft), Office of Solid Waste, U.S. Environmental Protection Agency, Washington, D.C.
4. P. Gorman and K. P. Ananth, 1982 (March), Trial Burn Protocol Verification: Performance Evaluation and Environmental Assessment of the Cincinnati Metropolitan Sewer District (MSD) Hazardous Waste Incinerator (unpublished draft), report by the Midwest Research Institute to the U.S. Environmental Protection Agency, Washington, D.C.
5. R. Esposito, 1982, The Burning Issue of Disposing of Hazardous Waste by Thermal Incineration, paper presented at the Air Pollution Control Association Meeting Conference, Neward, N.J., April 29-30.
6. D. Austin, R. Bastian, and R. Wood, 1982, Factors Affecting Performance in a 90 Million Btu/hr Chemical Waste Incinerator: Preliminary Findings, paper presented at the Air Pollution Control Association Conference on the Burning Issue of Disposing of Hazardous Waste by Thermal Incineration, Neward, N.J., April 29-30.
7. U.S. Environmental Protection Agency, 1981, *PCB Disposal by Thermal Destruction,* PBB2-241-860, Dallas, Tex.
8. L. G. Doucet, 1979, *Waste Incineration: Selection and Implementation of Industrial, Commercial and Institutional Systems,* Vanderbilt University, Nashville, Tenn.
9. L. Manson, and S. Unger, 1979 (Oct.), *Hazardous Material Incinerator Design Criteria,* EPA-600/2-79-197, U.S. Environmental Protection Agency, Cincinnati, Ohio.
10. R. H. Barnes, R. E. Barrett, A. Levy, and M. J. Saxton, 1979 (April), *Chemical Aspects of Afterburner Systems,* PB-298465, U.S. Environmental Protection Agency, Durham, N.C.
11. S. Calvert, J. Goldshmid, D. Leith, and D. Mehta, 1972 (Aug.), *Scrubber Handbook* vol. 1, *Wet Scrubber System Study,* EPA-R2-72-118a, U.S. Environmental Protection Agency, Durham, N.C.

12. D. Hitchcock, 1979, Solid waste disposal: Incineration, *Chemical Energy* **86**(11):185–194.
13. C. J. Wall, J. T. Graves, and E. J. Roberts, 1975 (April 14), How to burn salty sludges, *Chemical Energy* p. 77.
14. Personal communication with Phillip Beltz, Battle Memorial Interstate, July 6, 1982.
15. E. Bakke, 1975 (Feb.), Wet electrostatic precipitators for control of submission particles, *Air Pollution Control Association Journal* **25**(2):163–167. (See also *Code of Federal Regulations* Title 40, May 19, 1980.)
16. D. L. Hazelwood, and F. J. Smith, 1981 (March), *Assessment of Waste Fuel Use in Cement Kilns,* U.S. Environmental Protection Agency, Cincinnati, Ohio.
17. E. E. Berry, L. P. MacDonald, and D. J. Skinner, 1975 (June), *Experimental Burning of Waste Oil as a Fuel in Cement Manufacture,* EPS 4-WP-75-1, Environment Canada, Toronto.
18. L. P. MacDonald, D. J. Skinner, F. J. Hopton, and G. H. Thomas, 1977 (March), *Burning Waste Chlorinated Hydrocarbons in a Cement Kiln,* EPS 4-WP-77-2, Fisheries and Environmental Canada, Toronto.
19. Swanson Environmental, Inc., 1976 (Dec.), Peerless Cement Company, Detroit, Michigan.
20. B. Ahling, 1979 (March 16), *Destruction of Chlorinated Hydrocarbons in a Cement Kiln at Stora Vika Test Center,* Swedish Water and Air Pollution Research Institute, Stockholm.

5

Storage of Hazardous Waste

Roy Ball
ERM North Central, Inc.
Park Ridge, Illinois

James H. Johnson, Jr.
Howard University, Washington, D.C.

This chapter describes procedures for locating, sizing, operating, and closing hazardous-waste storage facilities. The reader should be aware of the extensive body of state and federal regulations that define and restrict these activities. Rather than duplicate the text of those regulations, we present material that addresses unregulated aspects of hazardous-waste storage and summarizes the regulatory requirements. In all cases it is essential that the reader refer to the most current text of the regulations (available from the *Federal Register* or the *Bureau of National Affairs Environment Reporter*) before finalizing any applications to state or federal authorities (also refer to chap. 1).

The most critical element in the success and cost of a storage facility is its location. The location will directly influence the construction cost and permit-ability of the site. The location will also affect, more indirectly, the operating and closure costs. The designer therefore must carefully consider several-score factors in the location process. There exists at present no official or widely used procedure for location analysis. In lieu of such consensus a locational procedure is presented herein that incorporates regulator guidance, the rating of abandoned-site hazard, and technical experience from hazardous-waste management. The procedure uses principles of indexing and scoring to develop a multiplicative-additive index for a proposed storage facility site. This ordinal index can be used to compare alternative sites, evaluate the purchase of an existing facility, and assess the "permit-ability" of a site.

LOCATING STORAGE FACILITIES

The criteria for locating storage facilities fall into three major categories: economic, regulatory, and social. Failure to satisfy the explicit or implied requirements of each category will generally result in an unsatisfactory facility—that is, one that cannot be permitted or operated. The locational criteria relevant to each category are described individually in this section.

Economic Location

Elaborate mathematical techniques for the optimal location of service and nuisance facilities have been developed for operations research. Many of these methods require computer facilities to calculate such locations, especially if multiple generators and disposers are to be integrated with many storage facilities. A simpler case, which is almost always that encountered in practice, is to locate one storage facility for a number (z) of generators and a smaller number (n) of disposal facilities. Even with the additional complexity that each generator will produce some amount of m components, the problem remains tractable as long as only one storage facility is to be located or, even more usual, one of several alternative sites is to be selected.

For example, let us assume the latter case—a small number of alternative storage facility sites are to be analyzed to determine which could be operated at the lowest total cost, regulatory and social constraints not considered. Let us further define the existence of z generators, each with varying amount of up to m components and the existence of n disposal facilities. Therefore, for the ith generator and the ith component, we may define W_{ij} as the amount (tons/month) of hazardous waste produced. We next need to define the cost of transporting the generated hazardous waste to the storage facility. Commercial hazardous-waste transporters typically provide prices in dollars per ton (or cubic yard) per mile. Let us therefore define C_{ij} as the dollars per ton per mile for transporting the ith component from the ith generator. In many cases C will be a function not of the generator location but only of the component properties—that is, containerized liquid, solid, semisolid, or gaseous wastes. There are cases, however, where access to the generator presents special difficulties or conditions, and there is the possibility of a higher cost due to those constraints; therefore C_{ij} rather than C_j is used.

Finally, the distance, measured according to the transporter's route, to the storage facility must be defined. For this example, D_i is the miles to a proposed storage facility location from the ith generator. The total cost therefore of transport to the storage facility is given in Eq. (5-1):

$$TC_{GS} = \sum_{i=1}^{z} \sum_{j=1}^{m} W_{ij} C_{ij} D_i \qquad \textbf{(5-1)}$$

The next cost area to be considered is the cost of transport from the storage facility to any of the n disposal facilities and the attendant cost of disposal (defined as A_{jk} in dollars per ton). For the jth component there will be a cost in collars per ton per mile for transport from the storage facility to the disposal facility, herein defined as C_{jk}. As with the transport to the storage facility, the amount of hazardous waste will be defined as W_{jk}, and the distance to the kth storage facility as D_k. The cost, therefore, of transport to the disposal facility and subsequent disposal is given by Eq. (5-2), with the constraint that, over time,

$$\sum_{i=1}^{z} \sum_{j=1}^{m} W_{ij} = \sum_{j=1}^{m} \sum_{k=1}^{n} W_{jk}$$

$$TC_{SD} = \sum_{k=1}^{n} \sum_{j=1}^{m} (W_{jk} C_{jk} D_k + A_{jk} W_{jk}) \qquad (5\text{-}2)$$

Clearly, an equation similar to Eq. (5-2) could be written for each component for each disposer, and the number of such combinations could become excessive. However, in practice, not all disposers receive all components of hazardous waste, and the matrix W_{jk} is normally rather sparse due to limitations on the number of waste components being handled at a given storage facility. Therefore, even with a large number (greater than 5) of disposal facilities under consideration, the number of cases is usually sufficiently small to allow the enumeration of each condition and direct comparison of potential cost of storage-disposal transport and subsequent disposal.

The third cost element is related to the cost of storage: The costs are typically in three groups. The first is the cost of actual storage of the material (B_j), which may be expressed in dollars per ton per month. The second group of costs is related to the receipt of shipments and their initial handling at the storage facility (O_{ij}), where O is the shipments of j received per month, and may be expressed in dollars per ton. The third group of costs is related to the preparation and dispatch of shipments from the storage facility to the various disposal facilities (P_{jk}), where P is the shipments of j sent per month, and may also be expressed in dollars per ton. The cost of these processes is summarized in Eq. (5-3).

$$TC_s = \sum_j \left(\sum_i O_{ij} \right) E_j + \sum_j \left(\sum_k P_{jk} \right) F_j + \sum_j \left(\sum_i W_{ij} \right) B_j T_j \qquad (5\text{-}3)$$

where T_j is the weight-averaged storage time associated with material j, and E_j and F_j are the costs of receiving and sending a shipment of j, respectively.

Because the cost provided by transporters and disposers is on a unit basis, there is normally little incentive to send the same component to more than one disposal facility. It is possible, under some circumstances, for the partitioning of a component among several disposal facilities to be economically attractive. In those relatively rare cases the degree of partitioning may be readily determined by simple linear programming techniques. In the majority of all cases, however, this is unnecessary, and the simple enumeration of the possible combinations and the calculation of the cost for each combination according to the above equations is more than sufficient for the economic selection of a storage facility location. These calculations can also be facilitated through microcomputer electronic spreadsheet software.

To further illustrate these concepts, we give an example. Four generators are assumed, located in Chicago (ORD), St. Louis (STL), Milwaukee (MKE), and St. Paul (MSP). Six alternative storage facility locations are to be evaluated: one at each generator, one at Madison (MSN), and one at Cedar Rapids (CID). Three disposal facilities are assumed: one at Chicago (ORD), one at Cincinnati (CIN), and one at Niagara Falls (BUF).

Matrix 1 in Table 5-1 lists the generator factors W_{ij} and C_{ij} and defines the j vector. Using the distance vector D_i for each storage facility location, we can calculate TC_{Gs} from Eq. (5-1). Matrix 2 lists the disposal factors W_{jk}, C_{jk}, and A_{jk}. Again, using the distance vector D_k for each storage facility location, we can calculate TC_{SD} from Eq. (5-2).

Matrix 3 lists B_j, E_j, and F_j. Using the generator and disposer load capacities (provided by the transporter) we can calculate

$$\sum_i O_{ij} \text{ and } \sum_k P_{jk},$$

respectively. Note that $T_j = P_{kl}$; TC_s is calculated using Eq. (5-3).

The cost data are summarized in Matrix 4, which indicates the optional location by simple enumeration.

Regulatory Location

The regulatory location of a storage facility is implicit in the permitting process. That is, facilities that cannot meet the regulatory standards will receive neither a construction permit nor an operating permit based on federal and state laws. (These constraints are discussed in greater detail in a later section). It is sufficient to say here that the considerations included in the regulatory location of a storage facility on the federal level include the following: physical security or isolation; separation from sources of ignition or reaction (for storage of ignitible or reactive wastes); avoidance of active seismic areas and floodplains; groundwater protection (does not apply to

Table 5-1. Cost Evaluation of Alternative Storage Facility Location

MATRIX 1: GENERATOR FACTORS

Generator Amount (W_{ij} = tons/month) and Cost Factors (C_{ij} = \$/ton/mile)

$j \backslash i$		*ORD* l	*STL* 2	*MKE* 3	*MSP* $4 = l$	W_{ij}
1	W	5	1	0	2	8
	C	12	15	—	15	
2	W	3	1	0	2	6
	C	12	15	—	15	
3	W	0	8	0	3	11
	C	—	20	—	25	
4	W	1	0	0	0	1
	C	16	—	—	—	
5	W	8	0	20	10	38
	C	40	—	50	40	
$m = 6$	W	0	2	10	10	22
	C	—	10	8	10	

Alternative Storage Facility Distance in Miles

Facility $\backslash i$	*ORD* l	*STL* 2	*MKE* 3	*MSP* $4 = l$	TG_{GS} (\$/month)
ORD	0	289	87	410	4,150
STL	289	0	376	630	9,310
MKE	87	376	0	337	3,306
MSP	410	630	337	0	6,734
MSN	140	375	77	272	3,951
CID	242	280	170	217	5,190

MATRIX 2: DISPOSAL FACTORS
(NA = not accepted at that site)

$j \backslash i$		*ORD* l	*CIN* 2	*BUF* 3	W_{jk}
1	W	NA	8	NA	8
	C	—	15	—	
	A	—	80	—	
2	W	6	0	0	6
	C	12	15	15	
	A	50	80	80	
3	W	11	0	0	11
	C	20	20	25	
	A	20	20	15	
4	W	NA	0	1	1
	C	—	15	15	
	A	—	60	40	
5	W	NA	NA	38	38
	C	—	—	12	
	A	—	—	70	
$m = 6$	W	10	10	2	22
	C	10	10	10	
	A	15	15	15	

Table 5-1. *(continued)*

Alternative Disposal Facility Distance in Miles

MATRIX 2: DISPOSAL FACTORS (continued)

Facility k	ORD 1	CIN 2	BUF 3	Transport	TC_{SD} disposed	($/month) total
ORD	50	289	522	3,395	+ 4,150 =	7,585
STL	289	340	716	5,396	+ 4,150 =	9,586
MKE	87	374	6 09	4,154	+ 4,150 =	8,344
MSP	410	692	917	7,681	= 4,150 =	11,871
MSN	140	430	670	4,785	= 4,150 =	8,975
CID	242	380	900	6,204	= 4,150 =	10,394

MATRIX 3: STORAGE FACTORS

j	B_j ($/ton/month)	E_j ($/shipment)	F_j ($/shipment)	W_{ij}
1	2	100	100	8
2	1	100	100	6
3	3	100	100	11
4	1	100	100	1
5	5	100	100	38
$m = 6$	3	100	100	22

Generator Load Capacity (tons/shipment) and Shipments/Months $(O_{ij})^a$

j \ i		ORD 1	STL 2	MKE 3	MSP 4 = 1	$\sum_j O_{ij}$	$\sum_j (O_{ij})E_j$
1	Cap.	5	5	10	20		
	O_{i1}	1.0	0.33	0	0.33	1.68	167
2	Cap.	5	5	10	20		
	O_{i2}	0.6	0.33	0	0.33	1.26	126
3	Cap.	20	20	20	20		
	O_{i3}	0	0.40	0	0.33	0.73	73
4	Cap.	10	10	10	20		
	O_{i4}	0.33	0	0	0	0.33	33
5	Cap.	1	1	1	1		
	O_{i5}	8.0	0	20.0	10.0	38.0	380
$m = 6$	Cap.	0	10	10	10		
	O_{i6}	0	0.33	1.0	1.0	2.33	233

$$\sum_j (\sum_i O_{ij})E_j = 1012$$

Table 5-1. *(continued)*

MATRIX 3: STORAGE FACTORS (continued)

Disposer Load Capacity (tons/shipment), Shipments/Month P_{jk}
and Storage Duration (T_j)

j	k	ORD *1*	CIN *2*	BUF *3 = n*	$\sum_k P_{jk}$	$(\sum_k P_{jk})F_j$	T_j	$(\sum_i W_{ij})B_jT_i$
1	Cap.	20	20	50				
	P_{1k}	0	0.4	0	0.4	40	2.5	40
2	Cap.	20	20	50				
	P_{2k}	0.3	0	0	0.3	30	3.3	20
3	Cap.	20	20	50				
	O_{3j}	0.55	0	0	0.55	55	1.8	60
4	Cap.	20	20	50				
	P_{4k}	0	0	0.02	0.02	2	50	50
5	Cap.	20	20	50				
	P_{5k}	0	0	0.76	0.76	76	1.3	247
$m = 6$	Cap.	20	20	50				
	P_{5k}	0.5	0.5	0.04	1.04	104	1.0	66
						307		417

MATRIX 4: COST SUMMARY

Facility	TC_{GS}	TC_{SD}	TC_s	TC
ORD	1,629	7,585	1,747	16,955
STL	5,435	9,586	1,747	24,116
MKE	1,233	8,344	1,747	18,869
MSP	6,734	11,871	1,747	23,825
MSN	2,278	11,871	1,747	23,145
CID	3,524	10,394	1,747	20,800

Note: Therefore, the preferred location (based on cost only) for the storage facility is Chicago, Milwaukee, or Madison.

[a]Note that O_{ij} must be greater than or equal to 0.33 to avoid excessive accumulation at a generator (RCRA regulations).

all storage facilities); potential cost of closure; ability to construct an impervious base and associated containment system; and ability to handle runoff and runon associated with the facility.

In general, any facility not located in a floodplain or seismic area is potentially capable of certification and permitting. Although many states add additional requirements such as minimum separation distance to existing wells, depth to bedrock, depth to groundwater, and allowable soil type, these are seldom implacable barriers to locating a storage facility.

Social Considerations

Any facility used in hazardous-waste management is invariably regarded by the public as a nuisance facility—that is one recognized as necessary but should be located elsewhere. The response, therefore, to the potential location of a storage facility is invariably negative and can only be converted into, at best, resigned neutrality by close coordination with local officials and scrupulous adherence to regulatory requirements. Location constraints of great importance to public neighbors, which are not usually explicit regulatory requirements, include such things as visual impact of the facility, dust, noise, and litter. A site that is, or can be, screened by natural elements such as plantings or other landscaping is generally tolerated far better than a facility in open view. Terrain and vegetation facilitating such screenings point to more initially favorable locations. Locating a faciilty near public institutions such as schools or hospitals is also a significant social criterion. Such nearness, in fact, is generally totally unacceptable to the affected public and may represent a serious deficiency in an otherwise suitable site. Even if a site is well screened and does not suffer from proximity to institutions or to established residential developments, it may still be effectively unacceptable if the access to the site suffers from exposed view or proximity. The perceived hazard of a facility to the public is often related to the route that the transportation equipment follows. If, for example, the site can only be accessed via local roads passing directly through small towns, residential areas, and/or adjacent to public institutions, the perceived risk of such transportation all but precludes acceptance of the site. If, on the other hand, the site is accessible from major primary or secondary roads and does not pass through residential or institutional areas, then the perceived risk is low and should not provide a barrier for site acceptability.

In general, therefore, this restricts the location of storage facilities to within approximately one mile of major primary or secondary highways. Even then, acceptability is marginal unless the access road passes through either undeveloped or industrial areas and is not in direct view of residential of public institutions.

Finally, a serious element of perceived risk arises if the transportation route in or out of the storage facility crosses a surface water stream of public importance (e.g., one that is used downstream as a major water supply). The perceived risk is the possibility of a transportation accident releasing hazardous materials directly into the water course without warning. Therefore, consideration must be given to developing alternative routes to a storage facility if such a situation exists. The perception of such an event as an unacceptable risk is not uniform but is serious enough to carefully consider it as a negative factor in locating a storage facility.

Site selection is discussed in greater detail in chapter 8.

FACILITY INDEX

It is recommended in this section that the criteria for evaluating alternative sites for storage facilities be developed from at least three categories: economic, regulatory, and social. As discussed in some detail in a later section, the preparation and negotiation of a permit for a storage facility may be a very time-consuming and costly process, and there is no guarantee that a permit will be obtained for a particular site. Given the cost and complexity fo the permitting process, therefore, it is not always prudent to attempt to permit only the most economically attractive site. The economic analysis is only operative in the event the site can be successfully developed and operated, which therefore requires that it receive both construction and operating permits from state and federal agencies. The overall evaluation of alternative sites should therefore also include consideration of the regulatory and social criteria. For this reason a multiplicative-additive index is proposed with the index for any potential site consisting of three scores—economic, regulatory, and social—each of which is determined as described below. Each score has been scaled such that 0 is the minimum and least desirable score, 100 is the maximum and most desirable score, and scores above 60 should receive the most consideration as feasible sites.

Economic Score

For n alternative storage facility locations let us define the total cost, TC, for the ith facility as follows, where the definitions are given in Eq. (5-1)–(5-3).

$$TC_i = (TC_{GS})_i + (TC_{SD})_i + (TC_s)_i \tag{5-4}$$

The facility economic score associated with that cost (S_E) may be defined as

$$(S_E)_i = 100 \times \frac{TC_{max} - TC_i}{TC_{max} - TC_{min}} \tag{5-5}$$

where TC_{max} and TC_{min} are the maximum and minimum values of TC_i, respectively, for $i \geqslant n$.

Regulatory Score

The ith-facility regulatory score (S_p) should consider both risk and exposure elements. The U.S. Environmental Protection Agency (EPA) has promulgated a scoring procedure in the National Oil and Hazardous Substances Contingency Plan that considers many such factors. Therefore, a modification of that scoring procedure, shown below, may be used to develop the regulatory score. The rating curves (subscore v. scoring factors) are provided in Appendix 5-1. The regulatory score is defined as

$$(S_R)_i = SS_{CR}(1/3(S_{GW}^2 + S_{SW}^2 + S_A^2))^{1/2} \tag{5-6}$$

where

$$
\begin{aligned}
S_{GW} &= \text{groundwater score} \\
S_{SW} &= \text{surface-water score} \\
S_A &= \text{air score}
\end{aligned}
$$

and

$$S_{GW, SW} = f(SS_R, SS_W, SS_T) \tag{5-7}$$

$$S_A = f(SS_W, SS_T) \tag{5-8}$$

where SS refers to subscore and the subscripts R, W, T, and CR refer to evaluation of routes for transport of spilled or leaked material, waste characteristics, target, and critical regulatory factors, respectively.

Social Score

The ith-facility social score $(S_S)_i$ includes consideration of the following factors: visibility (SS_V), access (SS_A), proximity to socially sensitive features (SS_P), and sensitivity of transportation routes vis-a-vis surface waters (SS_S). Appendix 5-2 contains the rating curves for this scoring procedure.

$$S_S = \left(1/4\,(SS_V^2 + SS_A^2 + SS_P^2 + SS_S^2)\right)^{1/2} \tag{5-9}$$

The facility index therefore includes consideration of criteria from each locational category. The index I_i is constructed as follows for the ith facility:

$$I_i = S_C\left(W_E(S_E)_i + W_R(S_R)_i + W_S(S_S)_i\right) \tag{5-10}$$

where W refers to the weights associated with each score such that $(W_E + W_R + W_S) = 1$. To avoid eclipsing of critical information, the score S_C has the following characteristics:

$$
\begin{aligned}
S_C &= 1 \quad \text{if } SS_{CR} = 1 \\
S_C &= 0 \quad \text{if } SS_{CR} = 0
\end{aligned}
\tag{5-11}
$$

The index is therefore constrained between 0 and 100, with index values above 60 associated with the most suitable facility locations. A facility location with an index of 0 should be eliminated from further consideration. Table 5-2 gives an example to illustrate the potential of the index.

Table 5-2. Indexing of Alternative Storage Facility Locations

Economic score TC_{max} = 24,116[a] and TC_{min} = 16,955[a]

Storage Facility	TC_i	S_E
ORD	16,955	100
STL	29,116	0
MKE	18,869	74
MSP	23,825	4
MSN	18,145	84
CID	20,804	46

Required Data for Regulatory Score

	ORD	STL	MKE	MSP	MSN	CID
Base data S_{GW}						
Depth to aquifer (ft)	300	50	300	300	100	400
Net precip. (in)	+3	+4	+3	+4	+2	-1
Logio perm. (cm/s)	10^{-4}	10^{-4}	10^{-4}	10^{-3}	10^{-4}	10^{-7}
Physical state of stored material	liquid					
Waste character						
Hazard (sax rating)	3					
Stored substances	halogenated					
Stored quantity [ton(s)]	86					
Targets						
Groundwater use	Comm.	Alt. Drink	Unusable	Sole Drinking	Comm.	Unusable
Dist. to wells (mi)	2	1	1	1	2	0.5
Pop. served	10^3	10^5	0	10^6	10^4	0
Base data S_{SW}						
Dist. to SW (mi)	10	0.5	1.5	5	0.25	10
1-yr 24-hr rain (in)	2.5	3.0	2.1	2.2	2.5	2.7
Terrain slope (%)	1	6	2	4	9	1
Facility slope	1	6	2	4	6	1

[a] Values are from Table 5-10.

(continued)

Table 5-2. (continued)

	ORD	STL	MKE	MSP	MSN	CID
Targets						
Water use	Ind.	Drinking	Ind.	Drinking	Recrea-tion	Not used
Pop. served	10^6	10^4	10^5	10^3	10^5	0
Dist. to wetland (mi)	10	0.5	10	0.5	0.5	10
Dist. to erd. species (mi)	10	10	1	0.5	0.25	10
Base data S_A						
Reactivity				4		
Compatibility (%)				50		
Targets						
Pop.	5×10^6	10^6	10^5	10^6	10^5	10^4
Dist. to pop. (mi)	2	4	2	4	1	4
Dist. to industry (mi)	0.25	0.25	0.25	1	1	1
Dist. to parks and prime ag (mi)	2	1	0.5	1	0.25	5

Calculation of Regulatory Score

		ORD	STL	MKE	MSP	MSN	CID
GW:	SS_{R_1}	100	30	100	100	65	100
	SS_{R_2}	45	40	45	40	50	65
	SS_{R_3}	25	25	25	0	25	100
	SS_{R_4}	0	0	0	0	0	0
	SS_R	54	25	54	48	41	73
	SS_{W_1}	0	0	0	0	0	0
	SS_{W_2}	65	65	65	65	65	65
	SS_W	20	20	20	20	20	20
	SS_{T_1}	65	35	100	0	65	100
	SS_{T_2}	55	20	90	20	45	85
	SS_T	57	23	92	16	49	88
	S_{GW}	39	19	46	25	34	50

SIZING STORAGE FACILITIES

Nowhere in environmental practice is the availability of a safety factor potentially more valuable than with a hazardous-waste storage facility. Therefore, it follows that great care should be taken in estimating the capacity that should be made available for each hazardous-waste component. The safety factor

Table 5-2. *(continued)*

Calculation of Regulatory Score (continued)

		ORD	STL	MKE	MSP	MSN	CID
SW:	SS_{R_1}	100	15	75	100	0	100
	SS_{R_2}	40	25	45	40	35	30
	SS_{R_3}	100	33	100	70	20	100
	SS_{R_4}	0	0	0	0	0	0
	SS_R	68	18	59	62	11	66
	SS_{W_1}	0	0	0	0	0	0
	SS_{W_2}	65	65	65	65	65	65
	SS_W	20	20	20	20	20	20
	SS_{T_1}	66	0	66	0	33	100
	SS_{T_2}	100	10	5	100	0	100
	SS_{T_3}	100	75	100	55	40	100
	SS_T	93	15	27	76	11	100
	S_{SW}	50	15	32	46	13	51
Air:	SS_{W_1}	0	0	0	0	0	0
	SS_{W_2}	0	0	0	0	0	0
	SS_{W_3}	65	65	65	65	65	65
	SS_W	26	26	26	26	26	26
	SS_{T_1}	10	35	30	35	10	50
	SS_{T_2}	100	75	100	55	40	100
	SS_{T_3}	25	25	25	50	10	100
	SS_T	25	40	40	40	15	63
	S_A	25	32	32	32	20	40
	SS_{CR}	1	1	1	1	1	1
	S_R	39	23	37	35	24	47

(continued)

applied to that capacity should be developed in a generous and conservative manner.

Although the reasons for this are presumably self-evident, they are worth repeating. The first constraint is that waste may not be legally stored at a generator planning off-site disposal for more than 90 days. Any shortfall, therefore, in storage capacity may necessitate direct disposal from the generator, which, on a short-term basis, will invariably be at a much greater cost than if the material is stockpiled and disposed of in larger lots from disposal facilities. Secondly, the manifest system requires commitment of

Table 5-2. (continued)

Required Data for Social Score

	ORD	STL	MKE	MSP	MSN	CID
Sight dist. (mi)	0.25	0.5	0.25	1	0.5	2
Captive pop.	10^2	10^4	10^2	10^1	10^4	10^1
Dist. to captive facilities (mi)	2	1	4	2	0.5	5
Dist. to surf. crossing (mi)	2	1	1	0.5	0.25	5

Calculation of Social Score

SS	10	25	10	100	25	100
SS_A	25	0	25	50	0	50
SS_P	100	25	100	100	10	100
SS_S	100	50	50	25	12	100
S_S	72	31	58	76	15	90

Calculation of Site Index

$$\text{Assume } W_E = 2W_R = 3W_S$$
$$0.55 = 2(0.27) = 3(0.18)$$
$$W_E + W_R + W_S = 1$$
$$S_C = 1$$

Facility	$W_E S_E$	$W_R S_R$	$W_S S_S$	I
ORD	55	11	13	79
STL	0	6	6	12
MKE	41	10	10	61
MSP	2	9	14	55
MSN	46	6	3	55
CID	25	13	16	54

Note: Therefore, Chicago (ORD) and Milwaukee (MKE) are the most favorable alternative sites.

a shipment to a particular disposal facility with the option to name an alternate. If, for some reason, both the primary facility and the alternate cannot accept the load, then the material must be returned to its origin. Here again, the availability of capacity at the storage facility allows a far greater liability and risk of holding the rejected material until either its properties are modified so that it becomes acceptable or until an alternative disposal facility is selected.

The principal basis for sizing, in terms of total capacity, follows from the presentation in the previous sections dealing with the economic location of

Table 5-3. Routing Matrix for *j* = 6

Disposal Facility	*i*	ORD 1	STL 2	MKE 3	MSP 4 = 1	W_{jk}
ORD *k* = 1		0	2	5	3	10
CID 2		0	0	5	5	10
BUF 3 = *n*		0	0	0	2	2
W_{ij}		0	2	10	10	22

storage facilities. Developing the information in the matrices W_{ij} and W_{jk} implicitly defines the minimum size of the storage facility for that component. Therefore, for any component *i,* a matrix similar to that in Table 5-3 (using data in Table 5-1) can be developed, where there are *k* rows and *i* columns. Any matrix element, therefore, represents the fraction of W_{ij} going to disposer 1. The marginal total of each column is W_{ij}, and the marginal total of each row is W_{jk}. The sum of either the row or column totals is the total amount of that component in the system. The final factor that must be defined is the frequency of shipments, both into and from the storage facility. This frequency is best defined in terms of hte optimal capacity as defined by the transporter. If, say, 200 tons per month of component *j* are produced from generator *i* and the economical load from the transporters is 22 tons per month from the generator to the storage facility and 50 tons per month from the storage facility to the disposal facility, then there will be 10 loads per month in and 4 loads per month out. The minimum storage capacity that must be provided is equal to the transport capacity from the storage facility to the disposal facility for component *j* . This must then be increased to allow for the lag time between the accumulation of a disposal load and the arrival of generator shipments, and the reliability of the disposal facility and the disposal transport equipment. In practice, these calculations are usually overridden by the considerations mentioned previously—that is, the economic and regulatory penalties as well as the liability considerations in the event storage capacity is insufficient. For most wastes prudence dictates that a minimum of six months capacity, in addition to that calculated as described above, be provided for a hazardous-waste component.

CONCEPTUAL DESIGN OF FACILITIES

Storage facilities are defined by federal regulations to be either containers, tanks, surface impoundments, or waste piles, and regulations have been pro-

Table 5-4. Federal RCRA [a] **Regulations** [b] **by Waste Type and Storage Process**

Storage Process \ Waste Type	General	Ignitible or Reactive	Incompatible	Wastes with Free Liquid and Autoleachable
All	264.1–4 264.10–16 264.30–37 264.50–56 264.70–77 264.110–120 264.140–151	264.17	264.17	—
Containers	264.170–178	264.176	264.177	—
Tanks	264.190–199	264.198	264.199	—
Surface impoundments	264.90–100	264.229	264.230	—
Waste piles	264.9–100	264.256	264.257	264.250

[a] RCRA = Resource Conservation and Recovery Act.
[b] Numbers are 40 CFR part and subpart.

mulgated by those categories. Once the types of materials (hazardous-waste components to be handled at the facility) are defined, then the first step in the conceptual design is to determine alternatives for management of the total hazardous-waste stream. These alternatives will basically consist of allocating the hazardous-waste material by certain physical and chemical characteristics and by storage process type.

In terms of differentiation between storage processes, hazardous wastes may be grouped by physical–chemical properties into the following types: waste containing free liquid; waste without free liquids; ignitible or reactive wastes; incompatible wastes; and nonautoleachable wastes. Table 5-4 indicates the major federal regulation sections by waste type and storage process.

In general, any waste type may be stored by means of any of the four storage processes. However, for a given waste or hazardous-waste stream there will be a storage-process configuration representing the minimum overall cost. The requirments of the federal regulations for each matrix element in Table 5-4 are described in more detail in what follows.

General Studies

All storage facility processes have general requirements described in federal regulations in seven categories: general facility standards; preparedness and prevention; contingency plan and emergency procedures; manifest system, recordkeeping, and reporting; groundwater protection; closure and post-closure; and financial requirements. These requirements do not necessarily represent a differential cost of one facility versus another but they are common to the operation of a hazardous-waste facility, in general. The requirements of the regulations will be examined in more detail in a later section. However, the information relating to the conceptual design of facilities is summarized here.

It is a general requirement that the owner or operator of a storage facility must be able to determine if the hazardous waste received matches the identity specified on the accompanying manifest. This must be done by a waste analysis plan. This implies that the conceptual design must allow for a staging area for transportation vehicles to hold pending completion of the waste analysis and approval of entry to the facility. Depending upon the waste, this holding time will be somewhere between 1 and 24 hours.

Site Requirements

The owner or operator is required to prevent physical contact with the waste by unknowing or unauthorized persons or livestock. This generally requires 24-hour surveillance and an artificial or natural barrier that surrounds the facility and has controlled entry. In addition, signs reading "Danger, Unauthorized Personnel Keep-Out" in English, and any other predominant language, legible from at least 25 feet must also be provided. This security must be carried on not only during the active life of the facility but potentially during its closure and post-closure care. Therefore a site with suitable natural barriers or an existing security system may represent a substantial savings over the life of the facility.

Waste-Related Requirements

In general, any liquid that comes into contact with hazardous waste through precipitation and runoff, runon, autogenerated leachate, or precipitation-induced leachate must be collected and suitably handled. One possibility is to treat the material and discharge it with provisions of the Clean Water Act (CWA). Another would be to solidify and/or encapsulate the liquid and dispose of it as a hazardous waste.

The conceptual design of the facility must also involve segregating ignitible reactive or incompatible wastes. Ignitible or reactive wastes must be separated and protected from sources of ignition, including flames, smoking, cutting and welding, hot surfaces, sparks, heat-producing chemicals, and radiant heat. Incompatible wastes, some examples of which are shown in

Table 5-5, must also be segregated. A general-purpose storage facility that uses tanks as its principal means of storage normally consists of many smaller tanks rather than one large tank, to allow for such segregation and/or protection.

Facility Requirements

Facilities must be designed to minimize the possibility of fire, explosion, or unplanned release of hazardous-waste constituents. This implies that operating areas must be distinctly segregated from areas of new construction, and that traffic systems must be isolated from the storage facilities so that the hazardous-waste transportation equipment could not, by accident, cause the release of hazardous materials. These considerations must also include the incursion of external traffic onto the storage facility property. Even accident conditions with relatively low probability must be considered and provided for in the layout of segregation and barrier design of the facility. The storage equipment must also be laid out so that water, at adequate volume and pressure, to supply hoses, foam-producing equipment, sprinklers, or water spray systems can be distributed over the facility. This requirement may be omitted only by demonstrating that no hazard posed by the waste-handling facility could require the equipment.

The layout of the facility must have sufficient aisle space to allow unobstructed movement of personnel, fire-protection equipment, spill-control equipment, and decontamination equipment to any area of facility operation in an emergency. This space must be available with the assumption that the normal transport vehicles are in operation on the site.

Other Requirements

An aspect of the contingency plan and emergency procedures that must be considered in the conceptual design of facilities is the requirement that the emergency coordinator must be able to assess possible hazards to human health that would result from a released fire or explosion. This implies that a control building must be included from which the emergency coordinator would be able to assess the extent and severity of the emergency and be able to monitor leaks, pressure buildup, gas generation, or ruptures in valves, pipes, or other equipment, as appropriate.

In terms of manifest information, it is necessary, in general, to be able to present an audit trail for all material accepted at the facility that indicates the source, the amount of material, the date accepted, the storage location, the date that material left the facility, and its ultimate destination. This implies, therefore, that the segregated areas, either by storage process or by

Table 5-5. Incompatible Materials

Group 1-A	*Group 1-B*
Acetylene sludge	Acid sludge
Alkaline caustic liquids	Acid and water
Alkaline cleaner	Battery acid
Alkaline corrosive liquids	Chemical cleaners
Alkaline corrosive battery fluid	Electrolyte acid
Caustic wastewater	Etching acid liquid or solvent
Lime sludge and other corrosive alkalies	Pickling liquor and other corrosive acids
Lime wastewater	Spent acid
Lime and water	Spent mixed acid
Spent caustic	Spent sulfuric acid

Potential consequences Heat generation; violent reaction

Group 2-A	*Group 2-B*
Aluminum	Any waste in Groups 1-A or 1-B
Beryllium	
Calcium	
Lithium	
Potassium	
Sodium	
Zinc powder	
Other reactive metals and metal hydrides	

Potential consequences Fire or explosion; generation of flammable hydrogen gas

Group 3-A	*Group 3-B*
Alcohols	Any concentrated waste in Groups 1-A or 1-B
Water	Calcium
	Lithium
	Metal hydrides
	Potassium
	SO_2Cl_2, $SOCl_2$, PCl_2, CH_3, $SiCl_3$
	Other water-reactive waste

Potential consequences Fire, explosion, or heat generation; generation of flammable or toxic gases

Group 4-A	*Group 4-B*
Alcohols	Concentrated Group 1-A or 1-B wastes
Aldehydes	Group 2-A wastes
Halogenated hydrocarbons	
Nitrated hydrocarbons	
Unsaturated hydrocarbons	
Other reactive organic compounds and solvents	

Potential consequences Fire, explosion, or violent reaction

Group 5-A	*Group 5-B*
Spent cyanide and sulfide solutions	Group 1-B wastes

Potential consequences Generation of toxic hydrogen cyanide or hydrogen sulfide gas

(continued)

Table 5-5   *(continued)*

Group 6-A	Group 6-B
Chlorates	Acetic acid and other organic acids
Chlorine	Concentrated mineral acids
Chlorites	Group 2-A wastes
Chromic acid	Group 4-A wastes
Hypochlorites	Other flammable and combustible
	wastes
Nitrates	
Nitric acid, fuming	
Perchlorates	
Permanganates	
Peroxides	
Other strong oxidizers	

Potential consequences   Fire, explosion, or violent reaction

Source: Hazardous Waste Management Law, Regulations, and Guidelines for the Handling of Hazardous Waste. California Department of Health, Sacramento, California, February 1975.

Note: The mixing of a Group A material with a Group B material may have the potential consequence noted.

waste properties, must be clearly discernible by the operating personnel and that the possibilities for mismanagement or transportation accidents or incidents in the facility must be minimized by the layout and conceptual design.

A very significant decision for the conceptual designer of a storage facility using surface impoundments or waste files is to carry out a design so that groundwater monitoring and protection either is or is not required. For surface impoundments this requires double-lining and the location of the impoundment, including its underlying layers, entirely above seasonal high-water tables. It also requires a leak detection system between the liners. For a waste pile this is only possible if the material to be handled in the waste piles does not contain free liquids, the pile is protected from runon, the pile is designed to control dispersal by wind, other than by wetting, and the pile will not autogenerate leachate. Here, the conceptual designer must limit the use of waste piles to only materials meeting these criteria.

The designer must also consider the requirements of closure and post-closure in the specification of processes, their layout capacity, and acceptance criteria. Closure, in general, requires the removal of all stored hazardous wastes and the decontamination of all facilities. The closure plan, unless approval is obtained otherwise, must involve removal of all waste within 90 days of receipt of the final volume of hazardous waste to be handled, and all activities must be completed within 180 days. The completion of closure must be certified by an independent registered professional engineer that the facility has been closed in accordance with specifications in an approved closure plan. The post-closure care, which normally must continue for 30 years (only for surface impoundments and waste piles), involves continuation of security and monitoring of the proposed use of the property to avoid threats to human health or the environment. As a final note, the designer must be aware of the financial requirements of the federal regulations and that alternative designs, however attractive in execution, that impose an extended post-closure burden may be less preferable than a somewhat more capital-intensive initial design that minimizes post-closure care. For this reason it is generally beneficial to make cost estimates over the life cycle of the facility on a present worth or equivalent annual cost basis.

DESIGN OF FACILITIES

The design of a storage facility is governed, in many ways, by regulatory requirements for information to accompany a permit application. The design must also carry out the intent of the conceptual design report. Finally, the design must provide suitable plans and specifications for bidding purposes. This section deals only with the first requirement since the other two are normal responsibilities of the designer.

Base Map

Many aspects of the facility (e.g., security, layout, access) can best be displayed on a scale drawing of the property. An early step, therefore, in the design process is the preparation of a base map. It will prove helpful to prepare the base map at a scale of 1 in. = 200 ft and to include the property within 1000 ft of the proposed facility boundary.

Base Map Overlays

It is necessary to prepare at least the following overlays to accompany the permit application: topographic map showing scale and date, 100 = yr floodplain area (if any), surface waters, surrounding land uses, wind rose, north arrow, facility legal boundary, access control, wells, buildings, utilities, drainage valves, and location of operational units. Contours must be at 2-ft intervals if the relief is less than 20 ft, 5 ft if over 20 ft, or large intervals appropriate to greater relief.

Operational overlays are described in the following list along with the required details of construction.

Containers

Dimensions and construction materials of barriers and containers
Drainage system

Number and volume of containers
Runoff controls
Container liquid removal to prevent overflow
Buffer and barrier systems for ignitible and reactive wastes

Tanks

Design standards for tank walls
Liner details
Dimensions of wall thickness and capacity
Process and instrumentation diagram
Feed, vent, and bypass controls
Buffer and barrier systems for ignitible and reactive wastes

Surface Impoundments

Dimensions and materials
Construction details
Liner materials and placement
Overtopping prevention system
Saturated zone location
Dike construction specifications

Waste Piles

Dimensions
Underliner details
Runon and runoff controls
Collectors and holding units for liquid management
Wind dispersal controls
Buffer and barrier systems for incompatible and reactive materials

Summary

Clearly, many details needed for permit applications are also needed to carry out the conceptual design and to prepare a bid package for construction. The designer is cautioned, however, that the same drawing may not be suitable for all three purposes. The EPA has distributed model permit applications for storage facilities. Simple prudence dictates that the application closely adhere to the approved model. For this reason, additional effort may be required to modify or redraw permit application drawings (which allow the model) so that they are suitable for conceptual implementation and/or bid preparation.

PERMITTING OF STORAGE FACILITIES

Consolidated Permit Program

Permitting of hazardous-waste storage facilities is governed by the consolidated permit program. There are two types of requirements: one for all environmental permits (the consolidated permit), another for RCRA facilities. Part 12 of the Code of Federal Regulations (CFR) describes the basic permitting requirements and administrative appeal process (under subparts A, B, E, and F).

The RCRA requires a permit for the "storage" of "hazardous waste" (as defined by 40 CFR 122.3). Application is required at least 180 days prior to construction; in no case, however, may construction begin before receiving an effective RCRA permit. Notice, however, that the EPA is not obliged to issue or even review a permit within that time. Although some states, most notably Michigan, have defined a limited time for review and denial of permit applications, no such limit is promulgated in federal regulations. The lead time, therefore, which may be 12 to 16 months, must be determined through discussions with the appropriate regulatory official.

The permit consists of two parts. Part A is essentially related to facility location and operation. The following list contains the information requirements (EPA Form 3510-1 and Form 3510-3).

Latitude and longitude

Facility location

Name, address, and telephone number of owner

Process descriptions

Storage Process	Process Code
Container	S01
Tank	S02
Waste pile	S03
Surface impoundment	S04

Process design capacity

EPA hazardous-waste codes (40 CFR subpart C or D) and estimated annual quantity

6" × 8" scale drawing of the facility showing
   property boundaries
   operational areas and name
   future operational areas
   approximate dimensions

Site photographs (aerial and/or ground)

Owner certification (general partner or vice president)

Part B requires more information; preparation cost and time to assemble the data are higher. A careful reading of 40 CFR 122.25 is essential to prepare this part: the following should be taken only as a guide in the preliminary states of preparation. Alternatively, the information here may be used to estimate the time and cost of preparation.

The following 20 categories of information are required by 40 CFR 122.25(a) as general information for all facilities.

1. General description
2. Chemical and physical analyses of wastes
3. Copy of waste analysis plan [40 CFR 264.13(b)]
4. Security procedures and equipment
5. General inspection schedule
6. Justification of waiver, if request, for 264 subpart C plan
7. 264 subpart D contingency plan copy
8. Contamination-prevention plan description
9. Ignitible, reactive, and incompatible waste-control plan copy
10. Traffic control procedures
11. Seismic and 100-yr floodplan location data
12. Training program outline
13. Closure and post-closure plan copy
14. Not applicable to new facilities
15. Closure cost estimate and financial guarantee
16. Post-closure cost estimate and financial guarantee
17. Insurance policy copy
18. Proof of coverage by state financial plan if applicable
19. 1 in. = 200 ft topographic map and details (see sec. 5.0)
20. Other information needed by the regional administrator

Assembling this information can be facilitated by obtaining a sample Part B application prepared by EPA. The current example, for the "Turkofile" facility, has examples of the general information described in the categories as well as examples of detailed storage facility information discussed later.

In any case the application preparer can anticipate two months of effort and a cost of $5000 to $20,000 per application, even more for more complicated sites.

Specific information is required for containers, tanks, surface impoundments, and waste piles. For containers, tanks, and surface impoundments the required information is

design drawings

runoff and runon control system description

containment and drainage system description

overflow control system description

ignitible, reactive, and incompatible waste buffer, barrier, and control systems descriptions

In addition, for surface impoundments, the type of liners and leak detection system must be described. For waste piles the control of wind dispersal must be addressed.

In addition to the permit example described here, the EPA has prepared checklists, similar to those used in the review process, for the applicant. The permit example, the checklist, the detailed regulations coupled with a review of previously submitted applications should be sufficient to avoid errors and omissions such that extensive revision resubmittal by the applicant is necessary.

The Decision-Making Process

The decision-making process used by EPA is neither simple nor short. The permitting, appeal, and nonadversary panel procedures are described in 40 CFR 124, subparts A, B, E, and F, and will not, due to their complexity, be summarized here. The process can be understood conceptually with the aid of three flowcharts promulgated with the regulations. These flowcharts, shown as Figures 5-1, 5-2, and 5-3, indicate the steps and reference the pertinent regulation sections. At present, if the application is complete and factual and the facility meets or exceeds all published criteria, then the applicant may never be exposed to any of the flowchart operations except permit application submission and receipt of the effective permit. The applicant must be aware, however, of the procedures in the event a more lengthy and intimate involvement becomes necessary.

(Text continues on page 348.)

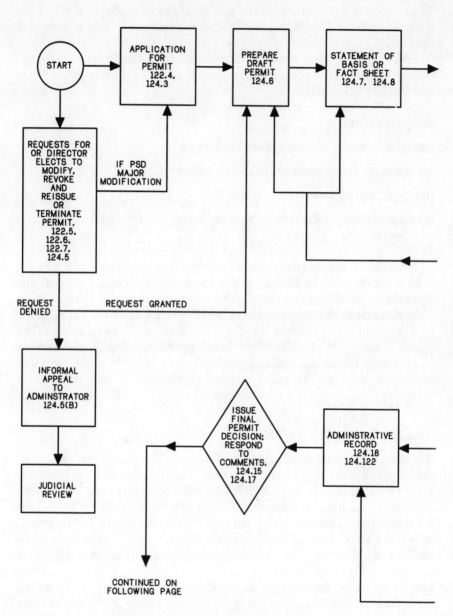

Figure 5-1. EPA permitting procedures.

CONTINUED ON
FOLLOWING PAGE

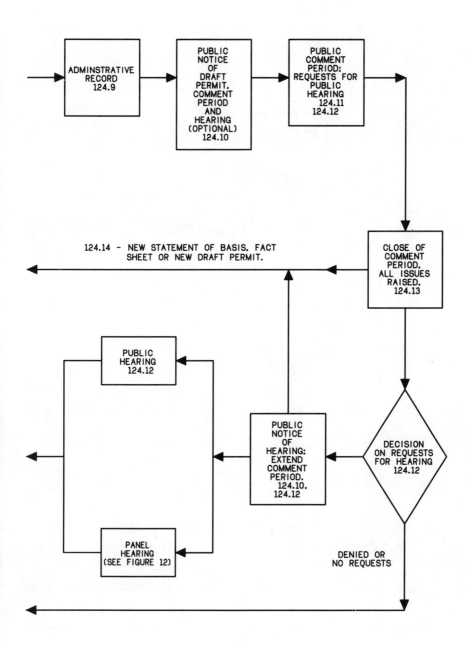

ADMINSTRATIVE RECORD 124.9

PUBLIC NOTICE OF DRAFT PERMIT, COMMENT PERIOD AND HEARING (OPTIONAL) 124.10

PUBLIC COMMENT PERIOD: REQUESTS FOR PUBLIC HEARING 124.11 124.12

124.14 - NEW STATEMENT OF BASIS, FACT SHEET OR NEW DRAFT PERMIT.

CLOSE OF COMMENT PERIOD, ALL ISSUES RAISED, 124.13

PUBLIC HEARING 124.12

PUBLIC NOTICE OF HEARING: EXTEND COMMENT PERIOD, 124.10, 124.12

DECISION ON REQUESTS FOR HEARING 124.12

PANEL HEARING (SEE FIGURE 12)

DENIED OR NO REQUESTS

344   R. Ball and J. H. Johnson, Jr.

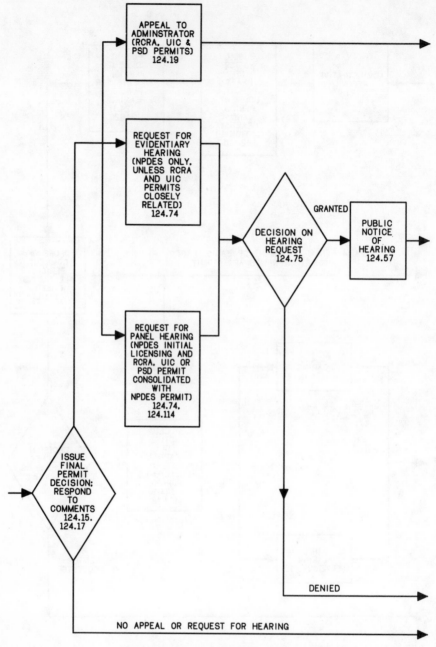

Figure 5-2. EPA appeal procedures.

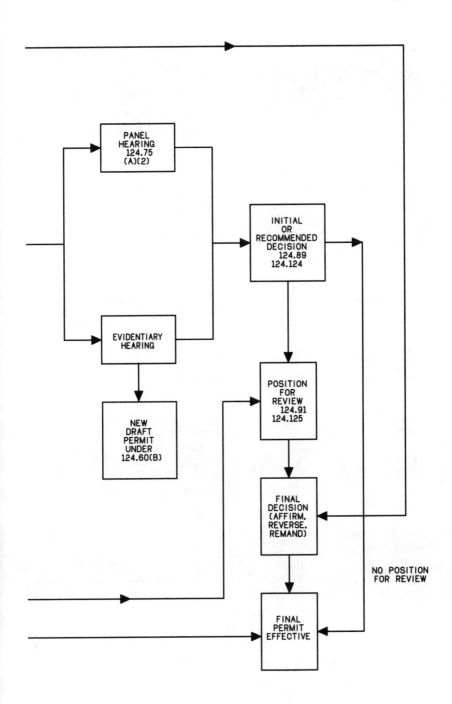

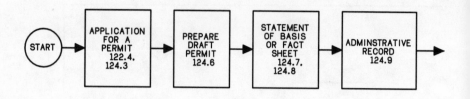

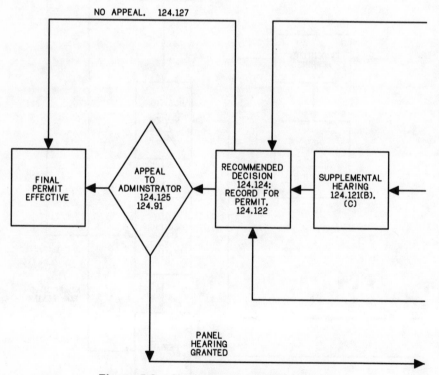

Figure 5-3.   Non-adversary panel procedures.

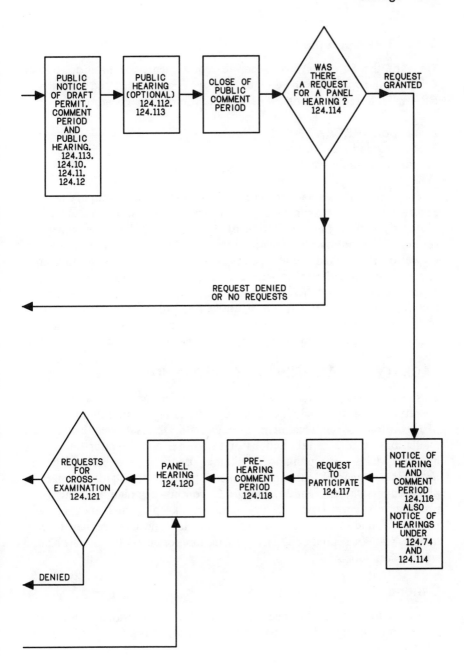

State Requirements

Many of the 50 states have separate requirements for hazardous-waste facilities. In some cases these requirements are as detailed and extensive as EPA's. State requirements, in general, fall into three categories:

1. addition to RCRA
2. state administers RCRA
3. state has parallel and separate requirements

The first two cases pose no special difficulties and may simplify the procedure due to the proximity of the reviewers. The third, however, may ultimately require the preparation of two permit applications with much common information but unrelated formats. The timing of the two permits may also be unrelated. Finally, some of the states have established a siting approval process prior to issuing a construction permit. Such requirements, entirely unrelated to RCRA, should be satisfied, if possible, before the costly and time-consuming RCRA permit application process is initiated. The only completely reliable source of information for current state requirements is the state department or division charged with hazardous-waste responsibility.

FACILITY MANAGEMENT AND OPERATION

This section deals principally with requirements for facility management and operation listed under 40 CFR 264, "Standards for Owners and Operators of Hazardous Waste Treatment, Storage, and Disposal Facilities." These regulations require certain actions of the owner and operator under nine headings, with corresponding regulations under each subpart. The ten headings are general, subpart A; general facilities standards, subpart B; preparedness and prevention, subpart C; contingency plan and emergency procedures, subpart D; manifest system, recordkeeping, and reporting, subpart E; groundwater protection, subpart F; use and management of containers, subpart I; tanks, subpart J; surface impoundments, subpart K; and waste piles, subpart L.

General

This subpart provides that regardless of any other provisions, enforcement actions may be brought if an imminent hazard, pursuant to Sec. 7003 of RCRA, exists.

General Facilities Standards

The owner or operator who receives hazardous waste from a foreign source must notify the regional administrator in writing four weeks in advance. Subsequent shipments do not require notice. A facility that receives hazardous waste from an off-site source (where the owner or operator is not the generator) must advise the generator in writing that he has the appropriate permits for the generator's waste.

Before storing any hazardous waste, a detailed chemical and physical analysis of a representative sample must be obtained. This analysis must provide sufficient information to store the waste in accordance with the provisions of part 264. The analysis must be repeated for a particular waste stream if inspection of the waste indicates a possible change or a lack of agreement with the manifest description. The waste plan must list the parameters for each hazardous waste and the rationale for selection of those parameters. In addition, the plan must list the test methods, the sampling methods, and the frequency of sampling.

As stated previously, the owner or operator must prevent unknowing entry of persons or livestock onto the active portion of the facility. The owner or operator can demonstrate to the administrator that physical contact with the waste or equipment within the active portion of the facility will not cause injury when other barriers or restrictions are unnecessary. Failing this demonstration, the facility must have either 24-hour surveillance of the active portion or an artificial or natural barrier completely surrounding the active portion, including controlled entry. In addition, a sign, legible from a distance of at least 25 feet, with the legend "Danger—Unauthorized Personnel Keep Out," in English and any other predominant language, must be posted such that the sign is visible from any approach to the active facility.

The owner or operator must inspect the facility to prevent the release of hazardous-waste constituents to the environment and to prevent threats to human health, particularly of employees. These inspections must follow a written schedule that identifies the types of problems to be checked, and the schedule should be based on the rate of possible deterioration of equipment and the degree of risk. These inspections must be at least as frequent as required for each type of storage facility, which is discussed under that heading. Finally, these inspections must be recorded in an inspection log or summary to be maintained for a minimum of three years. The records must include the date and time of inspection, the name of the inspector, notation of observation made, and the date and nature of repairs or other remedial actions.

Any personnel employed at the facility must receive a training program designed to insure that the facility personnel are able to respond to emergencies and operate emergency equipment. The training program must be completed within six months after date of employment or assignment to the facility, and the training must be reviewed annually. Compliance requires documents recording job title, name of employee, job description, description of introductory and continuing training, and documentation that training has taken place. Such training records must be maintained until closure of the facility, and the training record on any employee must be maintained for three years after the date the employee last worked at the facility.

The owner or operator must protect ignitible, reactive, or incompatible wastes from open flames and other heat sources. Smoking must be confined to specifically designated locations if ignitible or reactive wastes are handled. No smoking signs must be placed where the hazard exists. To further comply, the owner or operator must not mix waste or other materials unless documentation is available to indicate that such mixtures will not be ignitible or reactive.

Preparedness and Prevention

Emergency equipment, such as communications and alarms, fire-protection, spill-control, and decontamination equipment must be tested and maintained with an appropriate record of such testing and maintenance provided. The owner or operator must insure that any personnel involved in the handling of hazardous waste must have immediate access to alarm or communication, which could include visual or voice contact with another employee. Similarly, the owner or operator must maintain aisle space for emergency equipment at all times in all areas of the facility. This implies that the owner or operator must not only enforce safety work rules but also must be able to readily inspect the facility during its operation.

The owner or operator must arrange and document familiarization meetings with police, fire department, emergency response teams, and local hospitals. This familiarization must include the layout of the facility, the properties of the waste handled at the facility, and the type of injuries that could result from fire, explosion, and releases.

Contingency Plan and Emergency Procedures

The owner or operator must insure that an emergency coordinator is either on the facility premises or on call (with rapid availability), is familiar with the contingency plan and all aspects of hazardous-waste management, and has authority to carry out the contingency plan. The coordinator must activate alarms, notify local agencies, assess the characteristics of any released materials, and take steps to minimize the loss of materials or the continuation of emergency conditions. The owner or operator must further maintain sufficient documentation to show that all emergency equipment was fit for its use prior to and following the emergency. Finally, the owner or operator must submit a written report to the regional administrator within 15 days following the emergency.

Manifest System, Recordkeeping, and Reporting

The owner or operator, or his agent, must sign and date manifests received from the transporter, note discrepancies, provide the transporter with a copy, and within 30 days send a copy of the manifest to the generator. The storage facility copy must remain at the facility for at least three years from the date of delivery. Information on the manifest must be incorporated on a written operating record describing the quantity of hazardous waste received, method and date of storage, and location within the facility. The location and quantity of hazardous waste must be recorded on a map that is cross-referenced to manifest numbers. These records, and records required under previous parts, must be made available to EPA representatives upon request.

On March 1 of each year the owner or operator must submit an annual report to the regional administrator describing the hazardous-waste activities at the facility in terms of shipment, listed by EPA identification number, for each generator.

Groundwater Protection

For a storage facility nonexempt from groundwater monitoring requirements, the owner or operator must determine groundwater quality at each monitoring well at least semiannually and determine the groundwater flow rate and direction in the uppermost aquifer at least annually. If analysis indicates leakage from the facility, an elaborate monitoring program and corrective action program must be implemented. Because many storage facilities do not require groundwater monitoring and because the requirements for additional actions are extensive and complex if contamination is detected, these further requirements are not discussed in detail in this section. It is recommended that the facility owner or operator obtain the services of a qualified hydrogeological consultant to assist in the implementation of such actions in the event the routine monitoring described above indicates leakage. The extent and complexity of these measures further amplifies the recommendation in the section on conceptual design of facilities that the designer should seriously consider exempting the facility from groundwater protection by suitable design and thus avoid the requirements of subpart F.

Use and Management of Containers

Containers must have less than 1 in. of free liquid to be defined as empty (or meet future 40 CFR 261.7 requirements) and, therefore, exempt from the following requirements.

Container storage areas must be inspected at least weekly for signs of leakage or deterioration. If a problem is discovered, the container content must be transferred. Containers must be closed except during addition or removal of waste and must be constructed of, or lined with, materials that are unreactive and compatible with the waste.

The storage area must be underlaid by an impermeable base with sufficient storage capacity to hold 10% of the total container liquid volume (the liquid volume of the largest container, if greater) plus any runon and precipitation.

Ignitible or reactive waste must be stored at least 50 ft from the property line, and incompatible wastes must be separated with a barrier.

Tanks

Tank designs must be approved by the regional administrator with respect to design, foundation, structural supports, seams, and pressure controls. As with containers, the tanks must be constructed of, or lined with, materials that are unreactive and compatible with the waste. In addition, some positive control system to prevent overfilling is mandatory.

Overfilling-control equipment must be inspected daily. Other required daily data are tank pressure, temperature, and level (of uncovered tanks). The ground above and areas surrounding tanks must be inspected weekly for leakage. At some greater frequency, depending upon the potential rate of corrosion, tanks must be entered for interior inspections. Finally, precautions must be taken to avoid mixing incompatible and/or reactive materials in the same tank, with incompatible materials separated according to National Fire Protection Association (NFPA) standards.

Surface Impoundments

Surface-impoundment liners must be inspected weekly and after storms for evidence of deterioration or erosion. Levels must also be recorded, at least daily, so that sudden changes are discovered.

The requirements to avoid admixing incompatible or reactive wastes are most stringent with other types of storage processes due to the greater hazard. Therefore, unless the impoundment has been thoroughly decontaminated, no admixing is allowed.

Waste Piles

If the pile is under cover, does not autogenerate leachate, and is not subject to wind dispersal, there are no specific management requirements. Otherwise, weekly inspections are required for determination of runoff, runoff leachate control systems. Berms or other barriers are required in all cases to separate incompatible waste piles.

CLOSURE

Of all hazardous-waste facilities, storage facilities are the most straightforward in terms of closure and post-closure care. The requirements, however, are still both extensive and explicit.

General Requirements

The objectives of closure are to minimize the need for further maintenance and to prevent threats to human health and the environment. These objectives are realized through a written closure plan submitted with the permit application. The plan describes when and how the facility will be closed.

Within 90 days after receiving the final volume of hazardous waste, all wastes must be removed from the site, and all other closure activities must be accomplished within 90 additional days. The additional activities involve decontamination or disposal of all equipment and certification of closure by an independent registered professional engineer. In all cases the regulations allow for variances, delays, and modifications vis-a-vis the closure plan.

Specific Requirements

Containers and Tanks. For these processes closure is synonymous with removal of residues and equipment decontamination. Note that the owner/operator will become, therefore, a hazardous-waste generator for the closure period.

Surface Impoundments. The closure requirements are listed below:

1. Removal of waste residues and contaminated system components
2a. Elimination of free liquids (e.g., by solidification)
2b. Stabilization of remaining wastes to a suitable bearing capacity
2c. Cap the impoundment with an impermeable cover

If all residues are removed, post-closure care is not required. Otherwise, leak detection, groundwater monitoring, and runoff and runon controls are required.

Waste Piles. Removal of all residues and contaminated material relieves the owner/operator of post-closure care. If all contaminated materials cannot be removed, the facility must receive the same post-closure care as a landfill.

Summary. Post-closure requirements provide a strong incentive to design the facility to avoid them. The essential requirement is that all residues and contaminated portions of the facility be removable. Although it is desirable that the removal be accomplished in six months or less, the closure plan may justify a longer period if necessary. In all cases where removal is not complete, the property deed must record a notation that will notify any potential purchaser that the land was used to manage hazardous wastes. A description of the location, type, and amount of hazardous waste on the site must be filed with EPA as well.

APPENDIX 5-1
REGULATORY RATING CURVES

S_{GW} = groundwater score

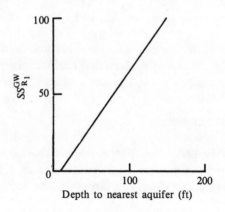

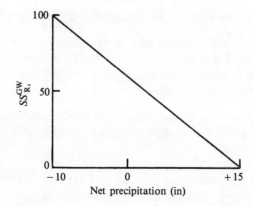

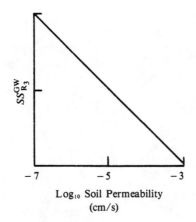

Log₁₀ Soil Permeability
(cm/s)

$$SS_R^{GW} = 0.4(SS_{R_1}^{GW}) + 0.2\sum_{i=2}^{4}(SS_{R_i}^{GW})$$

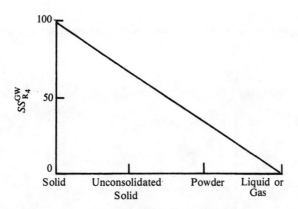

Physical State of Stored Material

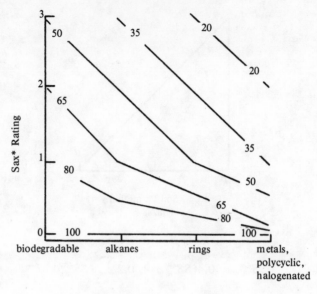

Stored Substance
(subscore contours $= SS_{w_1}^{GW}$)

*Sax, N.I., *Dangerous Properties of Hazardous Materials,* Van Nostrand Rheinhold Co., New York, 4th Ed., 1975.

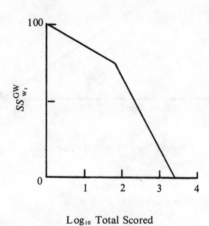

Log$_{10}$ Total Scored
Quantity (tons)

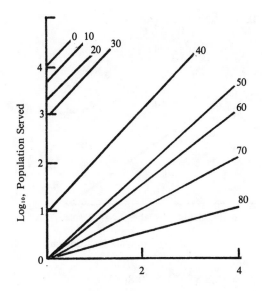

$$SS_{T_2}^{A} = SS_{T_3}^{SW}$$

Distance to population (mi)
(subscore contours $= SS_{T_1}^{A}$)

$$SS_T^A = 0.75(SS_{T_1}^A) + 0.15(SS_{T_2}^A) + 0.10(SS_{T_3}^A)$$

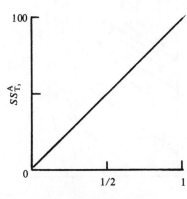

Commercial/Industrial

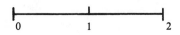

Parks and Prime Agricultural
Distance (mi)

$$SS^{GW}_{W_1} = 0.7(SS^{GW}_{W_1}) + 0.3(SS^{GW}_{W_2})$$

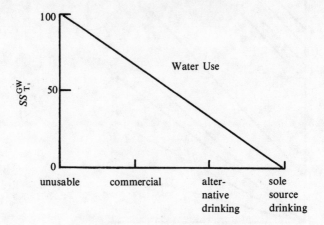

$$SS^{GW}_{T} = 0.2(SS^{GW}_{T_1}) + 0.8(SS^{GW}_{T_2})$$

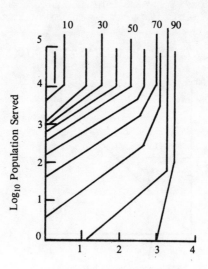

Distance to Nearest Well (mi)

(subscore contours = $SS^{GW}_{T_2}$)

$$S_{\text{GW}} = \{[0.5(SS_R^{\text{GW}})][1.0(SS_w^{\text{GW}})][2.0(SS_T^{\text{GW}})]\}^{1/3}$$

$$S_{\text{SW}} = \text{Surface-Water Score}$$

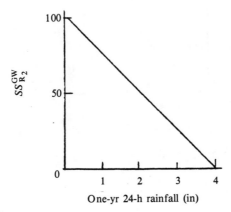

One-yr 24-h rainfall (in)

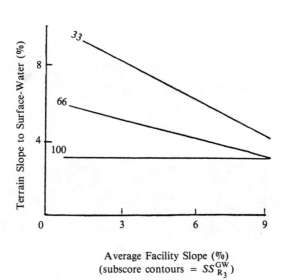

Average Facility Slope (%)
(subscore contours = $SS_{R_3}^{\text{GW}}$)

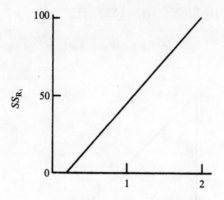

Distance to Surface Water (mi)

$$SS_{R_4}^{SW} = SS_{R_4}^{GW}$$

$$SS_R^{SW} = 0.4(SS_{R_1}^{SW}) + 0.2 \sum_{i=2}^{4} (SS_{Ri}^{SW})$$

$$SS_{W_i}^{SW} = SS_{W_i}^{GW}, i = 1,2$$

$$SS_W^{SW} = 0.7(SS_{W_1}^{SW}) + 0.3(SS_{W_2}^{SW})$$

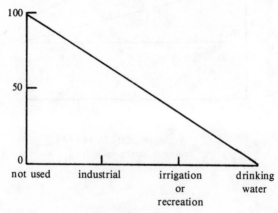

Surface-Water Use

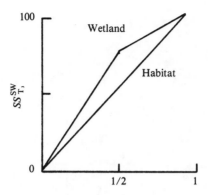

Distance to Wetland** or
Critical Habitat of En
dangered species*** (mi)

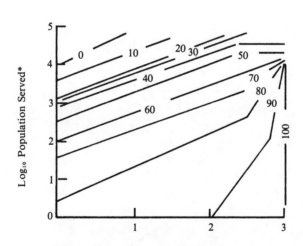

Distance to Surface-Water (mi)
(subscore contours = $SS^{SW}_{T_2}$)

$$SS^{SW}_{T} = 0.2(SS^{SW}_{T_1}) + 0.70(SS^{SW}_{T_2}) + 0.10(SS^{SW}_{T_3})$$

*Within 3 miles downstream as drinking water intake.
**40 CFR 230, Appendix A (use × distance for coastal).
***U.S. Fish and Wildlife Service.

$$S_{SW} = \{[0.5(SS^{SW}_R)] [1.0(SS^{SW}_w)] [2.0(SS^{SW}_T)]\}^{1/3}$$

$$S_A = \text{Air Score}$$

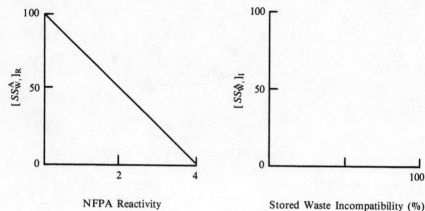

NFPA Reactivity   Stored Waste Incompatibility (%)

$$SS^A_{W_1} = \min\{[SS^A_{W_1}]_R, [SS^A_{W_1}]_I\}$$

$$SS^A_{W_3} = SS^{GW}_{W_2}$$

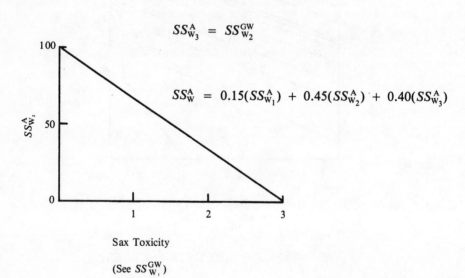

$$SS^A_W = 0.15(SS^A_{W_1}) + 0.45(SS^A_{W_2}) + 0.40(SS^A_{W_3})$$

Sax Toxicity

(See $SS^{GW}_{W_1}$)

APPENDIX 5-2
SOCIAL SCORE RATING CURVES

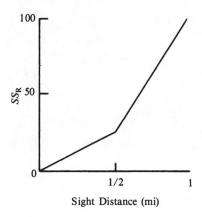

Sight Distance (mi)

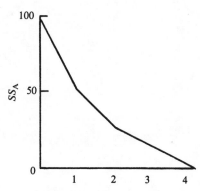

Log₁₀ Residential, recreational
or institutional population
exposed to facility access

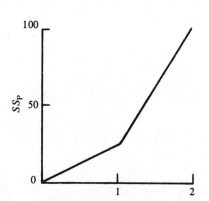

Distance to Residential,
Hospital, Health Care, or
Recreation Facilities (mi)

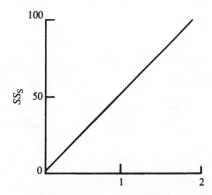

Distance from Surface-Water
Crossing to Downstream Population
(mi)

$$S_S = [1/4(SS_R^2 + SS_A^2 + SS_P^2 + SS_P^2 + SS_S^2)]^{1/2}$$

6

Land Disposal of Hazardous Wastes

Raymond C. Loehr
University of Texas, Austin

Hazardous-waste management involves prevention, treatment and disposal, and remedial action. Prevention includes both waste and volume reduction. One aim of any hazardous-waste management strategy should be to reduce, preferably to a minimum, the quantity of hazardous waste that must be managed. Hazardous-waste reduction can be achieved by changing production process conditions, changing chemicals used in a process, separating wastes at the source, and maintaining close control over spills and equipment breakdown.

Treatment and disposal techniques should render the wastes less hazardous and dispose of them in a manner such that environmental and human health problems do not occur. These techniques include chemical, physical, and biological treatment technologies to render the wastes less hazardous and high-temperature and land-application processes to dispose of the wastes and the residues from the treatment processes.

Remedial action is required for the hazardous wastes that already have been released to the environment as a result of improper disposal or poor management practices. Abandoned or inactive hazardous-waste sites are examples of locations where remedial methods are needed.

Treatment and disposal methods and remedial action methods both involve land to a large degree: the former because land application can be an economically and technically sound method for hazardous-waste treatment and disposal; the latter because remedial measures also require and understanding of the use of soil to control the pollutants improperly released.

Knowledge of land as a waste-management alternative is of prime importance for the management of hazardous wastes.

Treatment prior to the land disposal of hazardous wastes can be accomplished by one or more of the following:

Detoxifying the wastes

Separating and concentrating the hazardous constituents in a reduced volume

Fixing the waste in a matrix that inhibits leaching

Encapsulating and containing the waste to prevent leaching.

Fixation reduces the rate of release of toxic constituents. Chemical stabilization methods include those that are silicate, cement, lime, thermoplastic, or organic polymer based, and vitrification. The approach used depends upon waste characteristics and availability and cost of materials.

This chapter discusses those land-disposal methods in which hazardous wastes are intentionally placed or discharged on or into the land for the purpose of controlling the hazardous wastes in a technically and environmentally sound manner. These methods normally are part of an overall hazardous-waste management *system* that may include chemical, physical, or biological treatment as well as surface impoundments and storage prior to land disposal. Many of these other components are discussed in other chapters; this chapter focuses on land-disposal methods for treatment, disposal, and remedial action.

Land disposal of hazardous wastes represents the permanent placement of solid, liquid, semisolid, or contained gases in or on the land. It is expected that a portion or all of the wastes will be present at the site at closure. Unless the waste is totally and permanently contained, a portion of the wastes can be expected to migrate from the location at which they are placed. It is this migration of (a) components of the original waste or (b) decomposition or reaction by-products as runoff, leachate, or gaseous emissions that must be controlled at a land-disposal site.

Figure 6-1 illustrates some of the mechanisms of migration that can exist at a land-disposal site. As precipitation and surface water that run on to the site percolate through the disposal area, contaminants can be solubilized and carried to the water table where they are transported through the aquifer under the influence of the hydraulic head and groundwater flow. The contaminated leachate will exist as a plume because of incomplete mixing and diffusion in the saturated zone. Depending on the characteristics of the applied waste, the net infiltration rate, the distance to the water table, and the characteristics of the soil. Transport to and through the saturated zone can be slow, clay soils being a greater retardant than sands or gravels. If drink-

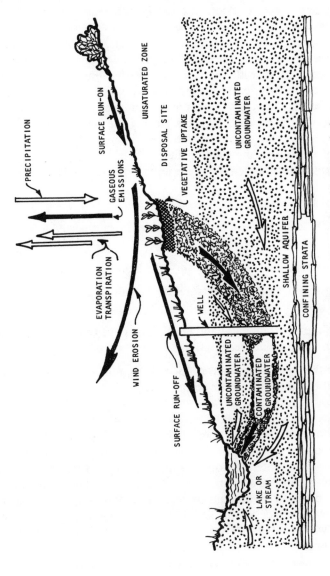

Figure 6-1. Contaminant transport from a land disposal site.

PRECIPITATION

SURFACE RUN-ON

GASEOUS EMISSIONS

UNSATURATED ZONE

DISPOSAL SITE

EVAPORATION TRANSPIRATION

VEGETATIVE UPTAKE

WIND EROSION

UNCONTAMINATED GROUNDWATER

WELL

SURFACE RUN-OFF

SHALLOW AQUIFER

UNCONTAMINATED GROUNDWATER

CONTAMINATED GROUNDWATER

CONFINING STRATA

LAKE OR STREAM

367

ing water or irrigation wells intercept the contaminated leachate or if the leachate enters surface waters, adverse environmental and public-health impacts can occur. Surface waters also can be contaminated by runoff from the disposal site.

Other adverse environmental impacts can occur. Air contamination at the site may occur by vaporization of volatile waste components, by gases emitted from the surface or within the site, and by wind-borne contaminated particles. In addition, any vegetation growing on the site may be contaminated by waste that may adhere to leaves and by uptake of constituents such as metals and other chemicals. Animals grazing the vegetation may have increased levels of waste constituents in their tissues and blood that may affect their progeny and the food chain.

Important factors in identifying the impact of contaminant migration are

Analysis of the wastes to be applied

Identifying reaction or decomposition by-products expected to occur at the site

Determining soil, geological, and surface characteristics of the site

Estimating the transport and fate of mobile waste constituents and by-products

Estimating the environmental and health impact of the mobile components if such components reach critical receptors

Estimating the type and amount of food chain, human health, and environmental exposure that may occur

A land-application–disposal site is designed and operated to avoid human health exposure and to minimize or eliminate migration of contaminants from the site. Emphasis is placed on approaches that reduce the possibility of contaminating surface water or groundwater, that control gaseous emissions and wind erosion, and that eliminate adverse food-chain impacts. These approaches normally involve one or more of the following: a natural or manmade impervious liner for the site, diversion of surface runon, incorporation of the wastes in the soil, and impermeable cover, and avoidance of food-chain vegetation on the site surface.

Key components of decisions related to land-disposal methods are waste and site characterization, environmental monitoring, relative risk assessment, evaluation of control and remedial action options, operation and management skills, and public acceptance. Environmental monitoring is needed both before and after land disposal is implemented to verify the design assumptions and criteria and to assure that environmental and human health problems do not occur. Even though a technically and environmentally sound land-disposal method can be designed and operated, public opinion and acceptance of the method frequently is the deciding factor.

CONTAMINANT CONTROL MECHANISMS IN SOIL

Land-disposal methods are designed to reduce or eliminate the migration of hazardous-waste constitutents or by-products from the disposal site. The soil characteristics at the site and the biological and chemical reactions taking place influence the mobility of constituents or by-products. In the soil, for instance, wastes that have a high or low pH can be neutralized, inorganic constituents can be converted to less mobile or toxic forms, and organic compounds can be degraded.

It is important to know the soil contaminant-control and removal mechanisms when deciding about the type of waste to be disposed on or in the land and when properly designing and operating land-disposal on or in the land and when properly designing and operating land-disposal sites. The physical, chemical, and biological contaminant-control and removal mechanisms occurring in soil are noted in Figure 6-2 and discussed here.

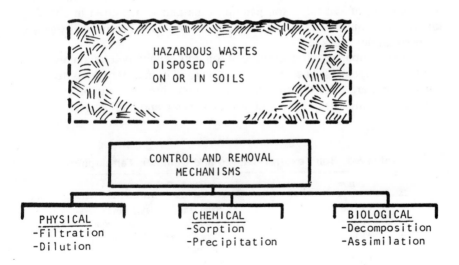

Figure 6-2. Contaminant control and removal mechanisms that can occur in soils.

Soil Characteristics

Soil consists of solids separated by pore spaces. The solid matter consists of mineral particles from rock fragments and organic matter from plant and animal decay. Microorganisms also are a component of the organic matter. In the top layer of soil, organic matter may compose only 1–10% of the solid matter. The pore spaces between the solids are filled with air or water and dissolved salts.

Water movement is a function of the nature and amount of pore space, whereas the mobility of waste constituents is related to water movement and the physical and chemical characteristics of the soil. Soil characteristics are variable in both space and time. The percentage of inorganic matter, for instance, increases with depth.

Soil texture refers to the size of individual mineral particles. Table 6-1 lists the basic textural classes, in terms of the amount of sand, silt, and clay in the surface layer. Other terms commonly used to describe soils are noted. By definition, sand must make up 70% or more of the weight of a sandy soil, and a clay soil must contain at least 35% clay. The textural terminology of the surface soils need not reflect conditions deeper in the soil profile.

Permeability, the ease with which air and water move through the soil, is determined by soil pore size. Individual pores can be a few microns to several millimeters in diameter. Generally, the finer the texture of the soil the slower the permeability. Silty clays have a very low permeability. Permeability is expressed in units of rate of water movement through a unit cross section of saturated soil per unit time and is measured in terms of distance per unit time (in. or cm/h). Table 6-2 relates soil permeability classes to permeability rates.

Another important characteristic of soil is its capability to retain dissolved ions and compounds. The net charge on clay and humus particles is negative. Positively charged ions (cations), such as ammonium, calcium, magnesium, and sodium, are attracted to and held by the clay and humus. The negative

Table 6-1. Soil Textural Classes and Common Terminology

Texture	Common Terminology
Coarse	Sand, sandy, loamy sand
Moderately coarse	Sandy loam, fine sandy loam
Medium	Very fine sandy loam, loam, silt loam, silt
Moderately fine	Clay loam, sandy clay loam, silty clay loam
Fine	Sandy clay, silty clay, clay

Table 6-2. Permeability Classes and Rates for Saturated Soils

Class	Permeability Rate [cm/h (in./h)]
Very slow	Less than 0.15 (0.06)
Slow	0.15–0.5 (0.06–0.2)
Moderately slow	0.5–1.5 (0.2–0.6)
Moderate	1.5–5 (0.6–2.0)
Moderately rapid	5–15 (2.0–6.0)
Rapid	15–50 (6.0–20)
Very rapid	Greater than 50 (20)

Source: Soil Survey Staff, 1971, *Guide for Interpreting Engineering Uses of Soils,* Soil Conservation Service, U.S. Department Agriculture, Washington, D.C.

charge on soil organic matter and, to a minor extent, on clay particles is pH dependent, with the net negative charge increasing with increasing pH.

As a result of this phenomenon, cations are retained by soils with high clay and humus content whereas negatively charged ions (anions), such as nitrate and chloride ions, are not retained but move with the water in the soil. The term "cation exchange capacity" (CEC) is used as a measure of the ability of a soil to retain cations. The greater the soil CEC, the greater the potential of the soil to retain negatively charged waste constitutents and the more effective the soil is for waste treatment and disposal. Soil CEC is expressed in terms of milliequivalents per 100 grams of soil (meq/100g) and is a function of the organic matter content and the type and amount of clay in the soil. The CEC of pure humus is about 200 meq/100g, respectively. The CEC of most soils ranges between 10 and 30 meq/100g.

Soil texture, permeability, and soil CEC are the major factors affecting the movement of waste constituents through soils. Other factors, however, affect the transport and fate of constitutents applied on or into the land. Basic soil science texts [1] and related documents [2,3,4] should be consulted for details on the above and other factors.

Physical

As wastewater moves through soil pores, large particles are removed by filtration. The larger the rate of water movement and the coarser the soil, the greater distance the particles will move.

Should any soil and applied waste erode from the site, particulate matter can be removed from the runoff by filtration as the runoff moves through surface vegetation. Such runoff, however, should be minimal at well-designed and well-operated sites.

Dilution of any leachate and runoff can occur as the leachate mixes with groundwater and the runoff mixes with other surface waters. Because hazardous-waste disposal sites are designed to contain rather than disperse the applied wastes, dilution is not an important contaminant-control mechanism. Filtration and dilution provide little net reduction in the hazardous nature of the applied wastes if they do not alter the chemical characteristics of the waste.

Chemical

Chemical reactions in the soil affect the mobility of dissolved ions or compounds. Thus some constituents are retained in the site indefinitely while others are more mobile. Adsorption and chemical precipitation are the most important chemical reactions governing the movement of constituents at a land-disposal site: CEC is an important adsorption parameter for metals.

Many of the potentially toxic metals, such as cadmium, nickel, zinc, and copper, are positively charged, and their retention by soil is related to the CEC. From the human health standpoint, cadmium has received the greatest attention because cadmium accumulation in the kidney can cause a chronic disease called proteinuria.

The relationship between soil retention of cations, pH, and CEC was included in criteria used for the application of municipal sewage sludge to agricultural land [5]. The use of CEC to indicate soil retention of cations should not imply that all metals are retained by the exchange complex of soils as exchangeable cations. Rather CEC is used as a single soil property that can be measured easily and is related to soil characteristics that may limit the mobility of metals.

Movement of organic compounds from land-disposal sites has not been defined. Where nonbiodegradable organic compounds are important to site operation, adsorption studies should be performed using the local soils for testing.

Biological

Organisms can alter waste constituents by organic matter decomposition and assimilation. Such biological processes can occur throughout a waste disposal site if the organisms are able to degrade the organics. The decomposition occurs by facultative or anerobic organisms. As the organisms degrade the waste, additional organisms are synthesized, and a small fraction of the carbon, nitrogen, phosphorus, and other elements in the waste is assimilated into the synthesized organisms.

Any plant, animal, or human pathogen in the applied wastes should not survive at a controlled hazardous-waste land-disposal site. Some pathogens could be in surface runoff if such runoff occurs immediately after the waste is applied. There should be little concern about pathogens at hazardous-waste disposal sites because such sites are designed to contain the applied wastes. Evaluations of a related type of disposal—the application of stabilized sewage sludges on agricultural cropland—have indicated no serious disease problems.

In general, soils have a high capacity for organic carbon decomposition. Hazardous wastes, however, contain organics that are not readily degradable. Organics such as the chlorinated hydrocarbons are relatively resistant to decomposition in soils. Polychlorinated biphenyls (PCBs) are the only group of organic compounds currently (1982) subject to regulations [5].

LANDFILLS

A landfill is a disposal facility where hazardous wastes are placed in or on the land. Landfills for hazardous wastes frequently are considered as a technology of last resort to be used after approaches to reduce or eliminate the hazards posed by the wastes have been evaluated or utilized. The intent is to bury or alter the wastes so that they are not an environmental or public-health hazard. Any soil cover must be greater than the depth of the plow zone so that subsequent use of the land will not return the landfilled wastes to the surface. Landfills are not homogenous and are usually made up of cells in which a discrete volume of the hazardous waste is kept isolated from adjacent cells and wastes by a suitable barrier. Examples of landfill cells are trenches and pits. Figure 6-3 is a schematic cross section of a hazardous-waste landfill.

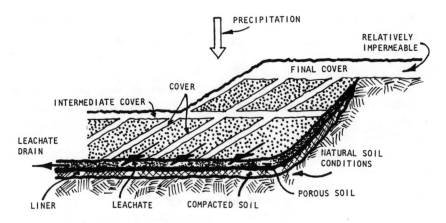

Figure 6-3. Schematic drawing of a hazardous waste landfill.

Barriers between cells or between the landfilled waste and the natural soil consist of a continuous layer of natural or man-made material that restricts downward or lateral escape of the hazardous waste, hazardous-waste constituents, or leachate. These barriers or liners can be compacted clay, soil, or plastic material with a very low permeability.

Landfilling relies on containment rather than treatment or detoxification for control of hazardous wastes. Technologically it is an unsophisticated method of containment. Landfilling is a common method of hazardous-waste management for both untreated wastes and the residues from treatment technologies. Landfills require careful construction, continuous maintenance and monitoring, and a high degree of management and technical attention.

For disposal of hazardous wastes in landfills, consideration should be given to the future protection of the environment. The Resource Conservation and Recovery Act (RCRA) requires that new secure landfills maintain their security for decades after closure. These are arbitrary time frames that do not adequately consider the long-term safe disposal of wastes. Many wastes will retain their hazardous characteristics for long periods of time, and much longer periods of security, perhaps hundreds of years, may be needed.

A "secure" landfill may not be secure over its lifetime. There are several basic causes of failure:

Operating methods may allow too much liquid to enter the landfill prior to closing.

Operating methods may result in cracks, punctures, and physical failure of liners.

Leachate collection systems may contain design or installation defects.

Consolidation over time may result in breaks in the liner or the cover material.

Solvents can affect the permeability of clay liners, causing the clay to shrink or crack.

Although there have been many instances of groundwater contamination resulting from landfills, they remain a key hazardous-waste-management strategy except in certain locations. The Netherlands, for example, has prohibited land disposal because of the high water table in that country. There is incomplete understanding of what happens in the ground and how best to avoid contaminant migration.

A hazardous-waste landfill is designed and operated to be a secure landfill used for permanent burial of the hazardous waste. The design and construction of such a landfill attempts to prohibit contact between the landfill contents and the surrounding environment. Migration (leakage and gaseous emissions) is acceptable only if it can be demonstrated that the rate and extent will not be harmful to man or the environment. The landfill must be structurally sound and not subject to flooding or external or internal displacements.

The primary concern at landfills is to prevent groundwater contamination. Design and management attention emphasize approaches to prevent formation of leachate and leachate migration. These approaches include eliminating free liquids (liquid waste should be dewatered or solidified before placement), diverting surface waters (runon), using relatively impermeable daily and final covers to minimize infiltration of precipitation, compacting wastes, using cells throughout the landfill, collecting and treating leachate, and monitoring groundwater.

The ideal hazardous-waste landfill is underlaid by hundreds of feet of impermeable clay in a nonseismic zone. Wherever possible the landfill should not be placed above a drinking-water aquifer. Such ideal conditions are not easily found, and creating a secure landfill meeting the above goals is difficult and expensive.

Plans for adequate records should be made. The location, dimensions, and contents of each cell in the landfill should be recorded. For future needs the location and dates of each type of waste material placed in the cells should be recorded.

Hazardous-waste landfills are subject to regulations developed by the Environmental Protection Agency (EPA) and states under the 1976 RCRA. The appropriate state agency should be contacted for specific information for the most current information on the design, construction, management, and permitting of hazardous-waste landfills.

Waste Piles

Landfills are part of a hazardous-waste-management system. Frequently storage and treatment, using waste piles and surface impoundments prior to landfills, also are part of such a system. A waste pile is a noncontainerized accumulation of solid, nonflowing hazardous waste used for storage or from which discharge to the land may occur. Waste piles are considered land-disposal facilities subject ot RCRA regulations, including those relating to groundwater and air-emissions monitoring.

If leachate or runoff from a hazardous waste pile occurs, then (a) the pile must be placed on an impermeable base compatible with the waste under conditions of storage, and (b) the pile must be protected from precipitation and runon. In addition, the design of a waste pile must prevent direct discharge of the pile constitutents to surface water during the life of the pile. Wind dispersal of the pile contents must be controlled. Liquids or wastes containing from liquids should not be included in the waste pile. If an impermeable base is not practical, a leachate detection, collection, and removal system may be required.

Unless otherwise authorized by the permit for the facility, at closure all hazardous wastes must be removed from the pile and managed as hazardous wastes. Any component of the containment system contaminated with hazardous wastes must be decontaminated or removed and managed as hazardous wastes. Waste piles that are closed with the waste left in place are subject to the same closure requirements as landfills.

Surface Impoundments

A surface impoundment is a man-made excavation, diked area, or natural topographic depression designed to hold an accumulation of liquid wastes or wastes containing free liquids. Examples include storage, settling and aeration pits, ponds, or lagoons. A surface impoundment must maintain enough freeboard to prevent overtopping by wave action, storms, or overfilling. A freeboard of at least 60 cm (2 ft) is recommended. Any point-source discharge from a surface impoundment is subject to regulatory control.

All earthen dikes must have a protective cover of grass, shale, rock, or comparable material to minimize wind and water erosion and to preserve their structural integrity. Dikes must have sufficient structural integrity to prevent massive failure without dependence on any liner system.

On closure, if the accumulated liquids, wastes, waste residues, liner, and equipment are determined to be hazardous wastes, such wastes must be managed as hazardous wastes. Surface impoundments that are closed with wastes left in place are subject to the same closure requirements as are landfills.

Conceptual Design

An initial step in developing a secure landfill is to determine the type and characteristics of the material to be placed in it. Information on the estimated waste volume, application rate, and physical and chemical properties of the materials expected to be landfilled is important to the actual design and management, to area and volume requirements, and to equipment needs. A list of hazardous wastes that will be permitted to be landfilled at the site should be prepared and used in subsequent design analyses.

Wherever appropriate, treatment and detoxification technologies that can reduce the volume and migration of hazardous wastes added to the landfill should be part of the hazardous-waste-management system. Thermal, chemical, physical, and biological treatment methods are available for this purpose. Encapsulation (enclosing hazardous wastes in an impermeable casing) and chemical fixation (solidification and stabilization) can be used to reduce the mobility of hazardous-waste constitutents in the landfill.

Segregation of wastes by type and chemical characteristics is essential to avoid mixing of incompatible wastes. Incompatible wastes may cause corrosion or decay of containment materials, explosions, fires, or toxic fumes, gases, or mists. Regulations usually require placing incompatible wastes in different cells in a landfill.

Many landfill configurations are possible to fit existing topographic conditions. Landfills can be built below original grade in flat terrain, in sloped and hilly areas, and in a valley. Some of the factors important to the design and construction of a secure hazardous-waste landfill are

Surface water runoff is intercepted and diverted from the site.

The integrity of intermediate and final cover soils is maintained.

Surface erosion is controlled.

Artesian pressures underlying the landfill are relieved.

Groundwater does not move laterally through the landfill.

Liners have been properly installed, and their integrity is adequate for the design period of the landfill.

Leachate collection and removal systems are considered.

Removed leachate is adequately treated prior to discharge.
Incompatible wastes are segregated.

The characteristics of the hazardous wastes to be landfilled are known and/or restricted.

Adequate consideration has been given to closure and post-closure management.

Proper environmental monitoring methods are utilized.

Liquid movement onto and through a landfill can be identified by liquid routing diagrams [6] using a series of lines and arrows to show the likely movement of liquids. Estimated quantities or percentages can be associated with the arrows to show the relative movement of the liquids and to identify the type of controls that may be necessary. Figure 6-4 provides an example of such a diagram.

Runon must be diverted away from the landfill. Closed portions of the landfill should be sloped and graded so that rainwater is not impounded. Terracing at reasonable intervals with proper limits on side slopes (generally 2:1 to 5:1) can avoid erosion and ponding.

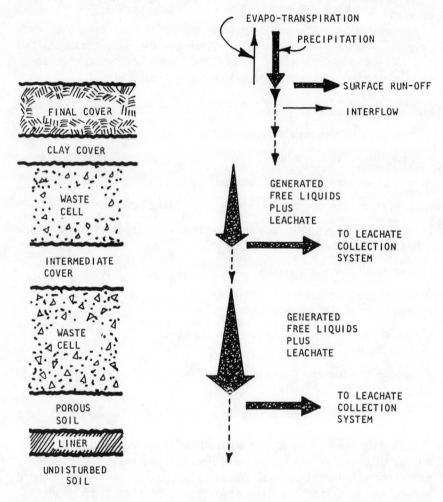

Figure 6-4. Schematic of a liquid routing diagram in a hazardous waste landfill.

Liners

Groundwater protection is a fundamental objective of a secure landfill. This can be accomplished by keeping water out of the landfill by one of the following means

Proper siting to avoid wetlands, floodplains, and high-groundwater areas

Diverting surface runon

Minimizing exposed waste surfaces

Avoid ponding of site precipitation

Proper use of intermediate cover material

Prompt covering and closing of inactive areas

Proper closure and post-closure management

Other methods are proper subsurface preparation, using a suitable liner, and leachate collection and treatment.

Proper subsurface preparation is important, but it varies with site conditions. Subsoils of high permeability should be sealed with natural or manmade materials to provide an unbroken barrier, thus preventing movement of liquids from the landfill. The depth and areal extent of subsurface preparation depends upon local soil conditions.

In addition to subsurface sealing, a liner may be required on the sides and bottom of the landfill. The liner must be installed and used in a manner that allows a reasonable prediction of its efficiency and useful life. Liners must be constructed of materials that have appropriate chemical properties and strength and of sufficient thickness to prevent failure from internal or external pressure gradients. Liner materials must resist failure because of failure contact with the wastes they are exposed to climatic conditions, stress and manner of installation, and stress of daily operation. The liner must also rest on a foundation or base capable of providing support and resistance to settlement or buckling.

Liners function in two ways: (a) They impede the flow of the pollutant and pollutant carrier; and (b) they absorb or attenuate suspended or dissolved constituents. A liner with low permeability is needed to impede pollutants. The absorptive or attenuative capability of a liner depends on the chemical composition of the liner and its mass.

Most liners incorporate flow-control and filtration mechanisms but to different degrees. Membrane liners are the most impermeable but have little absorptive capacity. Soils have a larger absorptive capacity but can be more permeable. However, the greater thickness of a soil liner can result in a low movement of pollutants through the soil liner.

Liners are materials constructed or fabricated by man and include soils and clays of low permeability either available at the site or brought to the site and compacted to reduce permeability and increase strengths. Liners can be classified in many ways (Table 6-3).

Due to their general availability, soils normally are considered as the first alternative for a hazardous-waste landfill liner. Soils used as liners should contain a relatively large proportion of fine particles. The proportion of clay-size particles (less than 2 μm) is an important criterion when selecting a soil to be compacted as a soil liner. Other factors include the gradation of the nonclay fraction and the physiochemical and mineralogical properties of the clay. Depending on these properties and the required saturated hydraulic conductivity, the acceptable proportion of fines in a soil liner cannot be less than 25% [7].

Contaminants in leachate can be attenuated when passing through soil liners by physical, chemical, and mechanical processes such as filtration, ion

Table 6-3. Classification of Liners for Hazardous-Waste Landfills

Type of Classification	Example
Construction method	
On-site construction	Compacted soil, sprayed on liner, raw materials brought to site and liner fabricated on-site
Prefabricated	Polymeric membrane Assembled on-site
Structure	
Rigid	Soil, soil cement
Semirigid	Asphalt concrete
Flexible (no structural strength)	Polymeric membranes, sprayed-on membranes
Materials and method of application	Compacted soils and clays
	Admixes (asphalt concrete, soil cement)
	Polymeric membranes (rubber and plastic sheetings)
	Sprayed-on linings
	Soil sealants
	Chemisorptive liners

Source: Adapted from U.S. Environmental Protection Agency, 1983 (March), *Lining of Waste Impoundment and Disposal Facilities,* rev. ed., SW-870, Office of Solid Waste and Emergency Response, Washington, D.C.

exchange, adsorption, and precipitation. The following soil properties are important in attenuating pollutants: clay content, pH, hydrous oxides (e.g., iron oxides), CEC, and surface area per unit weight of soil.

There is no such thing as an impermeable liner. Liquid is transmitted through all liners to some degree. Hazardous-waste landfills require liners with a very low permeability. Permeability is a numerical measure of the ability of a soil or other liner material to transmit a fluid. When water percolates through a soil, the permeability often is called hydraulic conductivity. Liquid movement through a liner can be determined from the following relationship [7]:

$$J = K(\Delta H) \tag{6-1}$$

where K is the permeability in cm/s, J is the volume of liquid passing through a unit cross section of soil per unit time (cm/s), and ΔH is the hydraulic gradient, that is, the rate of change of hydraulic head in the direction of flow (cm/cm). The validity of this approach has been questioned for very dense clay soils. However, the equation still is the main relationship used to express liquid movement through soils and liners.

The units of permeability (cm/s) can be put into practical perspective by identifying the relative rate of movement associated with different permeabilities (Table 6-4). Thus a soil with a permeability of about 1.5×10^{-3} cm/s (2.0 in.), suitable for slow rate infiltration of wastewater would permit liquid to move about 1500 ft/yr. A liner with a permeability of about 1.0×10^{-7} cm/s would permit liquids to move only about 0.1 ft/yr. A liner for a hazardous-waste landfill is recommended to have a permeability of about 1×10^{-7} cm/s or less. Fine sands have permeabilities in the range of 10^{-1} to 10^{-3} cm/s, clayey sands 10^{-3} to 10^{-5}, cm/s silt 10^{-5} to 10^{-7} cm/s, and clays 10^{-6} cm/s and less. Most undisturbed soils have permeabilities in the range of 10^{-3} to 10^{-7} cm/s [7]. Soils with more than 25–30% clay-size particles are

Table 6-4. Significance of Soil Permeability Values

Permeability Value (CM/s)	Equivalent Movement of Liquid and Associated Contaminants (ft/yr)
1.0×10^{-3}	1000
1.0×10^{-4}	100
1.0×10^{-5}	10
1.0×10^{-6}	1.0
1.0×10^{-7}	0.1
1.0×10^{-8}	0.01

concentrated in the lower range or permeabilities (10^{-5} to 10^{-7} cm/s).

Different clays have different permeabilities and CEC values. Although the capacity of a soil liner to attenuate leachate is a design consideration, the long-term effectiveness of the containment can only be verified by subsurface water and soil monitoring.

Information on site soil properties can be obtained from several sources, the primary being the Soil Conservation Service (SCS) of the U.S. Department of Agriculture. The SCS serves as the coordinating agency for the National Cooperative Soil Survey and cooperates with other government agencies, universities, and agricultural extension services to obtain and distribute soil survey information. Soil surveys include soil maps delineating the apparent boundaries of soil series, written descriptions of each soil series, information on chemical properties, and some engineering data. Specific site verification of soil properties at proposed hazardous-waste landfill sites should be performed.

All other factors being constant, a reduction in the void ratio of a soil results in a lower permeability. With compaction, several changes occur in a soil: a decrease of the effective area available for flow, a decrease in the median pore size, and a reduction in the void ratio. Thus compaction reduces soil permeability.

If existing soils are not adequate to serve as landfill liners, the soil permeability can be decreased by adding importing clays or polymeric materials. Clays such as montmorillonite or bentonite are available commercially and can be incorporated into existing soils to provide an adequate liner. The amount of additive should be such that the amended soil has a permeability of about 10^{-7} cm/s or less. The procedures to follow in preparing such a liner are

Select the most cost-effective soil additive.

Identify the mix that will achieve the desired structural characteristics.

Prepare the site by removing all roots, rocks, or other items that may penetrate the liner.

Grade the site to drain and collect leachate.

Incorporate the additive in the existing soil, generally to at least 30 cm (12 in).

Compact the mixture.

Saturate the amended soil with water. Clay liners can develop cracks if permitted to dry.

Cover the liner with a layer of permeable soil or sand for protection during landfilling and to facilitate leachate collection.

A number of admixed or formed-in-place liners have been used to impound water, although there is not extensive experience with such liners for hazardous-waste landfills. Possible admixes include asphalt concrete, soil cement, and soil asphalt. Example properties of such liners are noted in Table 6-5. Reactions of the landfilled material with these mixtures should be evaluated so that only compatible wastes are used with these liners.

Asphalt concrete is a controlled hot mixture of asphalt cement compacted into a uniform dense mass. Such liners are similar to asphalt concrete used to pave highways but have a higher percentage of mineral fillers and asphalt cement. Harder asphalts rather than softer paving asphalts are better suited as liners.

Asphalt concrete can be compacted to have a permeability of less than 1.0×10^{-7} cm/s. Thicknesses from 4 to 9 in. have been suggested [7]. Asphalt concrete liners may be placed with conventional paving equipment and compacted to the required thickness.

Although asphalts are resistant to water as well as some acids, bases, and organic materials, they are not resistant to organic solvents such as hydrocarbons. Asphalts are not effective liners for petroleum-derived wastes and solvents.

Soil cement is a compacted mixture of Portland cement, water and selected in-place soils. The mixture is a low-strength concrete with greater stability than natural soil. A fine-grained soil produces a soil cement with a permeability of about 10^{-6} cm/s (Table 6-5). There have been few studies to develop a soil cement with low permeabilities [7]. Coatings such as asphalt and coaltar have been added to reduce the permeability of the soil cement.

Soils with less than 50% silt and clay are suitable for soil cement. A high clay content impairs the ability to form a homogeneous material, reducing

Table 6-5. Properties of Soil Mixtures Used as Liners

	Asphalt Concrete[a]	Asphalt Concrete[b]	Soil Cement[c]	Soil Asphalt[d]
Thickness (cm)	5.6	6.1	11.4	10.2
Density (g/cm³)	2.39	2.52	2.17	2.23
Permeability (cm/s)	1.2×10^{-8}	3.3×10^{-9}	1.5×10^{-6}	1.7×10^{-3}
Compressive strength after 24 h (psi)	2230	2328	1323	184

Source: Tests on prepared specimens, unexposed samples, adapted from U.S. Environmental Protection Agency 1983 (March), *Lining of Waste Impoundment and Disposal Facilities,* rev. ed., SW-870, Office of Solid Waste and Emergency Response, Washington, D.C.

[a] 7.1 units asphalt and 100 units aggregate.
[b] 9.0 units asphalt and 100 units aggregate.
[c] 95 units soil, 5 units kaolin clay, 10 units type 5 cement, and 8.8 units water.
[d] 7.0 units SC-800 liquid asphalt and 100 units aggregate.

the efficiency of producing an impermeable layer [7]. the important factors in developing soil-cement liners are cement content, moisture content, and degree of compaction. Soil cements can resist moderate amounts of alkali, organic matter, and inorganic salts.

Soil asphalt is a mixture of on-site soil and liquid asphalt. A silty gravelly soil with 10-25 silty fines is the preferred soil type [7]. The permeability of soil asphalt after compaction varies with the percent compaction and the percent asphalt concrete (Table 6-5).

Sprayed-on linings have been used for water control in ponds, canals, and small reservoirs. Such liners are free of seams, but difficulties are encountered in eliminating pinholes in the liners. Other possibilities are air-blown and emulsified asphalt, urethane modified asphalt, rubber and plastics in liquid and latex form, organic polymers, and cationic salts. Many of the sprayed-on liners have an asphaltic base that may be incompatible with certain hazardous wastes.

The expected life of soil-mixture liners has not been firmly established, especially in hazardous-waste landfills. Such liners have been used at municipal solid-waste landfills for less than a decade. Their effectiveness in hazardous-waste landfills must be assured for many decades. Once a liner is in place and wastes are landfilled, liner failure cannot be easily detected or repaired. Therefore a combination of subsurface water monitoring, leachate collection, and a clay liner is suggested when constructing a hazardous-waste landfill.

Liners consisting of polymeric sheet (polymeric membranes) can be used for hazardous-waste landfills and waste storage impoundments. Such liners can have very low permeabilities and have found use in water reservoirs, sanitary landfills, and other waste facilities.

A wide variety of liner materials are being manufactured and marketed (Table 6-6). These materials have different physical and chemical characteristics, methods of installation, interaction with specific waste constituents, and costs. Some liners have fabric reinforcement. Detailed information on the applicability of a polymeric membrane should be obtained from the manufacturer.

Experience with these liners at hazardous-waste landfills is limited. Of concern is their ability to maintain integrity and impermeability over the life of the landfill. Subsurface water monitoring, leachate collection, and/or clay liners commonly are included in the design and construction of hazardous-waste landfills when polymeric membranes are used.

Laboratory and field experience with mixtures of wastes and liners permit an estimate of the compatibility of some of the liners (Table 6-7). The rationales for the noted estimates are [8] as follows:

Table 6-6. Polymers Used in the Manufacture of Lining Materials

Polymers	Trade Name	Thermo-plastic	Vulcanized
Butyl rubber	—	No	Yes
Chlorinated polyethylene	CPE	Yes	Yes
Chlorosulfonated polyethylene	Hypalon	Yes	Yes
Elasticized polyolefin	—	Yes	No
Elasticized polyvinyl chloride	—	Yes	—
Epichlorhydrin rubber	Herclor, hydrin	Yes	Yes
Ethylene propylene rubber	Epcar Epsyn, nordel royalene, vistalon	Yes	Yesd
Neoprene (chloroprene rubber)	Chemigum, hycar, krynac, paracril, NY syn	Yes	—
Polyethylene	—	Yes	No
Polyvinyl chloride	—	Yes	No

Source: Adapted from U.S. Environmental Protection Agency, 1983 (March), *Lining of Waste Impoundment and Disposal Facilities,* rev. ed., SW-870, Office of Solid Waste and Emergency Response, Washington, D.C.

Caustic petroleum sludge is alkaline and contains salts. Therefore soil sealants, admixed materials, and natural soils would be subject to attack. Such wastes may contain hydrocarbons that could attack asphalts and synthetics, with the possible exception of polyethylene and polypropylene.

Oil refinery sludge contains hyrocarbons, phenols, and heavy metals. Asphaltic and synthetic liners will not perform well.

Acidic steel pickling wastes contain high acid and salt concentrations and will attack soil-based liners. If high-temperature wastes of this type are added, some asphaltic and synthetic liners may not be suitable.

Electroplating sludges contain heavy metals and salts that attack soil-based liners. If these wastes also contain organic additives, asphalt and thermoplastic membranes may be inappropriate.

Toxic pesticide and pharmaceutical wastes may contain as much as 25% organics. Therefore only natural soil cement, and oil-resistant membranes are considered suitable.

The primary components of rubber and plastic wastes are oil, grease, acids, bases, and solids. Generally these are in low concentrations so that all liner materials would be suitable.

Table 6-7. Liner–Waste Compatibilities

Liner	Caustic Petroleum Sludge	Acidic Steel-making Waste	Electroplating Sludge	Toxic Pesticide Formulations	Oil Refinery Sludge	Toxic Pharmaceutical Waste	Rubber & Plastic Waste
Soils							
Compacted clay	P	P	P	G	G	G	G
Admixes							
Asphalt Concrete	F	F	F	F	P	F	G
Asphalt membrane	F	F	F	F	P	F	G
Soil asphalt	F	P	P	F	P	F	G
Soil cement	F	P	P	G	G	G	G
Soil bentonite	P	P	P	G	G	G	G
Polymeric Membranes							
Butyl rubber	G	G	G	F	P	F	G
Chlorinated polyethylene	G	F	F	F	P	F	G
Chlorosulfonated polyethylene	G	G	G	F	P	F	G
Ethylene propylene rubber	G	G	G	F	P	F	G
Polyethylene (low density)	G	F	F	G	F	F	G
Polyvinyl chloride	G	F	F	G	G	G	G

Source: U.S Environmental Protection Agency, 1983 (March), *Lining of Waste Impoundment and Disposal Facilities* rev. ed., SW-870, Office of Solid Waste and Emergency Response, Washington, D.C.

P = poor, combination should be avoided; F = fair, reasonable possibility, combination should be tested before use; G = good, probably satisfactory.

The information in Table 6-7 represents state-of-the-art estimates and should be used only as a general guide. Assessment of liner performance should be based on specific tests. In many cases combinations of liners must be tested before selecting acceptable liner materials. Field experience with liner integrity for hazardous-waste landfills is limited, and greater information on longevity and effectiveness is needed.

The estimated service life of a liner in a particular exposure condition is an important factor in selecting a liner material. For temporary holding situations such as impoundments and waste piles, a short life may be satisfactory. For a hazardous-waste landfill a very long service life is required. Physical, chemical, and biological failure of liners can occur. Principal causes of liner failure are:

Physical—Puncture, tear, differential, settling, thermal stress, hydrostatic stress, abrasion, cracking

Chemical—Solvents, hydrolysis, acids, bases, chemical oxidation

Biological—Microbial degradation

Physical failures are commonly related to subgrade preparation and movement, to operating conditions at the landfill, and to changing hydrostatic pressures. Chemical failures normally are related to the characteristics of the waste in contact with the liner.

Barrier Trenches

Liners are one way to create an impervious barrier between the waste in a landfill and the surrounding soil and groundwater. Another is to use impervious material to create clay cutoff trenches or slurry trench cutoff walls. Such trenches or walls frequently extend from the surface to underlying impervious geological strata (Fig. 6-5).

Barrier trenches can halt the movement of leachate from inactive and abandoned hazardous-waste sites because there are few options, other than excavating and moving the waste and contaminated soil to a secure landfill, for existing sites that are causing groundwater contamination. Compacted clay cutoff trenches usually are used to a maximum depth of 10-15 ft, and when excavation does not extend below the groundwater level. Slurry trench cutoff walls can extend to depths of 100 ft or more [9].

A slurry trench cutoff wall is constructed by excavating a narrow trench (about 2-3 ft wide) through the pervious soil into an impermeable layer. The sides of the trench are prevented from collapsing during excavation and prior to backfilling by maintaining the bentonite slurry 2-3 ft above ground level.

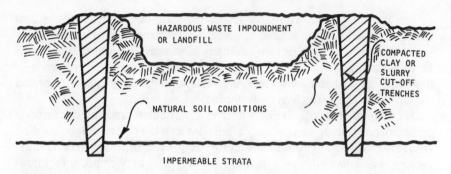

Figure 6-5. Schematic of barrier trenches for existing and inactive hazardous waste disposal sites.

The slurry deposits a filter cake on the walls of the trench that contributes to the low permeability of the completed cutoff. After excavation is complete, the trench is backfilled with material that forms the cutoff wall. The backfill mixture is prepared to have the required permeability characteristics and to be resistant to attack by the waste materials being contained.

The least expensive and most effective type of backfill material is a soil-bentonite clay mixture. The mixture can have permeabilities of 10^{-7} to 10^{-8} cm/s. Low permeabilities can be achieved by using large amounts of bentonite (about 4) with relatively coarse-grained soils (silts, sands, gravels) or by using small amounts of bentonite (0.5–1.5%) with clay soils [10].

Important design considerations for soil-bentonite cutoff walls include the connection details between the cutoff walls and the natural impermeable layer, and the mixture of soil and bentonite used for backfill. To obtain a low-permeability, waste-resistant backfill, a high percentage of fines must be used. A clayey sand or sandy clay containing 30–60% fines blended with a bentonite slurry usually is satisfactory for most waste-isolation applications [10].

Leachate Collection and Treatment

Collection.   A leachate collection and removal system at a hazardous-waste landfill must be installed and managed in a manner that allows its useful life to be attained. A leachate collection and removal system must be compatible with the characteristic of the leachate to be collected, of sufficient strength to resist collapse by pressures exerted by equipment used at the site and by the accumulated waste and covered materials, and capable of withstanding the hydrostic pressure.

A water balance at the site is desirable to determine the need for a leachate collection and treatment system as well as to determine its size. The volume of leachate produced at a landfill is a function of the amount of water flowing through the landfilled material and the liquid generated during waste decomposition and compression.

Precipitation and runon are important factors affecting the volume of leachate. External runoff should be diverted from the landfill site, and intermediate and final covers should help divert the precipitation that falls on the site. Figure 6-1 identified the factors influencing water infiltration and leachate volume, and Figure 6-4 illustrated a liquid routing diagram that can be used to estimate water movement and leachate volume. If there is no surface-water infiltration or groundwater moving through the landfill, the only leachate production will come from the landfilled wastes.

If a water balance indicates that appreciable leachate will occur at a landfill, a leachate collection system should be installed. The collection system will minimize the hydraulic head created at the bottom of the landfill and thus minimize percolation through the liner. As Eq. (6-1) notes, the volume of liquid passing through the liner directly is related to the hydraulic head.

A leachate collection system can consist of perforated pipes in a permeable media that discharge by gravity to a sump from which the leachate is pumped. The design approach used to size and space the collection system is similar to that used for drainage of agricultural soils.

Treatment. Once leachate has been collected, numerous alternatives exist for treatment and disposal [11]. Selection of a leachate treatment process is not simple. If the facility is not in operation, the quantity and characteristics of the leachate to be treated must be estimated. Other factors determining the type and degree of treatment are

Leachate characteristics—organic and inorganic concentrations

Is the leachate classified as a hazardous waste?

Discharge alternatives—surface waters, publicly owned treatment works, hazardous-treatment facility, land treatment

Treatment objectives—a function of discharge alternatives

Technological alternatives—a function of discharge alternatives

Technological alternatives—treatability studies

Cost of alternatives

Permit requirements

Table 6-8. Leachate Treatment Process Applicability[a]

Chemical Classification	Biol. Treat.	Carbon Adsorp.	Chemi. Precip.	Chemical Oxidation			Ion Exch.	Rev. Osmo.	Strip.	Wet Oxid.
				Alkal. Chlori-nation	Ozona-tion	Chem. Reduc.				
Alcohols	E	V		N	G,E[b]	N		V		
Aliphatics	V	V		N	P	N		V		
Amines	V	V		N	N	N				
Ammonia	G,E	N	N	N		N	G		G	
Aromatics	V	G,E	F	N	F,G	N		V		
Cyanide	F,G	N	N	E	E	N			N	
Ethers	G	V		N		N				
Halocarbons	P	G,E		N	F,G	G				
Metals	P,F	N,P	E	N		G	E	E	N	
PCB	N	E		N	E	N		E		
Pesticides	N,P	E		N	E	N		E		
Phenols	G	E		N	E	N		V		
Phthalates	G	E	G	N		N				
Polynuclear aromatics	N,P	G,E	R		G	N		E		E
TDS	N	N	N	N	N	N	E	E	N	N

Source: A. J. Shuckrow, A. P. Pajak, and C. J. Touhill, 1982 (Sept.), *Management of Hazardous Waste Leachate*, SW-871, U.S. Environmental Protection Agency, Office of Solid Waste and Emergency Response, Washington, D.C.

TDS = total dissolved solids; E = excellent performance likely; G = good performance likely; F = fair performance likely; P = poor performance likely; R = reported to be removed; N = not applicable; V = variable performance reported for different compounds in the class.

[a] A blank indicated that no data were available to judge performance; it does not indicate that the process is not applicable.

[b] Use of two symbols indicates differing reports of performance for different compounds in the class.

Operational needs—analytical testing, personnel training, safety, repairs

Post-closure considerations

The leachate characteristics depend on the nature of the landfilled wastes and on the stage of fermentation in the landfill. If the characteristics of the collected leachate indicate it is a hazardous waste, the leachate must be managed as a hazardous waste in accordance with the applicable permits and requirements. If the leachate is discharged as a point source to surface waters, it is subject to point-source permit limitations. If the leachate is discharged to a publicly owned teatment works, pretreatment requirements may apply.

The following list (adapted from [11]) gives potential leachate treatment possibilities:

Biological treatment	Distillation	Land treatment
Carbon absorption	Electrodialysis	Neutralization
Catalysis	Evaporation	Reverse osmosis
Chemical oxidation	Filtration	Solvent extraction
Chemical reduction	Flocculation and	Stripping
Chemical precipitation	sedimentation	Ultrafiltration
Dialysis	Ion exchange	Wet air oxidation

Their applicability is indicated in Table 6-8. The processes applicable at a specific site depend upon leachate volume, characteristics, treatability, and available discharge options. When selecting appropriate processes, consider residues, gases, and by-products that are formed during treatment and that may require additional control and management. Because leachates vary in composition, combinations of processes may be needed to achieve high levels of treatment. Table 6-8 relates leachate characteristics to the more applicable processes.

Closure and Post-Closure

Ultimately, hazardous-waste landfills reach the end of their useful life. In accordance with RCRA regulations, such landfills must be closed, and associated treatment and storage facilities decontaminated in a safe and environmentally acceptable manner. the landfill must be closed in a manner that minimizes the need for further maintenance and controls to the extent necessary to protect human health and the environment and post-closure escape of the hazardous waste or contaminated emissions. Such emissions include leachate, contaminated rainfall, and waste decomposition products.

States have guidelines or regulations to identify the steps to be followed in preparing a closure plan, closing a landfill, and providing for post-closure care. The general components of a closure plan are as follows:

Decontaminate and decommission any hazardous-waste treatment and storage facilities

Provide a final cover for the landfill

Control pollutant migration from the landfill via surface water, groundwater, and air

Establish a groundwater monitoring network for the required period of post-closure maintenance

Divert runon from the landfill

Prevent soil and wind erosion

Control surface-water infiltration and ponding at the closed site

Maintain any leachate collection, removal, and treatment system

Maintain any gas collection and control system

Maintain the integrity of the final cover and any liners

Note in a document examined in a title search, such as a deed, that the land has been used to manage hazardous waste and that its use is restricted

Restrict access to the landfill as appropriate for post-closure use

In developing a specific landfill closure plan, the following factors should be considered:

Type and amount of hazardous waste in the landfill

Mobility and expected migration of landfilled constituents and by-products

Site location, topography, and surrounding land use

Climate

Characteristics of the cover material

Soil and geologic profiles

Subsurface hydrology

The post-closure plan must provide care for an extended period of time after closure, generally for at least 30 years. This is an arbitrary time period that can be lengthened or reduced depending upon technical information related to pollutant migration that is acquired after closure. Properly closed hazardous-waste landfills can be used for purposes such as parking areas, open spaces, parks, ballparks, golf courses, and possibly light buildings. For such uses a satisfactory cover and post-closure monitoring of surface emissions is critical. The compatibilities of various features of a closed landfill and potential site uses are noted in Table 6-9.

On closure, the upper surface of the landfill should be sealed with soil or an impermeable layer of suitable material and graded to prevent ponding of surface water and to minimize erosion. Natural vegetation (grasses and weeds) is permitted to grow on the cover to reduce wind and surface erosion and to provide an aesthetic surface. Deep-rooted vegetation should be avoided since it can penetrate any impermeable barrier.

Landfill covers are vulnerable to soil erosion, unevenly settling fill material, and vegetation, animals, and human activity that can destroy the integrity and effectiveness of the liner. The closure cover thickness is important for reducing these vulnerabilities and is governed by one or more of the following: freeze and thaw effects; dry and wetting effects; trafficability needs; support requirements; gas migration control; infiltration; differential settlement; liner protection.

Table 6-9. Compatibility of Closed Landfill Components and Potential Site Uses

Component	Parking Areas	Open Spaces	Parks, Golf Courses	Light Buildings
Subsurface water control				
Leachate collection and removal system	DA	C	DA	C
Cutoff walls	C	C	C	C
Subsurface drainage	C	C	C	C
Surface water control				
Diversion of runon	C	C	C	C
Precipitation	DA	C	DA	DA
Intermittent grading	NC[a]	C	C	NC
Erosion control	DA	C	DA	DA
Air-emission control				
Gas control	C	C	C	NC
Other factors				
Access	C	C	C	C
Buffer area	C	C	C	NC
Surface settlement	NC	C	C	NC

Source: Adapted from A. W. Wyss, H. K. Willard, and R. M. Evans, 1982 (Sept.), *Closure of Hazardous Waste Surface Impoundments,* rev. ed., SW-873, U.S. Environmental Protection Agency, Office of Solid Waste and Emergency Response, Washington, D.C.

C = compatible; NC = not compatible; DA = may require design alteration.

[a] Not compatible with asphalt or cement-surfaced parking areas.

Generally a layered cover is provided to meet the multiple needs of a cover—that is, leveling of irregular waste surfaces, gas migration if needed, an impermeable layer, and soil for vegetative growth. Figure 6-6 illustrates the components of a layered cover. The buffer soil can be sand or a good grade of soil free of clods and rocks that may interfere with the leveling action. The buffer and the topsoil protect the impermeable layer of clay or membrane from punctures, tears, and cracks. The buffer soil smooths irregularities and provides basic support for surface traffic.

The topsoil should be capable of supporting a desirable vegetative cover and should not crack under drying or freezing conditions. The impermeable layer can be nature or off-site clays, synthetic membranes, or additives and cement mixed with soil. The various types of liners were discussed earlier in this section.

When gaseous emissions require control, a porous layer and vents can be designed into the cover (Fig. 6-7). The gas transmission layer can be a perforated pipe or a gravel layer above the buffer layer, or it can be a permeable buffer layer.

The objective of a final cover is to (a) prevent infiltration of rainwater, (b) prevent accidental dispersal of the landfill wastes, (c) minimize the need for maintenance during post-closure, (d) help provide for a beneficial use

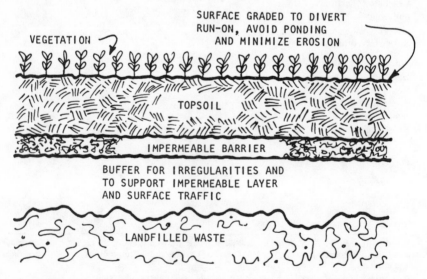

Figure 6-6. Schematic of layered hazardous waste landfill cover.

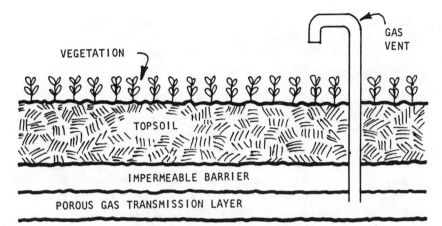

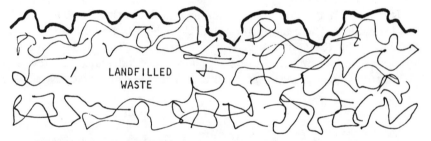

VEGETATION

GAS VENT

TOPSOIL

IMPERMEABLE BARRIER

POROUS GAS TRANSMISSION LAYER

BUFFER FOR IRREGULARITIES
AND TO SUPPORT UPPER LAYERS
AND SURFACE TRAFFIC

LANDFILLED
WASTE

Figure 6-7.   Schematic of a vented hazardous waste landfill cover.

after closure, and (e) protect human health and the environment. The final cover must be designed and constructed to

Minimize migration of liquids through the closed landfill

Require minimum maintenance

Minimize erosion of the cover

Maintain structural integrity by accommodating settling

A case-by-case approach is necessary to evaluate the appropriate type and thickness of hazardous-waste landfill cover material. The proper design, construction, and maintenance of a final cover is evaluated in terms of its ability to meet the above objectives. Detailed information is available on the evaluation of appropriate covers [12].

LAND TREATMENT

Land treatment of hazardous waste is a managed treatment and disposal process that involves the controlled application of a waste onto or incorporated into the soil surface. The objectives of land treatment are the biological and chemical degradation of organic waste constituents and the immobilization of inorganic waste constituents (Fig. 6-8). Land treatment also is a disposal process, because some of the original waste and certain waste by-products remain at the site at closure. Indiscriminate dumping of waste on land is not land treatment. Land treatment differs from landfills in that with land treatment, the assimilative capacity of the soil is used to detoxify, immobilize, and degrade all or a portion of the applied waste. Landfills are containments that store hazardous wastes and control the migration of the waste or by-products from the landfill sites. Liners are not required with land treatment.

The terms "land farm" or "land farming" also have been used to describe this technology. These terms are technically incorrect because the sites are not farmed in the agricultural sense of the term. To maximize the soil biological and chemical processes and to take advantage of the soil assimilative capacity, wastes are applied frequently. The surface traffic, the

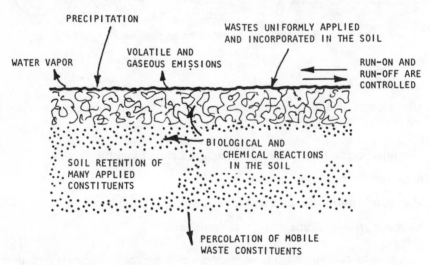

Figure 6-8. Schematic of reactions in a hazardous waste landfill.

incorporation of the wastes in the soil, and the possibility of crop uptake transmitting undesirable contaminants through the food chain normally preclude the growing of agricultural crops on the site during active use. Thus the term "land treatment" is more descriptive and appropriate for this treatment and disposal technology. The degradative and removal processes in the soil are used to treat the applied wastes.

When used for hazardous-waste treatment, the basic philosophy of land treatment is that it is a waste-management practice for wastes treatable in a soil system. Hazardous waste should not be placed in or on a land-treatment site unless the waste can be made less hazardous or nonhazardous by the reactions occurring in or on the soil.

Hazardous-waste land treatment is the controlled application of hazardous waste on or into the aerobic soil horizon, accompanied by continued monitoring and management in order to alter the physical, chemical, and biological characteristics of the waste and to render the waste less hazardous. Land treatment is a widely used practice for the treatment and disposal of nonhazardous and hazardous wastes.

Land treatment can result in the neutralization of wastes with high or low pH values and in the conversion of inorganic constituents to a less mobile or toxic form. However, land treatment is best used for those wastes that are biologically degraded or chemically stabilized. The greater the degree of treatment a waste undergoes in a soil, the more acceptable the waste is for land treatment. A waste that contains components that are degraded, neutralized, made less mobile, and/or made less toxic in the soil is an ideal candidate for land treatment.

Although filtration and dilution occur when waste are land applied, such mechanisms are not acceptable for the treatment of hazardous wastes if they are the only or primary mechanisms that occur. Filtration and dilution provide little net reduction of hazards if they do not alter the chemical state of the waste.

The hazardous wastes that are applied can be liquids, semisolids, or solids. Waste constituents or by-products can be transmitted to groundwater as leachate, to surface waters by erosion and leachate interflow, and as gaseous emissions. Some of the waste remains in place at closure. The design of a land-treatment system minimizes such transmission and losses.

Advantages of land treatment are (a) relatively odorless process, (b) waste application can be repeated safely at frequent intervals, and (c) minimum energy is used to dispose of the wastes. Well-designed and well-operated land-treatment facilities should have low short- and long-term liabilities relative to other available disposal methods. Land treatment has relatively low initial and operational costs, low energy consumption, and a low pollution potential. With proper management land treatment can be an environmentally sound and technically feasible method of hazardous-waste management. With sound controls over application rates and the type and volume of wastes applied to the land, land treatment sites can be reclaimed for productive uses after closure.

General Operating Requirements

The following approaches are involved in operating a land-treatment facility:

Runon should be diverted from active portions of the site.

Runoff from active portions of the site should be collected and disposed on in an environmentally sound manner.

Periodic analysis of the applied waste should be conducted.

Records of the application dates, rates, quantities, and location of the applied wastes should be kept.

The use of the site for food-chain crops should be avoided.

An unsaturated zone monitoring plan must be implemented.

A closure and post-closure plan must be prepared and implemented at the proper time.

Runon must be diverted to avoid excess water entering the site, creating anaerobic or ponding conditions, increasing erosion, or increasing the quantity of contaminated runoff that may require treatment.

Runoff from the active portions of a site must be collected and, if determined to be a hazardous waste, must be managed as a hazardous waste. If it is not a hazardous waste, it should be monitored and analyzed to identify proper treatment and disposal methods. Not all the runoff collected at a land-treatment site will require treatment prior to discharge. Field experience suggests that if the waste is thoroughly incorporated into the soil, contamination of runoff can be minimized or prevented. Careful study of the need for continuous runoff treatment will result in considerable cost savings. Intermittent treatment of runoff only during certain periods should be considered. Runoff treatment may be required only after the beginning of a storm, immediately following waste application, or in the period after waste application but prior to soil incorporation.

If the collected runoff is discharged as a point source to surface waters, it is subject to a state point-source discharge permit. The appropriate state agency should be contacted to determine discharge limits and permit requirements.

The applied wastes should be analyzed periodically for conventional parameters and those constituents that cause the waste to be defined as hazardous. Analysis results also permit researchers to verify the design and operating application rates, and to predict degradation and stabilization rates of the applied wastes.

Records should be kept of where and when wastes are applied, the rates of application, and the types of waste applied. These accounts are useful for maintaining a history of the site, for closure and post-closure plans, and for monitoring results.

Due to uncertainties of transmission of hazardous-waste constituents through the food chain, and because the use of a land-treatment site for the growth of food-chain crops is not necessary to meet national or regional food supplies, food-chain crops should not be grown on a hazardous-waste land-treatment site. These crops must not be grown unless it can be demonstrated that hazardous-waste constitutents that cause human or animal health problems will not be transferred to the food portion of the crop by plant uptake or direct uptake. If such constitutents are transferred, the crops should not be ingested by food-chain animals if the constitutents occur in concentrations greater than those grown on nonhazardous waste sites under similar conditions in the same region. These demonstrations should be based on field evaluations at the specific hazardous-waste land-treatment site. Information relating to demonstrations and the regulations governing the growth of food-chain crops on such sites are identified in state and federal [13] documents.

Unsaturated zone monitoring will detect migration of the applied waste and waste by-products. As a minimum, the monitoring should include soil cores and soil pore water samples. The number of samples to be taken and analyzed is a function of the hazardous-waste constitutents expected to migrate, soil type and permeability, climatic conditions, and waste application rate and frequency.

When the site is no longer used, it must be closed and properly managed after closure to control subsequent migration of waste constituents from the site and to control post-closure use to those functions compatible with the prior history of the site.

Conceptual Design

General. Land treatment is a system consisting of the soil, the atmosphere, and any vegetation that may grow at the site. The soil is the key component of the system and the treatment that occurs is within this matrix. Hazardous wastes are applied to the soil at controlled rates. The soil and waste are mixed to maximize waste biological and chemical reactions. Figure 6-9 identifies some of the reactions that are involved.

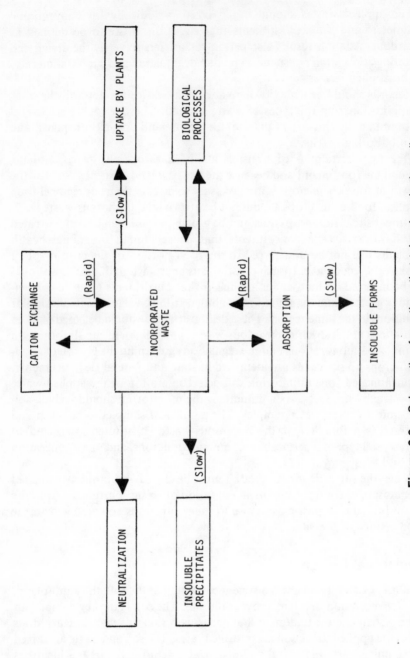

Figure 6-9. Schematic of reactions that take place in the soil.

Degradation of organics results from indigenous microbes and other biological forms such as protozoa, mites, and earthworms. Volatile constituents and gases from biological and chemical reactions can be emitted to the atmosphere. Major chemical reactions include adsorption, precipitation, ion exchange, and gas transfer. The soil also functions as the ultimate sink for immobile, nondegradable wastes and waste by-products (Fig. 6-8). Mobile constituents will move with water percolating through the site.

Atmospheric conditions affect the reactions in the soil by controlling the water content and temperature of the soil. Surface winds increase transfer of oxygen to the soil as well as volatile compounds, CO_2, and other gaseous by-products from the soil. Wind also affects the moisture content and heat balance of the soil. Solar radiation can enhance photooxidation of waste constituents exposed at the soil surface.

Vegetation serves a secondary function to protect the surface, retard erosion, and maintain the integrity of the site. Vegetation will increase water loss to the atmosphere by evapotranspiration. Any waste constituents taken up by the vegetation will be returned to the soil as the vegetation decays, unless the vegetation is harvested.

The soils at a land-treatment site should be permeable and tillable, such as loams, sandy loams, or clay loams. Such soils permit aerobic conditions to be maintained after waste application, result in drainage of high rainfall events, and allow relatively easy incorporation of the waste.

Limiting Constituent Concept. Nondegradation of the site is an important and primary concept for land treatment. The concept implies that wastes shall be applied to the site at such rates and over such limited time spans that no land is irreversibly removed from potential use [14]. Such uses include food-and non-food-chain crops, forests, open space, development, and parks.

Because site nondegradation cannot always be guaranteed as a project is designed and implemented, monitoring must be part of a land-treatment system to determine performance and to extend or shorten the life of the site as appropriate. The design of a land-treatment system involves identifying the waste constituents that limit the feasibility and life of a site, and evaluating methods, such as pretreatment, or process changes to modify the limit.

Designed properly, land-treatment systems and application rates can be determined for specific industrial wastes and for many hazardous wastes. With some wastes the resulting land areas and controls may be large, thus land treatment will not be cost-effective. Decisions to use land treatment for industrial wastes are a function of economic rather than technical feasibility.

The land-limiting constituent [14] or limiting-parameter [4] concept is based upon the fact that soil has an assimilative capacity for inorganic and organic constituents. As the amount of waste applied to the soil increases, the maximum assimilative capacity of a parameter can be exceeded. The constituent that results in the lowest application rate is the limiting constituent and results in the largest land area being required. Use of this land area means that the other waste constituents are applied at a conservative rate that would not cause the other constituents to reach levels of environmental concern.

An example may illustrate this concept. In Figure 6-10 assume that parameter I is the nitrogen concentration in the leachate, and that this concentration is not to exceed 10 mg/L nitrate nitrogen. Also assume that parameter II is the cadmium concentration not to be exceeded in the surface layer of soil, and that parameter III is the concentration of a specific synthetic organic chemical not to be exceeded in the water percolating to the groundwater.

In this illustration nitrogen in the applied waste is the limiting constituent because the maximum concentration of environmental concern is reached at the lowest application rate (R_1). With a given amount of waste to be applied per unit of time, the lowest application rate results in the largest land area. Thus the land area required to avoid any environmental concern occurs as a result of the nitrogen in the applied waste. If the waste were applied at the R_1 rate, neither cadmium nor the synthetic organic chemical would reach concentrations of environmental concern in this example.

A variety of parameters can limit waste-application rates. Examples include nitrate leached to groundwater, cadmium in food-chain crops, synthetic organic compounds in surface, groundwater, and crops, salts that inhibit seed germination, and metals that may be phototoxic to crops. Limits for such constituents can be found in state water quality criteria and in regulations for food-chain crops [13]. Because food-chain crops are rarely grown on hazardous-waste land-treatment sites, the important short-term, limiting constituents at such sites are water quality related (i.e., constituents that impact surface and groundwater quality).

However, the long-term limiting constituents may relate to food-chain crops because the intent of the nondegradation concept is not to preclude such use after closure of the site. Monitoring of the soil and subsurface water will identify if and when limits to such use are being approached. Reduction in application rates or ceasing application can preserve the future of the site long before the limit is exceeded.

In Figure 6-10, parameter I is the limiting constituent in the waste being applied and results in the acceptable application rate being R_1 (quantity of waste applied per unit area being required, because with a specific amount of waste to be treated in a period ot time, Q, the active land area, A_1, needed is Q/R_1.

If the limiting parameter or constituent is eliminated through pretreatment, industrial process changes, or other methods, parameter II becomes

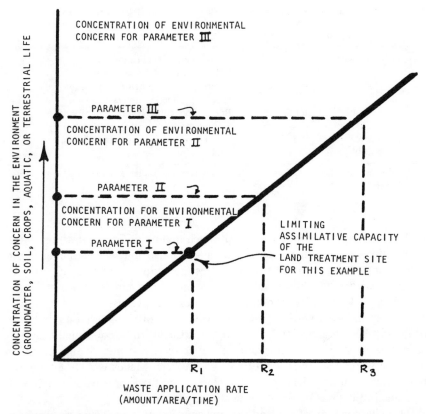

Figure 6-10. Schematic of land limiting concept or limiting parameter concept for land treatment.

the limiting parameter, and the maximum acceptable application rate is R_2. The required land area, A_2, is now smaller. The need for and value of pretreatment is a function of the relative difference in land areas. There are several methods that can be used to relieve specific limiting constituents (Table 6-10). Because most hazardous wastes applied to land-treatment sites are not liquid, hydraulic limitations rarely limit the land area at such sites.

With hazardous wastes the limiting constituents usually are synthetic organics, metals, and nitrogen; with nonhazardous sludges the limiting constituents are usually soil permeability and nitrogen. Salts can be a limiting constituent in arid regions. The rate of degradation of organics usually is so large that biodegradable organics rarely are a limiting constituent. Nonbiodegradable organics will accumulate in the soil and, unless they leach in unacceptable concentration, will not be a limiting constituent.

Table 6-10. Methods to Modify Land-Limiting Constituents

Constraint	Modification
Nitrogen	Reduce before application
	Enhance losses after application
Metals	Remove or reduce before application
	Modify production processes
Synthetic organics	Remove or reduce before application
Pathogens	Remove or reduce before application
Salts	Remove or reduce before application
Soil permeability	Underdrain [a]

[a] Not used at a hazardous-waste land-treatment site if a hazardous-waste leachate or highly contaminated leachate will result.

The land-limiting concept is an approach that identifies a constituent that limits the use of land treatment as a hazardous-waste-management technology. The approach also identifies that constituent that, if reduced or removed, would reduce the necessary land area and system costs.

Table 6-11 provides another illustration of the use of the limiting-constituent concept with a hazardous waste. In this example one or more of the synthetic organics is the limiting constituent, thus requiring the largest land area to avoid exceeding one of the concerns—water quality criteria, drinking-water standards, or food-chain impacts. Metals would be the next limiting constituent if the constraint due to the synthetic organics were relieved by reducing their concentration in the applies waste.

The determination of the land-limiting constituent is made on a site-specific basis to select the parameter or class of constituents requiring the largest land area. A site and waste-specific analysis is necessary because soil, waste, climate, and other factors vary from site to site. The type and cost of pretreatment options to reduce or eliminate land-limiting constituents also should be determined and compared to the relative land areas needed for the limiting constituents.

The major constituent classes to consider in determining the limiting constituents are those that degrade biologically in the soil (i.e., biodegradable organics); are relatively immobile and nondegradable and therefore accumulate in the soil (i.e., metals, refractory organics); and are mobile, nondegradable, and impact surface and groundwater uses (i.e., salts, synthetic organics).

Figure 6-11 illustrates the response that occurs when these major constituent classes are added to a land-treatment site. At a land-treatment site wastes are applied intermittently to avoid overloading the site and to permit degradative and detoxification reactions to occur. Figure 6-11(a) illustrates what happens to degradable compounds. Application rates are limited by

Table 6-11. Illustration of the Limiting Concept

Waste Constituent	Application Rate Limitations	Estimated Land Area Needed
Nitrogen	Nitrate groundwater	⇒
Synthetic organics	Surface-water quality criteria Drinking-water standards Food-chain impact	⇒
Metals	Food-chain impact Water quality criteria or standards	⇒
Water	Soil permeability	⇒
Biodegradable organics	Maintenance of anaerobic conditions Surface plugging	⇒

the amount of material that can be easily applied and incorporated and that will not inhibit the microorganisms involved in the biodegradation.

To have adequate biodegradation, sufficient nutrients must be available in the soil. Nitrogen, phosphorus, and potassium may have to be added to the soil if the hazardous wastes are deficient in these compounds. After each application, the organics will be degraded. The rate of degradation depends on the temperature and the ease of degradation. The degradation normaly is approximated by a first-order rate reaction. When the organic compounds are reduced to above 15–20% above background, another application can occur. Soil monitoring and experience dictate when the subsequent applications should occur.

Immobile and nondegradable constituents are retained and accumulate in the soil. With intermittent applications a stepwise accumulation occurs [Fig. 6-11(b)]. The assimilative or limiting capacity is based upon acceptable limits for accumulation. Examples of such limits are cumulative amounts of cadmium, plant phytotoxic limits of metals or organics, or inhibiting concentrations of salts. When the appropriate limits are reached, application of the waste must cease. Monitoring for specific constituents can verify and modify actual application rates and the expected life of the site.

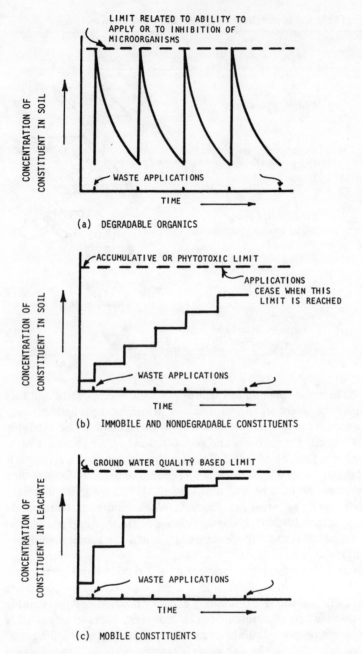

Figure 6-11. Changes in soil characteristics due to waste applications at a land treatment site.

Limits for mobile constituents are based on not exceeding groundwater quality standards. Mass balances that include precipitation, infiltration, and gaseous emissions and solubilization rates of constituents can be used to determine proper application rates. With repeated waste applications the leachate flushes out the original soil water, and the mobile constituents reach the groundwater-quality-based limit as shown in Figure 6-11(c). If the initial application rate is determined correctly and no changes take place in the soil-waste system, the groundwater limits should not be exceeded with continued waste applications. An example of a groundwater-based limit is the 10-mg/L nitrate nitrogen standard set for drinking water. Drinking-water standards for other compounds are available from state and federal agencies.

Information and experience from a number of disciplines, such as soil chemistry, soil analysis, agronomy, hydrology, agricultural engineering, environmental engineering, environmental science, and microbiology may be needed to determine the land-limiting constituents and to design the system. Accurate analyses of representative waste samples are important to determine the constituents or classes of constituents that should be evaluated as a limiting constituent. Table 6-12 lists the waste characteristics to be determined. This is only a partial list because there may be other compounds of concern in the applied waste. Chemicals likely to be present in the wastes are candidates for analysis. The parameters in Table 6-12 are related to the general classes noted in Table 6-11 and to the impact that the wastes may have on soil characteristics and post-closure site use.

Table 6-12. Waste Characteristics to Determine Land-Limiting Constituents and to Design Land-Treatment System

Classes	Characteristics
General	Volume, dry solids, moisture content, pH
Nitrogen	Organic nitrogen, ammonia nitrogen, nitrate nitrogen
Nutrients	Phosphorus, potassium, sulfur
Metals	Copper, lead, nickel, zinc, cadmium
Ions	Sodium, chloride, calcium, magnesium, carbonate
Synthetic organics	Those relevant to the industry and waste-production processes
Biodegradable organics	A measure of oxygen demand (BOD, COD, oxygen uptake rate)
Other	Substances causing a waste to be listed as hazardous

Information and approaches are available to determine the limiting constituents of a particular waste [4, 14, 15, 16]. Critical constituents for which little assimilative capacity information exists should be evaluated by using laboratory or field experiences. Conditions in the experiments should be as near as possible to those that will occur at the site.

It provides a standard methodology for considering important site-specific soil data, climatic conditions, waste characteristics, and environmental concerns.

It provides a direct relationship between waste characteristics and land assimilative capacity.

The methodology is useful for regulatory agency personnel as well as for the design engineer, because all facets of concern are considered.

The methodology is based upon scientific fundamentals rather than empirical, black-box approaches that cannot be extrapolated due to variations in soil, climate, and waste characteristics.

Attention can be focused on management techniques, site selection, and pretreatment aspects that improve the assimilative capacity of the site and avoid environmental and public-health problems.

Monitoring programs can be focused on the critical parameters, such as the limiting constituents.

Pretreatment.   The characteristics causing a waste to be classified as hazardous may be among those that limit the use of land treatment. However, having identified the limiting constituent or constituents, researchers can direct their attention to those constituents that, if reduced, will decrease the needed land area. The costs of pretreatment must be balanced against the costs of the reduced land area when determining if pretreatment is cost effective.

Pretreatment is any end-of-pipe or in-plant process change that alters the characteristics of the hazardous waste. Pretreatment is used to reduce the volume of wastes to be treated, to reduce constituents limiting the use of land treatment, and to detoxify the wastes. The limiting-concept approach identifies which constituents should be removed by pretreatment and how volume and constituent reduction will affect needed land area and related factors. The extent of removal is determined by economic factors.

Many physical, chemical, and biological end-of-pipe technologies can be used to pretreat hazardous wastes prior to land treatment. Descriptions and details of the processes are available in many texts and documents [14, 15, 17, 18] and are also described in chapter 3. In-plant controls are specific to the particular industry and production process.

Most end-of-pipe technologies are separation and concentration processes that produce a clarified and purified treated liquid and a concentrated solution, slurry, or solid. One of the results is volume reduction. Many end-of-pipe processes do not detoxify, degrade, or selectively remove hazardous-waste constituents. Table 6-13 summarizes end-of-pipe treatment processes and their potential use for pretreating hazardous wastes prior to land treatment.

System Design and Management

A land-treatment system is a dynamic system in which the waste, soil, climate, and management interact. Previous sections have focused on the basic soil and waste interactions and on the conceptual design. This section focuses on the general system components, operation, and management. The components of a land-treatment system are

Active application site areas

Future application site areas

Buffer areas

Waste storage areas

Sufficient areas near waste storage and the application site for emergency procedures if needed

Runon diversion

Runoff retention ponds with emergency spillway

Roads to bring waste to the site and to apply the wastes

Equipment washing and storing areas

Wastes to be treated are applied intermittently to each unit of land. This permits waste degradation and soil interactions to occur and aerobic soil conditions to be restored prior to the next application. Wastes may be applied during one season of the year or at more frequent intervals. The actual intervals are determined by the degradation rate of the applied material and the need to empty storage units and dispose of wastes. Because a land-treatment site is not used for agricultural purposes, frequent applications are commonly used, which minimize the need for excessive storage between applications.

The land-treatment area can be subdivided into smaller areas to which the waste is applied sequentially, one subarea after another during a season,

Table 6-13. End-of-Pipe Pretreatment for Hazardous Wastes

Method	Heavy-Metal Removal	Organic Removal	Destruct. of Organics	Volume Reduct.[a]	Applies to[b]	Comments
Biological						
Activated sludge	Some	Yes	Yes	Little	L	Biological processes are sensitive
Trickling filters	Some	Yes	Yes	Little	L	to toxic compounds
Ponds and lagoons	Little	Yes	Yes	Little	L	
Anaerobic digestion	Little	Yes	Yes	Little	L, SL	
Composting	None	Yes	Yes	Yes	SL, S	
Chemical						
Carbon absorption	Possible	Yes	No	Little	L	
Calcination	Some	No	Yes	Yes	SL, S	
Catalysis	No	No	Yes	No	L	
Chlorinolysis	No	No	Yes	Some	L	Converts chlorinated hydrocarbons
Dialysis	Some	Some	No	Some	L	Separates salts
Distillation	No	Yes	No	Some	L	
Electrodialysis	Possible	No	No	No	L	
Hydrolysis	No	No	Yes	No	L, SL	
Ion exchange	Yes	Possible	No	No	L	Recovers inorganic salts
Liquid-liquid extraction	No	Yes	No	No	L	
Neutralization	No	No	No	No	L, SL, S	
Ozonolysis	No	No	Yes	No	L	
Photolysis	No	No	Yes	No	L	
Reverse osmosis	Some	Some	No	Little	L	
Physical						
Centrifugation	Some	Some	No	Yes	L, SL	
Evaporation	No	Some	No	Yes	L, SL	
Filtration	Some	Little	No	Little	L, SL	
Flotation	Little	Yes	No	Little	L, SL	Dissolved air, chemicals may help
Sedimentation	Some	Some	No	Little	L, SL	Chemicals can be used
Stripping	No	Yes	No	No	L	Air and steam can be used
Ultrafiltration	Some	Some	No	No	L	

Source: Adapted from U.S. Environmental Protection Agency, 1983 (April), *Hazardous Waste Land Treatment*, rev. ed., SW-867, Office of Solid Waste and Emergency Response, Washington, D.C., Chap. 5.

[a] The adjectives relate to the reduction of initial waste volume that occurs; many of the processes concentrate separated constituents; i.e., activated sludge concentrates metals and organics in the settled sludge, and flotation concentrates organics in the float. Such processes may reduce the volume of hazardous waste to be handled.

[b] L = liquids; SL = slurries and sludges; S = solids.

until a soil constituent limit is reached or monitoring indicates that the desired groundwater quality will be degraded. At such time the site is closed. Alternatively, each subarea can be used until the limits are reached, the subarea closed, and application transferred to the next subarea.

The land-treatment site should have adequate buffer areas between the site and potential visual and olfactory receptors. A buffer area minimizes aesthetic and odor problems. Many land-treatment sites are on the property of the industry producing the waste, and aesthetic problems may be minimal. The buffer zones also should permit adequate area for possible emergency procedures. Such procedures include waste removal, cleanup, and treatment.

Waste storage areas may exist either where the wastes are generated or at a land-treatment site. The storage volume will be determined by the application frequency, which is influenced by climatic conditions. Adequate buffer zones should be around the storage areas for aesthetics, odor control, and aerosol drift.

Water control is important at the land-treatment site because the site should be isolated hydrologically. Runon diversion structures and runoff retention ponds must be included at the site. The runon diversions can be grassed waterways or open ditches designed to divert the upgrade surface drainage. Every retention pond should have an emergency spillway for overflows. Because the runoff can contain hazardous-waste constituents and conventional pollutants, environmental damage may result from pond failure. The spillway should be vegetated to prevent erosion and pond failure in the event of an overflow and, if possible, drain to another retention pond or to a land-treatment subarea. Figure 6-12 illustrates the components of a land-treatment site.

Runoff retention ponds are designed to accommodate specific precipitation events, such as the 24-h, 25-y return-period storm, and to have an adequate freeboard for extreme events. The ponds act as flood-control reservoirs and should be drawn down between storm events. Collected runoff volume can be reduced by natural evaporation and by pumping, siphoning, draining, or irrigating the pond contents to adjoining land.

The collected runoff can be treated by several methods. If the industry operates a wastewater treatment plant, the runoff can be pumped to such a plant on a controlled basis so that the treatment plant is not hydraulically or otherwise overloaded. The collected runoff also can be applied to the existing land-treatment site in a controlled manner. As a last resort, an on-site treatment system can be constructed for discharge to surface waters.

Equipment washing facilities should be provided at the site so that no hazardous waste leaves the site on the equipment. Adequate roads and equipment storage areas also should be provided. Fences may be needed to prevent public access to the site.

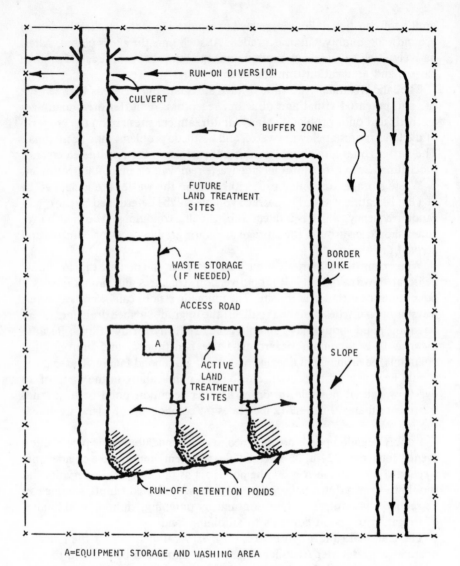

A=EQUIPMENT STORAGE AND WASHING AREA

Figure 6-12. Schematic of land treatment site components.

The waste characteristics will determine the appropriate application methods. Liquids can be surface applied by irrigation techniques. Semisolid wastes can be spread over the site by trucks equipped with flotation tires, or they can be subsurface injected. Wastes too thick to pump and low-moisture solids can be spread over the surface of the land. Surface-applied wastes should be mixed with the soil by plowing, rototilling, or discing as soon after spreading as possible so that contaminated runoff is minimized. Surface spreading is not appropriate for wastes with volatile constituents that cause environmental problems or that endanger those working at or adjacent to the site.

Uniform application is important to continuous and proper use of the site. The most appropriate application equipment is based upon

Ability to handle the solids content of the waste

Minimizing any risk to operator health and safety and to the environment

Minimizing operator contact with the waste

A long service life

Not being adversely affected (i.e., corroded) by the waste

Reasonable capital and operational costs

Erosion and dust control must be incorporated in site management plans. Dust can occur during discing and mixing operations and from roads during dry periods. Dust from the site can contain some fraction of the applied hazardous waste, and it should be suppressed. Both wind- and water-erosion control are needed during site use and post-closure.

Closure and Post-Closure

At some point the soil assimilative capacity of the land-treatment site is reached, or the site is no longer needed and must be closed. A closure and post-closure plan must be prepared and approved by the permitting agency. Closure and post-closure operations should

Control the migration of contaminants to the groundwater

Control release of any contaminated runoff

Control release of any airborne contaminants

Avoid update of contaminants by food-chain crops

The factors that must be considered in developing closure and post-closure plans and operations are [13]:

Type and amount of hazardous waste and other waste constituents applied to the site

Mobility and expected rate of migration of contaminants

Site location, topography, and surrounding land use

Climate, including amount, frequency, and pH of precipitation

Geological and soil profiles at the site, subsurface hydrology, and soil characteristics, including CEC, total organic carbon, and pH

Unsaturation zone monitoring

Type, concentration, and depth of migration of hazardous-waste constituents in the soil as compared to background concentrations

Collection and appropriate treatment and disposal of site runoff, erosion control, runon diversion and monitoring of soil, soil pore water, and groundwater characteristics should continue during the post-closure period. Access to the site should be restricted to appropriate post-closure uses.

When a land-treatment site is closed, a final cover should be installed and an appropriate vegetative cover should be grown, and contaminated soils must be removed to a secure hazardous-waste landfill.

If the land-limiting constituent is properly identified and if the land-treatment system is properly designed and operated, the land-treatment site should be closed before any soil constituents reach concentrations of concern. Therefore there should be no contaminated soils that require removal and disposal.

If there is active decomposition of the applied organics, application of any cover material should be delayed until adequate stabilization has occurred. Depending on the degradation rate, the waiting period may be several months to a year or longer. Fertilizer and lime to adjust the soil pH to greater than 6.5 may be added prior to closure to increase the organic degradation rate, provide nutrients for the vegetation, and reduce mobility of metals.

Practical Application

Land treatment of industrial wastes has been practiced for many years. The following discussion illustrates results that have been obtained.

Chlorophenols have been degraded by the microorganisms in soil and sediment [19]. Under aerobic conditions at 0 °C and 4 °C o-chlorophenol, p-chlorophenol, and 2, 4-dichlorophenol were degraded.

Halogenated hydrocarbons are particularly interesting because of their general persistence in the environment. Therefore the adsorption, mobility, and degradation of hexachlorobenzene (HCB) and polybrominated biphenyls (PBB) in soils have been studied [20]. Both PBB and HCB were strongly adsorbed by soil materials, with HCB being adsorbed to a greater extent than PBB. The adsorption capacity and mobility of PBB and HCB were correlated with the organic carbon content of the soil materials. In a soil incubation study both PBB and HCB persisted for up to six months with no significant microbial degradation. Because of their low solubility in water and strong adsorption, these compounds resist movement with water through soils. They are, however, highly mobile in organic solvents. Because PBB are not degraded, leached, taken up by plants, or volatilized, they will be a long-term component of the soils they are applied to. The same conclusion was indicated as true for HCB if the applied wastes are well covered to avoid volatilization.

Organic solvents are used in most industries and are commonly found in industrial wastewaters and sludges. The response of the soil microorganisms to organic solvents in land-applied wastes depends upon the amount added and the nature of the chemical species. When the application rates are low, there is little or no adverse response. At higher application rates, both short- and long-term effects are noted. Data related to the decomposition rates, the growing plant response, and the critical soil concentration are needed to identify more precisely the design relationships for such solvents. Information on the decomposition and environmental losses of organic solvents is available [21] and can help identify proper land-application rates.

Oils, oily sludges, and petroleum industry wastes have been applied to land in many locations for a long time. Perhaps the greatest amount of experience related to the land treatment of industrial wastes has originated from these studies. As early as 1970, published reports identified land spreading as a nonpolluting method of disposing of oily wastes. Soil microbes were able to oxidize and decompose a large quantity of petroleum hydrocarbons under a wide range of soil and environmental conditions [22]. At proper oil application rates, the eventual effect on the soil was considered to be an improvement of physical and chemical soil properties such as an increase in organic content, porosity, and moisture-holding capacity.

Detailed analyses of the intensive land treatment of six oils at three locations (Pennsylvania, Oklahoma, and Texas) have been conducted [23]. At these locations a single application of 0.003 m^3/m^2 was made. The concentration of the oils in the soils decreased significantly at all locations, with the average reduction ranging from 48.5 to 90%, depending on the type of oil and the location. All classes of compounds degraded, the more polar hydrocarbons degrading more slowly. Significant increases in hydrocarbon-utilizing microorganisms were noted in all of the treated plots. At the loading rates used the concentrations of residual oils or their oxidation products were high enough nine months after application to inhibit plant growth.

Concern over oil spills and waste-management approaches in northern climates spurred investigation of oil degradation in cold climates. In a three-year study [24] oil was applied to replicate plots at an application rate of 7.3 L/m². Fertilizer was added to some of the plots. The initial oil treatment killed all surface plant growth. Revegetation did occur and was fastest on the fertilized plots. In the plots there was a rapid increase in bacteria with a rapid disappearance of n-alkanes and a continuous loss of saturated compounds. The study indicated that microbial activity can accelerate the degradation of oil spilled on northern soils if the environment is modified to support microbial growth.

In a comparable study Prudhoe Bay crude oil was added as 0, 3, and 6% of the weight of soil [25]. Disturbance of the soil and fertilization hastened oil mineralization. The study was conducted at 10 °C, and the time required for doubling of the indigenous aerobic population was two days. Fertilization further reduced the doubling time.

Hydrocarbon residues from crude-oil storage tanks also have been successfully degraded in Canadian soils [26]. Experimental plots were treated with fertilizers, and the residual sludge was applied at 41.1 tons/ha, the equivalent of 9.45 tons/ha of hydrocarbons. The oil content of the soil immediately after application was 1.45%. The experiments were conducted over a two-year period. About 50% of the total applied hydrocarbons were degraded in this period. About 55% of the saturated compounds, 50% of the mono-aromatics, 57% of the diaromatics, 44% of the polyaromatics and polar compounds, and 11% of the high-molecular-weight compounds were degraded during the study period.

Experience with the land treatment of oily wastes and biological sludges from the petroleum industry has shown [27] that this technology can be a viable treatment and disposal method for these wastes. High oil-removal efficiencies have been experienced at all such facilities. The organics have been degraded, refractory hydrocarbons have been retained within the zone of incorporation, and the metals were immobilized. Conservative waste-application rates for the site, appropriate monitoring, and a conscientious group of operators are keys to the success of the effort.

For example, extensive investigations have indicated that land treatment is a sound way to manage the treatment and disposal of petroleum refinery wastes. At the Conoco refinery in Oklahoma [28], wastes were applied to the plow zone of the soil to produce about 5% oil in the soil on a weight basis. The added waste contained 16% oil, of which 76% was organic carbon (TOC). The plots received supplemental fertilizer. Over 20 months of exposure, an oil loss of 54 to 67% was found, and a significant portion of the soil had been transformed into microbial mass of partially degraded organic matter. No leaching of oil, TOC, or heavy metals was observed.

Studies to identify the factors influencing the biodegradation of refinery wastes [29] found that the soil moisture content had an effect on the biodegradation rates of API Separator waste from a petrochemical plant and a petrochemical refinery. The rates decreased when the soil became too wet or too dry. Temperature also affected the biodegradation rates. The rate at 10°C was about half of the maximum found at 30°C. Adjustments in the C:N ration to about 10:1 increased the rate of biodegradation. Microbial analysis revealed that both hydrocarbon-utilizing bacteria and fungi were present in the soils.

The quantity of hydrocarbon emissions from land-treatment sites has been assessed [30]. The volatility of the applied material was found to be the most important factor that affected the rate and amount of emissions. From 0.01 to 3.2 weight percent of the oil refinery sludges applied to the soil was vaporized. For the applied wastes this was the equivalent of 0.1 to 3.5% of the applied oil. The highest emission rates occurred within 30 min after the application of the sludge. Subsurface injection of sludges appeared to be the preferred method for minimizing emissions from land-treatment operations.

In summary, the above and other studies indicate that for many wastes, including those identified as hazardous wastes, land treatment can be a technically and economically feasible and environmentally sound treatment and disposal method. Determining proper application rates, using the land-limiting constituent concept, and having good operation are important to the success of this technology.

SITE SELECTION

The feasibility of landfills and land treatment for the management of hazardous wastes depend on factors such as the type, amount, and characteristics of the wastes, laws and regulations, public perception and acceptance, and soil and site characteristics. This section focuses on important characteristics for selecting a landfill or land-treatment-system site.

The objective of landfills and land treatment is to contain and detoxify hazardous wastes. To do this, sites and soils must

Minimize migration of hazardous or conventional contaminants from the site to surface water, groundwater, or atmosphere

Minimize risk to workers and adjacent public as the wastes are handled, stored, or applied at the site

Permit reasonable closure and possible subsequent use of the site after closure

Minimize direct contact with the public, domesticated animals, and important wildlife

Careful site selection is important to the ultimate use and feasibility of the facility. Once the landfill or land-treatment site is operating, little control over natural processes or public activities is possible.

Approach

The design and management of landfills and land-treatment systems are site specific. The technical and economic feasibility of such methods depends on the topography, soils, climate, and hydrology of the site, on transport distances from the waste source to the site, and on current and projected land uses for the available land.

Considerable information exists that can be used for site characterization and selection, but much of it is in a format and in disciplines not commonly used by design engineers. Design engineers must locate and use the available information and acquire additional information that may be needed.

The site selection approach can be considered to have at least three major components (Fig. 6-13):

General evaluation—problem definition and initial solution evaluation

More detailed analysis of options

Final site selection and design

The general evaluation screens out sites and pretreatment alternatives that clearly are not feasible. The more detailed analysis closely evaluates the site and soil characteristics, includes economic and political considerations, and narrows site options. Final site selection and design can involve detailed economic, technical, and political evaluations of specific sites.

The general evaluation is concerned with defining technical feasibility; for example, is it worth considering these methods in greater detail, or are other methods clearly more appropriate? Figure 6-14 identifies items considered in this evaluation. The general area requirements are identified after consideration of waste characteristics, characteristics of possible sites, and possibilities to modify waste characteristics to reduce the needed land area. When the general land area needs and location are identified, the approximate costs can be estimated. Costs and pertinent factors such as site ownership and obvious public reactions can be used to judge whether a landfill or land treatment continues to be a possible hazardous-waste disposal option. If a positive

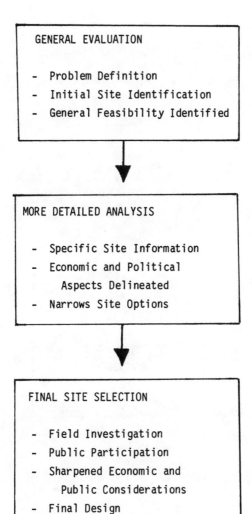

Figure 6-13. Schematic approach to landfill or land treatment site selection.

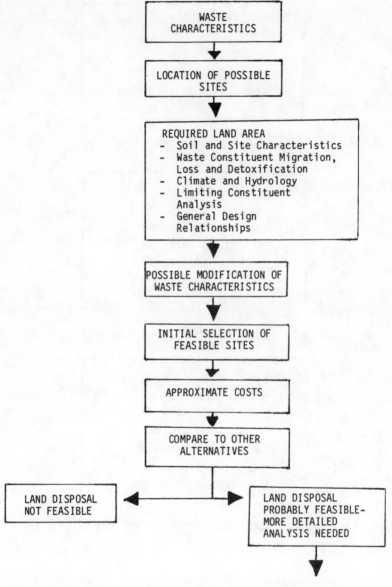

Figure 6-14. Schematic of items considered in general site selection evaluation.

conclusion is reached, a more detailed technical, economic, and political analysis of the more feasible sites can be made.

The selection of landfill and land-treatment sites is not likely to involve an extensive number of sites. Because of the constraints related to the site (e.g. minimize pollutant migration), as well as public and environmental risk, there may be only a few potential sites for evaluation. For many hazardous-waste-generating industries, such sites may be on their own property.

The more detailed analysis is initiated after land disposal has been estimated to be potentially feasible. This analysis is a cost-effectiveness evaluation of the possible sites and land-disposal options. It must evaluate critical site characteristics of each of the candidates, identify the land-disposal method to be investigated for each candidate site, define the detailed design requirements for each land-disposal method and candidate site, develop a cost-effectiveness evaluation for each alternative, encourage public participation throughout the planning stages, determine whether land disposal continues to be feasible, and identify a site or sites for intensive field investigations. The analysis use available published information, although field checks may be desirable.

The public should be involved actively in both the detailed analysis and final site selection steps. Public participation and understanding is crucial to the final selection of a land-disposal site as well for the implementing and operating such a site. The economic and environmental impacts on the adjacent community should be addressed, and public concerns such as odor, groundwater contamination, health effects, and site closure and ultimate use should be discussed. One or more public hearings are usually required before a site or permit is approved by a governmental agency, even for sites on industry-owned land.

Land Use and Political Considerations

Unacceptable areas based on land use and political considerations should be screened out in the general evaluation. The screening procedure involves using available land-use information and discussion with local and regional government officials.

Land-use information varies widely in terms of specificity. Local and regional agencies can provide such pertinent information. An excellent example is the New York State Land Use and Natural Resource Inventory (LUNR), which tabulates statewide land-use data derived from the interpretation of aerial photographs. The LUNR inventory divides land into 13 main categories (agricultural, recreational, residential, etc.) and 130 subcategories (vineyards, golf courses, low-density residential, etc.). Transparent overlays of the LUNR information are available for use with 7.5 min U.S. Geological Survey (USGS) topographic maps.

Few states have developed as detailed an evaluation of existing land use as New York State has. However, the same type of information available in LUNR can be developed from aerial photographs by experienced personnel. Topographic maps also can provide information regarding residential densities, recreational areas, and surface-water location.

The major land-use restrictions relate to creating a public nuisance or a public or environmental risk. Sites near areas with high population densities may be unacceptable.

Actual or projected use of the land at or adjacent to a potential site does not impact on the technical aspects of site selection. Such uses do impact the economic, political, and social aspects of site selection. This latter aspect may pose important constraints on the use of a possible site. Zoning and related land-use restrictions should be evaluated as part of the site selection process.

Ideally, a land-treatment operation should be located at the site of or immediately adjacent to the waste generator. If hazardous wastes must be transported to an off-site facility via public roads, rail systems, or other means, the transporter must comply with state and federal regulations concerning the transport of hazardous wastes.

Political boundaries can be constraints in selecting sites. Questions can arise regarding whether one municipality can own land within the jurisdiction of another municipality. One political jurisdiction may not permit hazardous waste from another to be disposed of within their boundaries. Some communities may prohibit the transport of hazardous waste through their jurisdiction. Whenever possible, it is best for an industry to dispose of its hazardous wastes within its boundaries.

Site Characteristics

An ideal land-disposal site is rare, and many factors must be considered in choosing a site. Some of these are

Land use

Hydrology: surface, subsurface

Geology

Soil properties: physical, chemical, hydraulic

Topography

Climate: wind, temperature, moisture

Clearly unacceptable site characteristics [15] for a hazardous-waste land-disposal facility are the following:

Land use
   areas formerly used for landfills
   areas contaminated with persistent residues from past chemical spills or
      waste-treatment processing

Hydrology
   aquifer recharge zones
   flood-prone areas
   wetlands
   seasonally high water tables
   near private or community water supply wells or reservoirs

Geology
   bedrock outcrops
   irregularities such as fissures or faults overlying groundwater

Soils
   thin soil above groundwater
   saline soils
   highly permeable soils above
   shallow groundwater
   soils with extreme erosion potential

Topography
   overly steep slopes
   broken terrain

Climate
   extremely wet or cold conditions

Hydrology.   The hydrology of a site is important because it determines the movement of water and associated contaminants. Both surface and subsurface hydrology must be considered.

The important hydrologic parameters are those affecting the movement of a liquid through or over a soil or causing it to evaporate. Applicable data for many soils are available from USGS surveys, soil surveys, and serial photos. Each source varies according to the level of detail, the type of information, and the quantitative nature of the information available.

The USGS has regional reports available, including well logs of water levels and quality and drillings logs detailing surface and subsurface geology, depth to water, and saturated thickness. The geographic areas covered by these reports are very extensive, and considerable variance from the tabulated data may be found locally. Information in these reports may be of limited use, although it can serve to define groundwater quality for an area and local geologic conditions. In addition, extensive information may be available for local areas detailing the lithology, porosity, horizontal and vertical permeability, transmissibility, and water quality of an area.

Other information is available in modern (post-1956) soil surveys at either the soil association or soil series level. The soil series analysis provides a more accurate database, but soil association analysis can be adequate when large land areas must be evaluated.

The soil association analysis involves rating the major individual soil series units composing the soil association. Care must be taken to ensure that the soil series unit naming the association makes up the major part of the association. A soil association may be named after two or three soil series units, but the soil series units may compose only 30% of the soils in the association. Estimates of the percent occurrence of the soil series contained within a soil association will be described in the soil survey.

When rating soil suitability, one must identify the predominant slope from the soil survey. Depth to bedrock, depth to water table, and most limiting permeability of soil series also are reported. The parent material of geology of the series is reported in the description of the series.

Modern soil surveys are not available for many areas of the United States and aerial photos may provide the necessary information on hydrogeology and soil features. Although quantitative information is difficult to define from these photos, trained individuals can distinguish many soil and hydrogeologic features that affect site selection.

Landforms that can be identified from aerial photos and that may be suitable for a land-disposal site are noted in Table 6-14. Other geological features that should be identified during the site selection process include

Depth of subsurface soils

Bedrock outcrops

Type of bedrock

Fissures, faults, fractures, crevices, joints, caves, springs, sinkholes, seeps, limestone cavities, or other irregularities in the bedrock

Important soil characteristics affecting surface and subsurface hydrology include depth to bedrock or gravel, depth of the seasonal water table, thickness of soil horizons, soil texture, permeability of each horizon, pH of each horizon, soil salinity, and the shrink–swell potential of the soil. The depth of soil to bedrock, gravel, or an underlying formation affects the potential for groundwater contamination.

Table 6-14. Landforms Identifiable from Airphotos that May Be Suitable for Land-Disposal Sites

Landform	Comment	Suitability for Land-Disposal Site
Moraine	Highly heterogeneous material	Detailed soil analysis required
Floodplain	Frequent flooding causes environmental concerns	Exclude
Drumlins	Limited areas and steep slopes	Exclude
Filled valleys	Heterogeneous material, dependent on parent material and climate	Detailed soil analysis required
Delta	Frequent flooding causes environmental concerns	Exclude
Beach ridges	Limited area, heterogeneous conditions, high ground water	Exclude
Coastal plains	Highly heterogeneous material	Detailed soil analysis required
Tidal flats	Frequent flooding causes environmental concerns	Exclude
Sand dunes	Wind erosion, unstable landform	Exclude

Source: J. R. Ryan and R. C. Loehr, 1981 (Nov.), *Site Selection Methodology for the Land Treatment of Wastewater,* Special Report 81-28, U.S. Army Corps of Engineers, Cold Regions Research and Engineering Laboratory, Hanover, N.H.

Shallow soils (1.5–2 m or less), especially over karst formations, and those classified as sand (2-mm particles or greater) have a high potential for transmitting soluble hazardous constituents to groundwater. The depth of soil to the seasonal water table (the highest standing water table) should be a minimum of about 2 m to prevent contamination of the water table with untreated waste and to allow sufficient soil aeration for microbial degradation of organic compounds. A complete soil series classification including number of horizons, classification, and thickness should be obtained for any potential site. Sieve analysis for soil texture should be included in the soil series classification.

The permeability of each horizon or zone should be known or determined, because high permeabilities result in rapid movement of water and associated contaminants. High-salinity soils can reduce the effectiveness of biological treatment of wastes and may cause soils to crust, resulting in excess runoff. The shrink–swell potential, especially in certain clay soils, can increase water infiltration through the surface soils due to deep cracks during dry seasons.

Water table data is needed to position upgradient and downgradient monitoring wells and to determine if the water table is too near the surface. Because local water table depths and gradients may not be readily available, observation wells may need to be installed in and around a potential site. Sampling frequency at these observation wells should account for seasonal changes. If care is taken in locating and properly installing these initial observation wells, they could be used for future groundwater monitoring.

With the above data, a map of the potential site showing groundwater depths, seasonal water tables, flow directions, and groundwater quality can be prepared. Existing public and private wells, springs, and other water supplies that may be influenced by water moving from a land-disposal facility also should be noted on the map, even if the location of such supplies is not within potential site boundaries. Equally important is the identification of septic tanks and other point and nonpoint sources of potential surface-water and groundwater contamination.

Soil Properties.   Soil properties requiring evaluation depend on the waste constituents and the type of land-disposal method. Soil parameters that are commonly measured and can be potentially significant are noted in Table 6-15.

Measurement of soil physical properties should occur as part of the detailed analysis and field investigations. Data collected from soil profile descriptions can be used to prepare a detailed soil map of the site. This map will help locate sampling points as well as help prepare a management scheme for the site. A detailed site topographic map should be developed from a survey of selected grid points so that land grading requirements and surface and subsurface drainage can be evaluated.

Table 6-15. Commonly Measured Soil Properties Relevant to Land-Disposal Sites

Physical Properties	Chemical and Biological Properties	Hydraulic Properties
Soil description	CEC[a]	Infiltration rates
Topography	Exchangeable cations[a]	Aquifer pump tests
	pH	Drainable porosity
	Electrical conductivity[a]	Hydraulic conductivity
	Organic Matter %	
	Nutrients (N, P, K)	
	Boron[a]	
	P adsorption[a]	
	Base saturation	

[a] Need for analysis should be based on an evaluation of waste characteristics.

Where needed, a grid system is used to determine topographic elevations and prepare soil descriptions. Soil cores should be extracted and described at various locations along the topographic grid, especially at locations where soil surveys indicate boundaries between soil types or where changes in slope occur.

Profile descriptions of each soil boring should include profile depth, profile boundary, texture when moist, structure, degree of mottling, presence of carbonates, depth to groundwater, and percentage and type of coarse fragments (greater than 7.5 cm). The descriptions should be prepared by a soil scientist. Personnel from the local soil conservation service office may be able to assist. The descriptions then can be correlated to descriptions of known soil series that occur in the area. Once the field data are collected, detailed soil and topographical maps can be developed, and the type of hydraulic tests and the location of sampling points for chemical and hydraulic parameters can be determined.

The steps to take in acquiring specific field soils data [31], especially for land-treatment system sites are as follows:

Step 1
Prepare soil maps from air photos, soil survey, and topographic map overlays
Determine approximate soil boundaries, grid spacing, and location of
    monitoring points

Step 2
Lay out grid and determine elevation of individual grid points
Conduct soil borings and describe soils at designated grid points. Note vegetation and farming practices in grid area
Install temporary groundwater monitoring wells

Step 3
Develop topographic maps of soil surface elevations, impermeable horizon
    elevations, groundwater elevations, and depth to bedrock
Analyze soil descriptions and correlate to known soil series
Develop soil map and indicate location of sampling units for each soil type
Determine chemical and hydraulic parameters to be measured

Step 4
Collect samples for chemical analysis
Conduct hydraulic tests at selected grid points
Conduct deep borings to determine subsurface geology

Certain soil chemical parameters are measured routinely. The need for knowledge of other parameters will depend on the concentration of various constituents in the applied waste.

At land-treatment sites percent organic matter and pH should be measured in all site investigations to provide background data for subsequent monitoring. If pH corrections are required, base saturation should be analyzed to determine the lime requirement of a soil.

Although deep soils having relatively uniform physical and chemical characteristics are found occasionally, most soils are characterized by distinct horizons that differ in texture, water retention, permeability, CEC, and chemical characteristics. Knowledge of these characteristics is important to limit subsurface movement of leachate from a landfill or a land-treatment site.

At a land-treatment site most of the biological activity and the waste decomposition is accomplished in the zone of incorporation, which may range from 15 cm to 30 cm. Therefore characterizing this horizon is very important. Lower horizons will influence the rate of downward water movement and may remove waste constituents or their degradation products. The permeability of lower horizons also influences the amount of water remaining in the surface horizon following rainfall, irrigation, or waste application. For example, a change from an average permeability in the upper soil layer to a lower permeability in the lower horizons produces greater amounts of water retained in the soils of greater permeability. This may give wetter conditions than otherwise expected and a high-perched groundwater table. Thus, it is important to detemine the pertinent characteristics of all horizons affected by the land-disposal method.

At sites with deep homogeneous soils, the predominant flow of water is vertical, and the long-term soil infiltration capacity should equal the saturated vertical soil hydraulic conductivity. However, in situations where an impermeable horizon exists, the predominant subsurface water movement will be in the horizontal direction. The hydraulic gradient determines the rate of water movement. Monitoring wells and groundwater levels will identify the gradient.

Topography.   The configuration of the landscape determines the pathway and rate of surface-water movement and often the subsurface water movement. The mode of waste transportation to the site also requires an evaluation of the topography.

The topography of the area adjacent to a potential site should also be evaluated. A level site adjacent to higher sloping land can receive considerable surface runoff and subsurface flow. Surface and subsurface hydrologic considerations can help evaluate such conditions and the need for and type of surface-water diversion systems.

Topography also affects the erosion potential of a site. Minimizing erosion avoids loss of surface cover at landfills and the soil-waste mixture at land-treatment sites. Erosion can be controlled by soil and water conservation measures such as diversion terraces and grassed waterways.

Climate. Climatic conditions can influence the selection of a land-disposal site because of how they affect the transport of waste constituents from a site. Winds impact site selection because they can transport waste constituents, aerosols, suspended particulates, and material released by an accident or explosion. Wherever possible, hazardous-waste land-disposal sites should be downwind of population centers and ecologically sensitive areas.

Temperature and moisture conditions affect the ability to use a site, the microbial degradation rates, and the leaching and volatilization potential. For land-treatment sites, seasonally wet conditions can restrict access to the fields and produce anaerobic conditions. The size of storage areas is influenced by temperature and precipitation patterns and intensity. Field and site drainage systems, runoff and runoff control, and the type of equipment used are all affected by climatic conditions. Table 6-16 gives climatic conditions and effects they may have on the selection and use of a land-disposal site.

Summary

Compared to other types of hazardous-waste management, land-disposal methods depend heavily on site-specific conditions. Both quantitative and

Table 6-16. Effect of Conditions on Land-Disposal Sites

Condition	Effect
Biodegradation	
Temperature	Indirect—controls soil temperature and microbial populations and activity
Precipitation-Evapotranspiration	Indirect—controls soil moisture that affects soil aeration and available water
Waste application	
Temperature	Direct—cold: increases waste viscosity, decreasing ease of handling; hot: increases waste volatility
	Indirect—cold: controls soil temperature that can limit soil workability and waste degradation
Precipitation-Evapotranspiration	Indirect—wet: can inhibit field accessibility and enhance the waste leaching hazard
Winds	Direct—transport of volatiles and particles
Atmospheric stability	Direct—surface inversions can lead to direct contact of volatile waste constituents and the public
Location	
Winds	Direct—potential public hazard

Source: Adapted from U.S. Environmental Protection Agency, 1983 (April), *Hazardous Waste Land Treatment,* rev. ed., SW-874, Office of Solid Waste and Emergency Response, Washington, D.C.

qualitative information must be used in selecting suitable sites. The approach described in this section uses available information for a general evaluation of the technical feasibility of landfill and land treatment and for narrowing the possible site options. Specific data for a detailed analysis and for system design comes from field investigation.

Land disposal of hazardous waste will continue to be a method that causes public concerns and controversy. The selection of a site must involve

A detailed understanding of site and soil characteristics

Identification of water control and migration and loss of waste constituents

An analysis of the impact on surface and groundwater, environment, and public health

Public understanding and participation

If a competent assessment of site characteristics occurs and that minimization of public and environmental risk will result, the selection of a land-disposal site depends primarily on factors such as public perception, existing and projected land-use patterns, and political will.

MONITORING

A monitoring program is an essential component of a land-disposal system. It has the following objectives:

Corroborate the initial design assumptions and criteria

Establish performance of the system

Verify long-term use

Identify changes in soil and water characteristics

Demonstrate groundwater and surface-water protection

Meet regulatory requirements

Provide for public-health protection

Establish public, industry, and regulatory confidence in the system

Identify that detoxification and/or containment of the hazardous wastes does occur

Indicate acceptable closure methods and need for post-closure monitoring

The details of a specific monitoring program are a function of the characteristics of the waste, the type of system (landfill or land treatment), soil characteristics, and climate. When developing a monitoring program, the researcher should collect and analyze samples of the waste, any runoff water, the groundwater, any vegetation grown on or near the site, soil at the site, especially in the zone of incorporation for a land-treatment facility, and soil pore water in the unsaturated zone below the landfill and below the zone of incorporation at a land-treatment site.

At a landfill site the design of the landfill focuses on containing the hazardous constituents of the wastes. At a land-treatment site the land-limiting or limiting-parameter concept is a key design parameter. Thus, the *prime* objective of a monitoring program is to monitor the hazardous-waste components likely to be mobile under the conditions at a landfill or land-treatment site, and those parameters identified as limiting or potentially limiting at a land-treatment site.

A *secondary* objective is to monitor parameters that can affect the mobility of hazardous-waste components (such as pH), that may be indicative of contaminant migration (such as chloride or nitrate), and that may identify overall system performance. These latter parameters are monitored with less frequency than those that are part of the prime objective.

Nonrelevant parameters and sampling at unnecessarily frequent intervals should not be monitored. Changes in soil and groundwater characteristics normally are slow, and sampling frequencies can be four times per year or less. A suggested monitoring program for a land-treatment site is given in Table 6-17.

As a minimum a monitoring program should be able to detect the migration of hazardous-waste constituents and provide information on the background concentrations of such constituents in similar but nearby untreated soils. This latter information permits changes in characteristics and potentially adverse trends to be identified.

The following sections summarize general monitoring and analytical aspects of a monitoring program for wastes, groundwater, runoff, soil, and the unsaturated zone. Both EPA and state regulations should be consulted for the details of monitoring programs that may be needed to acquire necessary permits and for operation of a landfill or land-treatment facility.

Waste

Routine sampling and analysis of the wastes to be contained in a landfill or to be applied to a land-treatment site are needed. How often these are done depends on variations in the quantity and quality of the hazardous waste. Adequate samples should be obtained to determine the consistency and

Table 6-17. Possible Monitoring Program for a Land-Treatment Site

Item Monitored	Purpose	Sampling Frequency	Parameters Measured
Waste	Determine changes	Quarterly if continuous stream; each batch if intermittent generation	Rate and capacity-limiting constituent, plus those with 25% of being limiting
Soil in zone of incorporation	Determine degradation pH, nutrients, and rate- and capacity-limiting constituents	Design life: Frequency of sampling during operation and closure: <2 yr quarterly; 2–10 yr semiannually; >10 yr annually	Rate and capacity-limiting constituent, those within 25% of being limiting, pH, nutrients, and residual organics
Soil below zone of incorporation	Determine movement of metals and soluble constituents	Design life: Frequency of sampling during operation and closure: <2 yr quarterly; 2–10 yr semiannually; >10 yr annually	Metals that limit the loading capacity, those within 25% of being limiting, plus slowly mobile organics
Vegetation	Phytotoxic constituents and food-chain hazards	Annually or at harvests	Metals; nutrients if needed for diagnosis
Runoff water	Soluble or suspended constituents	As required for NPDES[a] permit; post closure: 1/2, 1, 2, 4, 8, 16, and 30 yr after closure	Discharge permit or background parameters plus organics
Unsaturated zone water	Determine mobile constituents	Quarterly during operation and closure	Potentially mobile constituents
Groundwater	Determine mobile constituents	Semiannually during operation and closure: post closure: 1/2, 1, 2, 4, 8, 16, and 30 yr following closure	Potentially mobile constituents

Source: Adapted from U.S. Environmental Protection Agency, 1983 (April), *Hazardous Waste Land Treatment*, rev. ed., SW-874, Office of Solid Waste and Emergency Response, Washington, D.C.

[a] National Pollutant Discharge Elimination System.

Table 6-18. Contaminants Defining EP Toxicity Characteristics

Contaminant	Maximum Concentration (mg/l)
Arsenic	5.0
Barium	100.0
Cadmium	1.0
Chromium	5.0
2, 4-D	10.0
Endrin	0.02
Lead	5.0
Lindane	0.4
Mercury	0.2
Methoxychlor	10.0
Selenium	1.0
Silver	5.0
Toxaphene	0.5
2, 4, 5-TP silvex	1.0

Source: General Services Administration, 1982, Identification and listing of hazardous waste, *Code of Federal Regulations* Title 40, Protection and the Environment, Part 261.24.

uniformity of the waste. Waste sampling techniques should assure that the samples are representative.

If the waste is generated from a routine, continuously run process, weekly samples can be composited and analyzed quarterly. If a production process change occurs that could impact the characteristics of the waste, the wastes should be sampled and analyzed as soon as possible. If the waste is land-filled in batches or if nonroutine batches are to be applied to a land-treatment site, the waste should be fully characterized prior to each application.

Analyses of hazardous wastes that are to be disposed of in a landfill or treated by land treatment must be accurate and up-to-date. The analyses should be repeated when there is no reason to believe that the process of operation generating the waste has changed. For a land-treatment facility the waste should be analyzed for the substances that cause a waste to exhibit the extraction procedure (EP) toxicity characteristics. These substances and the concentrations that define EP toxicity are listed in Table 6-18. In addition, the waste should be analyzed for substances that caused the waste to be listed as a hazardous waste, and that limit or potentially limit the land area and waste-application rate.

Groundwater

A groundwater monitoring program should identify whether the landfill or

land-treatment site complies with or violates the groundwater quality provisions of the facility's permit. Groundwater monitoring is required during
the active life of the facility and, if hazardous wastes or decomposition
products remain after closure, also during the post-closure period.

Groundwater and unsaturated zone monitoring are complementary. An
increase of potential contaminants detected in leachate (unsaturated zone
monitoring) can help identify potential problems and provide time for
remedial action. Although such monitoring can provide valuable information
on the effectiveness of waste contamination and attenuation processes in the
soil, it is not a substitute for groundwater monitoring.

Wells are the primary tools used to monitor groundwater contamination.
The monitoring system should consist of an array of wells strategically placed
to identify groundwater movement. A sufficient number of wells to
characterize the potential contamination of the groundwater quality should
be used. One group of wells should be located at the downgradient edge of
the land-disposal site. Another group should be located further downgradient
from the disposal site in the direction of decreasing static head. These two
groups of wells are needed so that changes in quality of the groundwater
can be determined as it flows through the aquifer away from the site. Additional monitoring wells should be installed upgradient (in the direction of
increasing static head) from the disposal site.

The upgradient wells provide an estimate of the direction and head potential of groundwater flow. If the hydrology of the site is complicated and includes several aquifers or several flow directions within a given aquifer, many
wells may be required to adequately characterize the movement of potential
contaminants from the disposal site.

The upgradient wells, or those installed in areas not affected by the disposal
site, provide samples indicative of background water quality, whereas the
downgradient wells identify movement of any contaminants from the site.
The selection of the monitoring well locations must be justified in terms of
the hydrogeology at a particular site.

The exact number of monitoring wells needed at a site depends on soil
type, complexity of the groundwater hydrogeologic system, waste
characteristics, and site history. Monitoring wells may be needed at more
than one depth. Mobile waste constituents first enter the upper portion of
the groundwater and are diluted as they move with the groundwater flow.
Thus, samples from the upper portion of the groundwater can better indicate
contamination than samples from the lower portion.

Monitoring wells must be cased in a manner that maintains the integrity
of the bore hole. The casing must be screened or performated and packed
with gravel or sand to enable sample collection at depths where groundwater
flow exists. The space above the sampling depth must be sealed with cement

grout, bentonite, or other suitable material to prevent surface contamination of samples and groundwater.

To reduce the cost and analytical burden of groundwater sampling and analysis, only those parameters generally used to establish groundwater quality or that are especially unique to the applied waste should be measured. Monitoring these indicator parameters can establish background water quality and detect the entry of hazardous-waste-constraints constituents into the groundwater. There is no purpose in analyzing for a large number of chemical parameters that may be released from a site until there is reason to believe that they have been released.

Thus broad indicators such as chloride, iron, manganese, phenol, sodium, sulfate, pH, or an easily identified parameter found in the landfilled or land-treated waste can be used to indicate if a potential problem is occurring. These organic parameters are generally used to characterize groundwater suitability for various uses. Specific conductance and two general measures of organic contamination, total organic carbon (TOC), and total organic halogens (TOX), also can be used as indicators. The parameters TOC and TOX identify if organic contaminants are leaching from the disposal site. Such parameters are found in uncontaminated groundwater in only low concentrations if they can be detected at all.

A statistically significant change in these parameters from background conditions indicates that inorganic and/or organic substances are being introduced into the groundwater at the disposal site. The methodology to sample and analyze these parameters is available. If monitoring these parameters indicates that contamination is likely or has occurred, a more detailed sampling and analysis program can be initiated.

Because groundwater monitoring data should be statistically valid, the reported value of a parameter should be based upon at least four replicate measures. Mean and variance information can be obtained by pooling the replicate measurements from the background wells during the year.

The elevation of the groundwater surface at a monitoring well should be measured each time a sample is obtained. If a wide seasonal variation in groundwater elevation is noted, such elevations should be measured frequently, such as every month, to establish the specific groundwater elevations, head differential, and direction of flow.

Unless there is reason to suspect a rapid change in groundwater characteristics, the monitoring wells do not have to be sampled more often than quarterly. Once background conditions are established, the frequency of sampling may be reduced to perhaps semiannually. If an adverse change in characteristics appears likely to occur, or is occurring, more frequent sampling will be necessary to substantiate the change, its rate, and its implications.

Runoff

Both runoff from the disposal site and from surface waters surrounding the site, if lateral movement of waste constituents is expected, should be part of the monitoring program. Analyzing runoff samples is necessary to determine if the runoff can be discharged and to satisfy any point-source discharge conditions if the facility has such a permit.

The runoff and adjacent surface waters should be monitored during the active life of the site as well as during closure. Reasonable sampling of these waters during post-closure (such as quarterly) can help determine when the closure period can end.

Unsaturated Zone

Liquids from a landfill may bypass the leachate collection system and the bottom barrier and move toward the groundwater. At a land-treatment site water added to the soil by precipitation, by irrigation, or by the applied waste will pass through the soil and incorporated waste and may transport mobile waste constituents or by-products through the unsaturated zone to the groundwater.

Groundwater monitoring does not provide an early warning of such movement. Unsaturated zone monitoring—that is, collection and analysis of the percolate as it passes through the unsaturated zone—can measure the concentration and type of contaminants moving toward the groundwater and provide early warning of potential problems. Various types of lysimeters (pressure, vacuum, and trench) as well as vacuum extractors have been used for this purpose [15]. Sample collection and analysis is possible only during periods when there is liquid percolation.

Soil

Soil monitoring, especially at a land-treatment facility, serves several purposes: (a) to determine the degradation rate of constituents in the applied wastes; (b) to verify estimated system performance; (c) to estimate possible site uses after closure; and (d) to indicate post-closure monitoring needs. Samples of the soil immediately after waste incorporation and at periodic intervals should be taken and analyzed for specific waste constituents, particularly those that were identified as the limiting constituents.

Samples of the lower soil horizons should be taken, using cores or borings, to indicate contaminant migration and because soils are not homogeneous. Soil sampling is done on a random basis, several subsamples being composited to obtain one sample for analysis. Samples should be taken for each major soil type in the land-treatment site.

Each soil sample should be thoroughly mixed. After mixing, a representative sample should be taken, air dried, and analyzed. With some samples it may be easier to air dry the subsamples before mixing. High-temperature oven drying can volatilize some organic constituents and must be avoided.

Summary

A sound, well-designed monitoring program is an integral part of any land-disposal system for hazardous wastes. Such systems are designed, constructed, and operated in an engineered, controlled manner to avoid contamination of surface and groundwaters and to avoid an adverse impact on adjacent property and environment. Proper monitoring will verify that the applied wastes are being contained or degraded and do not represent a public-health or environment threat.

Reasonable approaches, suitable equipment, and analytical methods exist to develop the necessary monitoring program. It is equally important to analyze the monitoring data, to note trends that occur, and to identify and implement appropriate remedial action if adverse-health or environmental effects begin to appear. A monitoring program is worthless and a waste of money if the data are not evaluated and used for better site operation and to verify or improve design procedures.

REFERENCES

1. N. C. Brady, 1977, *The Nature and Properties of Soils,* 9th ed., Macmillan, New York.
2. W. E. Sopper, E. M. Beaker, and R. Bastian, 1982, *Land Reclamation and Biomass Production with Municipal Wastewater and Sludge,* Pennsylvania State University Press, University Park, Pa.
3. A. L. Page, T. L. Gleason, J. E. Smith, I. K. Iskandar, and L. E. Sommers, 1983, *Utilization of Municipal Wastewater and Sludge on Land,* University of California, Riverside.
4. R. C. Loehr, W. J. Jewell, J. D. Novak, W. W. Clarkson, and G. S. Griedman, 1979, *Land Application of Wastes,* vol. 1 and 2, Van Nostrand Reinhold, New York.
5. General Services Administration, 1982, Criteria for classification of solid waste disposal facilities and practices, *Code of Federal Regulations,* Title 40, Protection of the Environment, Part 257.
6. C. A. Moore, 1983 (April), *Landfill and Surface Impoundment Performance Evaluation,* SW-869, rev. ed., U.S. Environmental Protection Agency, Office of Solid Waste and Emergency Response, Washington, D.C.
7. U.S. Environmental Protection Agency, 1983 (March), *Lining of Waste Impound-*

ment and Disposal Facilities, rev. ed. SW-870, Office of Solid Waste and Emergency Response, Washington, D.C.

8. A. W. Wyss, H. K. Willard, and R. M. Evans, 1982 (Sept.), *Closure of Hazardous Waste Surface Impoundments,* rev. ed., SW-873, U.S. Environmental Protection Agency, Office of Solid Waste and Emergency Response, Washington, D.C.

9. D. J. D'Appolonia, 1980, Soil-betonite slurry trench cut-offs, *Journal of the Geotechnical Engineering Diversion, ASCE 106* (April): 399–417.

10. D. J. D'Appolonia, 1981, *Slurry Trench Cut-Off Walls for Hazardous Waste Isolation* (reprint), Engineered Construction International Inc., Pittsburgh, Pa.

11. A. J. Shuckrow, A. P. Pajak, and C. J. Touhill, 1982 (Sept.), *Management of Hazardous Waste Leachate,* SW-871, U.S. Environmental Protection Agency, Office of Solid Waste and Emergency Response, Washington, D.C.

12. R. J. Lutton, 1982 (Sept.), *Evaluating Cover Systems for Solid and Hazardous Waste,* rev. ed., SW-867, U.S. Environmental Protection Agency, Office of Solid Waste and Emergency Response, Washington, D.C.

13. General Services Administration, 1982, Interim status standards for owners and operators of hazardous waste treatment, storage and disposal facilities, *Code of Federal Regulations:* Title 40, Protection of the Environment, Part 265, Subpart M-Land Treatment.

14. M. R. Overcash, and D. Pal, 1979, *Design of Land Treatment Systems—Theory and Practice,* Ann Arbor Science Publishers, Ann Arbor, Mich.

15. U.S. Environmental Protection Agency, 1983 (April), *Hazardous Waste Land Treatment,* rev. ed., SW-874, Office of Solid Waste and Emergency Response, Washington, D.C.

16. G. W. Leeper, 1978, *Managing the Heavy Metals on the Land,* Marcel Dekker, New York.

17. Metcalf and Eddy Inc., 1979, *Wastewater Engineering—Treatment, Disposal, and Reuse,* McGraw Hill, New York.

18. U.S. Environmental Protection Agency, 1981, *Treatability Manual* (revised), vols. 1–4, EPA-600/2-82-00-e, Industrial Environmental Research Laboratory, Cincinnati, Ohio.

19. M. D. Baker, C. I. Mayfield, and W. E. Inniss, 1980, Degradation of chlorophenols in soil, sediment and water at low temperature *Water Research* **14**:1765–1771.

20. R. A. Griffin, and S-F. J. Chow, 1980, Disposal and removal of halogenated hydrocarbons in soils in *Disposal of Hazardous Waste,* Proceedings of the Sixth Annual Research Symposium, EPA-600/9-80-010, Municipal Environmental Research Laboratory, Cincinnati, Ohio, pp. 82–92.

21. D. Pal, M. R. Overrash, and P. W. Westerman, 1977, Plant-soil assimilative capacity for organic solvent constituents in industrial waste in *Proceedings of the 32nd Purdue Industrial Wastes Conference,* Ann Arbor Science Publishers, Ann Arbor, Mich. pp. 259–271.

22. G. K. Dotson, R. B. Dean, W. B. Cooke, and B. A. Kennar, 1970, Land spreading, A conserving and non-polluting method of disposing of oily wastes, in *Proceedings of the 5th International Water Pollution Research Conference,* Paper II-36 Pergamon Press, pp. 1–15.

23. R. L. Raymond, J. O. Hudson, and V. W. Jamison, 1976, Oil degradation in soil, *Applied and Environmental Microbiology* **31**:522-535.
24. D. W. S. Westlake, A. M. Jobson, and F. D. Cook, 1978, in situ degradation of oil in a soil of the boreal region of the Northwest Territories, *Canadian Journal of Microbiology* **24**:254-260.
25. T. E. Loynachon, 1978, Low-temperature mineralization of crude oil in soil, *Journal of Environmental Quality* **7**:444-500.
26. P. E. Cansfield, and G. J. Racy, 1978, Degradation of hydrocarbon sludges in the soil, *Canadian Journal of Soil Science* 339-345.
27. American Petroleum Institute, 1983 (June), *Land Treatment Practices in the Petroleum Industry,* Washington, D.C.
28. R. L. Huddleston, and J. D. Meyers, 1979, Treatment of refining oily wastes by land farming, *Water-1978 A.I.C.E. Symposium Series,* pp. 327-339.
29. K. W. Brown, K. D. Donnelly, J. C. Thomas, and L. E. Duel, 1981, Factors influencing the biodegradation of API separator sludges applied to soils, in *Land Disposal: Hazardous Waste,* Proceedings of the Seventh Annual Research Symposium, 81-002b, Municipal Environmental Research Laboratory, Cincinnati, Ohio, pp. 188-199.
30. R. G. Wetherold, D. D. Rosebrook, and E. W. Cunningham 1981, Assessment of hydrocarbon emissions from land treatment of oily sludges, Proceedings of the Seventh Annual Research Symposium, EPA-600/9-81-002bn, *Land Disposal: Hazardous Wastes,* pp. 213-223.
31. J. R. Ryan, and R. C. Loehr, 1981 (Nov.), *Site Selection Methodology for the Land Treatment of Wastewater,* Special Report 81-28, U.S. Army Corps of Engineers, Cold Regions Research and Engineering Laboratory, Hanover, N.H.

7

Hazardous-Waste
Leachate Management

Alan J. Shuckrow
C. J. Touhill
Andrew P. Pajak

*Baker/TSA Inc.
Beaver, Pennsylvania*

Leachate generated by water percolating through a hazardous-waste disposal site could contain significant concentrations of toxic chemicals. Proper leachate management is essential to avoid contamination of surrounding soil, groundwaters, and surface waters. This chapter provides guidance on available management options for controlling, treating, and disposing of hazardous-waste leachates. Its primary focus is upon leachate treatment-process selection.

Available data on leachate characteristics at existing disposal sites and identification of major leachate management options are presented in order to provide a framework for process selection. In addition, the potential applicability of various unit treatment processes is assessed. This information is used to enumerate those factors influencing treatment-process selection. Using selected hypothetical and actual leachate situations, we suggest a systematic approach for attacking hazardous-waste leachate problems.

LEACHATE CHARACTERISTICS

The type of unit processes used for hazardous-waste leachate treatment depend on leachate pollutant characteristics. Presumably, in a planned disposal operation safeguards will be engineered that minimize dilution of the leachate due to percolation of precipitation, runoff, or flowthrough of extraneous water sources such as groundwater. Moreover, the collection system will intercept the leachate before migration from the site, and dilution can occur. Thus, leachate is envisioned as a concentrated solution of chemicals representative of soluble or leachable materials contained in the disposal site. Leachate from existing sites, constructed prior to implementation of Resource Conservation and Recovery Act (RCRA) regulations, might be more dilute because of infiltration of extraneous water.

Ideally, for existing and planned disposal sites, leachate characteristics could be predicted using previous experience from documented sites. Unfortunately, there are deficiencies in the existing database. For example:

Very little data on actual hazardous-waste leachate exist. Most available leachate composition data pertain to sanitary landfills.

Reported information is such that it often is difficult to distinguish between leachate and contaminated groundwater wherein some dilution has occurred.

Most available composition data on contamination associated with hazardous-waste disposal sites pertain to surrounding ground- and surface waters.

Composition is highly variable from site to site, at different sampling locations within a site, and at a given location over time.

Analytical testing is difficult and costly in a complex hazardous aqueous-waste pollution matrix. These factors serve to limit the database. In addition, analytical errors and interferences also may contribute to some of the variability.

Because of the analytical complexity and expense, "complete" characterizations are nonexistent.

There is a general lack of information regarding the physical characteristics of sites.

Thus, the existing database is characterized primarily by its incompleteness and variability.

Despite these deficiencies, available information does give insight into leachate characteristics at actual hazardous-waste disposal sites. Moreover, the available information can be used to provide guidance for selecting and evaluating leachate treatment technologies.

The following data summarize available characterization information on leachates and contaminated ground-and surface waters associated with existing hazardous-waste disposal sites. The latter categories were included because they represent the preponderance of the database and because they provide information on the types of compounds that have been associated with previous disposal operations.

In conjunction with the U.S. Environmental Protection Agency, (EPA) technical resource document, *Management of Hazardous Waste Leachate* [1], characterization data were compiled on leachates and contaminated ground- and surface waters in the proximity of 30 sites containing hazardous wastes. Tables 7-1 and 7-2, respectively, summarize the more frequently occurring inorganic and organic contaminants reported. Only those contaminants that appeared at four or more sites are listed in the tables.

Note that the tables list more contaminants than those addressed by RCRA regulations. Leachate treatment processes must deal with a broader spectrum of compounds than those listed in RCRA as acutely hazardous, hazardous, or toxic. That is, treatment processes must be designed to deal with hazardous constitutents in the matrix in which they occur. Moreover, it is likely that effluent from a leachate treatment process will have to meet requirements in addition to RCRA regulations (e.g. National Pollutant Discharge Elimination System [NPDES], pretreatment).

Further insight into leachate characterization may be obtained from the data in Table 7-3, which summarize a survey of ground- and surface-water quality in the vicinity of 43 industrial waste disposal sites (landfills and impoundments). Note that this information, while representing more sites, is less detailed than that in Tables 7-1 and 7-2.

Table 7-1. Inorganic Contaminants Reported

Contaminant	Pollutant Group[a]	Conc. Range Reported	No. Sites Reported
As	H,P	0.011–< 10,000 mg/L	6
Ba	H	0.1–2,000 mg/L	5
Cd	H,P	5–8,200 µg/L	6
Cl	S	3.65–9,920 mg/L	6
Cr	H,P	1–208,000 µg/L	7
Cu	P	1–16,000 µg/L	9
Hg	P	0.5–7.0 µg/L	7
Ni	H	20–48,000 µg/L	4
Pb	H,P	1–19,000 µg/L	6
pH	C	~ 3–7.9 (pH scale)	7
Se	H,P	3–590 µg/L	4
Suspended solids	C	< 3–1,040 mg/L	4
TKN[b]	C	> 1–984 mg/L	4

Source: A. J. Shuckrow, A. P. Pajak, and C. J. Touhill, 1981, Management of Hazardous Waste Leachate, SW-871, U.S. Environmental Protection Agency, Cincinnati, Ohio.

[a] C = conventional pollutants (from Clean Water Act); H = hazardous (RCRA list); P = priority pollutant; S = section 311 compound.

[b] TKN = total Kjeldahl nitrogen.

Table 7-2. Organic Contaminants Reported

Contaminant	Pollutant Group[a]	Conc. Range Reported	No. Sites Reported
Benzene	H,P,S,T	<1.1–7,370 µg/L	5
Chlorobenzene	H,P,S,T	4.6–4,620 µg/L	5
Chloroform	H,P,S,T	0.02–4,550 µg/L	4
COD	C	24.6–41,400 mg/L	6
1,2-dichlorethane	H,P,S,T	2.1–4,500 µg/L	5
1,1-dichlorethylene	H,P,S,T	28–19,850 mg/L	5
Dichloromethane	H,P,S,T	3.1–6,570 µg/L	4
Ethylbenzene	P,S	3.0–10,115 µg/L	4
Phenol	H,P,S,T	<3–17,000 µg/L	4
TOC	C	10.9–8,700 mg/L	8
Toluene	H,P,S,T	<5–100,000 µg/L	7
1,1,1-trichloroethane	H,P,T	1.6 µg/L–590 mg/L	5
Trichloroethene	H,T	<3–84,000 µg/L	4
Trichloroethylene	H,P,S,T	<3–260,000 µg/L	4

Source: A. J. Shuckrow, A. P. Pajak, and C. J. Touhill, 1981, *Management of Hazardous Waste Leachate,* SW-871, U.S. Environmental Protection Agency, Cincinnati, Ohio.

[a] C = conventional pollutants (from Clean Water Act); H = hazardous (RCRA list); P = priority pollutant; S = section 311 compound; T = toxic (RCRA list).

Table 7-3. Characterization of Hazardous Leachate and Groundwater from 43 Landfill Sites

Pollutant	Conc. Range (mg/L)	Typical Conc. (mg/L)	No. Sites Detected
As	0.03–5.8	0.2	5
Ba	0.01–3.8	0.25	24
Cr	0.01–4.20	0.02	10
Co	0.01–0.22	0.03	11
Cu	0.01–2.8	0.04	15
CN	0.005–14	0.008	14
Pb	0.3–19	—	3
Hg	0.0005–0.0008	0.0006	5
Mo	0.15–0.24	—	2
Ni	0.02–0.67	0.15	16
Se	0.01–0.59	0.04	21
Zn	0.1–240	3.0	9
Light organics	1.0–1000	80	10
Halogenated organics	0.002–15.9	0.005	5
Heavy organics	0.01–0.59	0.1	8

Source: Based on data from Geraghty and Miller, Inc., 1977, *The Prevalence of Subsurface Migration of Hazardous Chemical Substances at Selected Industrial Waste Land Disposal Sites,* EPA/530/SW-634, U.S. Environmental Protection Agency, Washington, D.C.

Even though the database presented has deficiencies, it does provide guidance in formulating treatment alternatives, provided that data are used cautiously, recognizing that leachate from secured landfills may have higher concentrations.

MANAGEMENT OPTIONS

The full range of hazardous-waste leachate management options could begin with manufacturing process modifications and extend through leachate treatment–disposal options. Such a span of options is illustrated conceptually in Figure 7-1. Because the focus of this chapter is primarily upon the last stage—leachate treatment–disposal—the other options are reviewed only briefly in light of leachate generation and treatment.

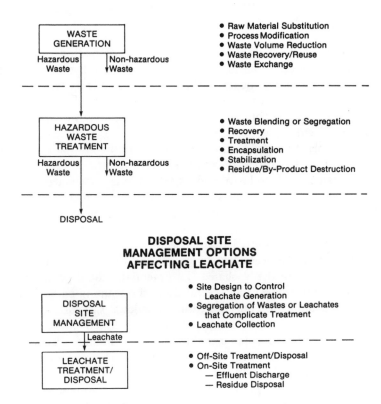

Figure 7-1. Disposal site management options: Effect on leachate generation.

The objectives of the waste-generation option are the following:

1. Eliminate hazardous wastes by substituting less troublesome raw materials
2. Reduce waste generation through source reduction, process modifications, waste segregation, and material recovery and reuse
3. Exchange wastes with entities capable of using them in their production processes
4. Delist materials in question

The hazardous-waste treatment option involves processing the wastes before disposal to reduce or eliminate their hazardous properties or to reduce their potential leachability. Such treatment could be accomplished by

1. Detoxificating the entire waste stream
2. Concentrating hazardous constituents in a reduced-volume waste stream that can be further treated, detoxified, destroyed, or reused
3. Fixating waste in a matrix that inhibits leaching
4. Encapsulating the waste to prevent leaching

In the disposal-site-management option, leachate formation can be limited or controlled by means of liners, covers, and other liquid-diversion techniques (see chap. 6). It also may be desirable to dispose of certain types of wastes in separate cells at the site or to exclude others altogether. Such an approach could be used to avoid combining wastes that ultimately would complicate leachate treatment.

Leachate collection and storage systems are an integral part of disposal site management. Regardless of measures adopted to limit leachate generation, a leachate is likely to be formed, especially in areas where precipitation exceeds evaporation and/or at sites used for disposal of liquid-containing hazardous wastes. Effective collection permits an equalizing and storage capability that will reduce overloading and avoid possible reduction in subsequent treatment-process efficiency. Collection and storage may also allow for a reduction in the necessary equipment cost by providing for periodic or batch treatment of the leachate. This potentially could permit a mobile unit to treat the leachate from several sites on a rotating basis, increasing treatment unit utilization and decreasing individual site cost.

The final option, leachate treatment–disposal, which entails processing leachate to render it acceptable for discharge or ultimate disposal, is influenced by precedent options. Thus, it is important to have a thorough understanding of all the various options available. However, as a practical matter, site operators may not have full control over some of these options, and it is expected that leachate will be generated at most sites.

The following section addresses leachate management separately because of its importance relative to the purposes of this chapter.

LEACHATE MANAGEMENT

Off-site treatment–disposal of leachate refers to treatment–disposal at a facility not associated with the landfill or surface impoundment operation. Primary off-site treatment–disposal alternatives include publicly owned treatment works (POTWs), hazardous-waste treatment–disposal facilities, and industrial waste-treatment facilities. For several reasons it is expected that off-site treatment will be feasible only in a few cases. In most instances, neither POTWs nor industrial waste-treatment facilities will be available at reasonably distances, nor will they be technically capable of accepting hazardous leachate while satisfying their permit requirements. It also is unlikely that such facilities will assume the potential liabilities associated with accepting a hazardous leachate that is expected to vary in composition and quantity. Therefore, stringent pretreatment requirements probably would be imposed, making on-site treatment a necessity.

On-site hazardous leachate treatment can be used to accomplish either pretreatment of the leachate with discharge to another facility for additional treatment before disposal, or treatment complete enough to meet direct discharge limitations. Pretreatment processes will be dictated by the capabilities of the subsequent off-site facility. Objectives of pretreatment could be to

Equalize leachate quality and quantity fluctuations and provide short-term storage

Adjust pH to within acceptable limits for discharge to a POTW

Reduce concentrations of toxic components to acceptable levels for discharge to a POTW

Remove hazardous constituents so that a portion of the leachate can be judged nonhazardous (the hazardous or nonhazardous fraction could be shipped to off-site treatment)

Reduce the volume of leachate transported off-site

Complete treatment, on the other hand, should produce an effluent suitable for discharge to surface water or groundwater. Thus, the major difference between complete on-site treatment and pretreatment is likely to be the extent of the treatment. That is, the treatment technologies are essentially the same, but the extent of application will differ, depending upon effluent objectives.

Another possible approach to on-site leachate management is leachate recycling. This approach involves the controlled collection and recirculation of leachate through a landfill for the purpose of promoting rapid landfill stabilization. The precise mode of operation of leachate recycling is poorly understood, since it only recently has been investigated in sanitary landfill simulations.

LEACHATE TREATMENT TECHNOLOGIES

This section summarizes an extensive assessment of various unit processes relative to their applicability to treat hazardous-waste leachate [2]. Because leachates vary widely in composition and often contain a diversity of constituents, it is likely that process trains composed of several unit treatment technologies will be needed to achieve high levels of treatment in the most cost-effective manner.

Although research [2] currently is underway to define better performance and design criteria for hazardous-waste leachate treatment technologies, actual full-scale applications are few. Activated carbon adsorption and chemical coagulation–precipitation are the only technologies known to have been used in large-scale applications. Extensive experience with sanitary landfill leachate treatment has only limited applicability for hazardous wastes. On the other hand, many technologies used to treat industrial process wastewaters containing hazardous constituents may be applicable because of similarities in chemical components and discharge goals. Thus, technologies considered in the assessment included those used to treat hazardous-waste leachate, sanitary landfill leachate, and industrial process wastewaters.

Based upon results of the technology assessment [2] referred to above, workers judge unit processes to fall into three general categories:

1. Processes judged to have *demonstrated,* broad potential for full-space leachate treatment applications
2. Processes judged to be *promising* for limited or specialized treatment, or where less full-scale experience exists
3. Processes believed to have *minimal applicability* for hazardous-waste leachate treatment.

Processes forming the demonstrated and promising technologies together with their major areas of application are listed in Table 7-4.

In addition to treated effluent, most leachate treatment processes generate sludges, brines, gaseous emissions, or other by-product streams containing hazardous constituents requiring disposal. Disposal techniques available are

Table 7-4. Technologies Applicable to Hazardous-Waste Leachate Treatment

Category	Unit Process	Major Application
Demonstrated	Biological treatment	Soluble biodegradable organics and nutrients
	Chemical precipitation	Soluble metals
	Carbon adsorption	Soluble organics, especially toxics and refractories
	Density separation	Suspended solids, chemical precipitants, oily materials
	Filtration	Suspended solids and precipitants
Promising	Chemical oxidation	Cyanide and organics
	Chemical reduction	Hexavalent chromium
	Ion exchange	Inorganics, especially fluoride and total dissolved solids
	Stripping (air)	Ammonia nitrogen
	Membranes (reverse osmosis)	Total dissolved solids
	Wet oxidation	High-strength or toxic organic aqueous streams

hazardous-waste landfill, hazardous-waste treatment facility, hazardous-waste incinerator, deep-well injection, or land application. The technique chosen depends upon residue characteristics and economics.

No historical cost data exist on the use of unit processes for hazardous-waste leachate treatment. Consequently, one must rely on information based upon municipal and industrial water and wastewater treatment experience to develop cost estimates. Such information, however, should not be applied directly in developing cost estimates for hazardous-waste leachate treatment. Nevertheless, it reflects the best available information and with care can be used to make approximate cost comparisons among leachate treatment alternatives.

LEACHATE TREATMENT-PROCESS SELECTION

Selection Methodology

Because hazardous-waste leachate treatment is an emerging area still in its infancy, the methodology suggested herein should be used as a guide rather than a prescription. Ideally, process selection should be based upon treatability studies (laboratory or pilot scale) using actual leachate. This is recommended for several reasons:

1. Published hazardous-waste leachate-treatment performance data are rare. In the absence of treatability studies, inferences must be drawn from other laboratory experimental studies, and industrial and municipal water and wastewater treatment experience.
2. Lacking previous experience and/or treatability data, there is no guarantee that high levels of treatment can be achieved.
3. It is likely that a combination of several unit processes will be needed to deal with the complex leachate matrix. Arriving at the optimum system is unlikely without treatability studies.
4. The complex leachate matrix may not behave like other wastewaters, thus affecting design and operating criteria (e.g., chemical dosage requirements), and invalidating extrapolations from other experiences.
5. Capital investment and especially operation and maintenance costs are likely to be greater per unit volume treated than for municipal or industrial wastewater. However, costs will be difficult to estimate without treatment experience. Investment in a costly, unproven system that may not meet the required treatment objectives is imprudent.

In spite of these considerations, treatability studies may not be possible at new disposal sites or existing sites where leachate has not appeared or its quality is expected to change greatly. Thus, a more theoretical approach (at least to conceptual design of a treatment system) with greater dependency on published data must be taken.

A general methodology that can be used to select a leachate-treatment process is shown in Figure 7-2. This methodology revolves around the question of existence of a leachate treatability study. The lower portion of the flowchart applies to cases where a leachate exists and can reasonably be used in treatability studies. The upper portion addresses the case where leachate treatability studies cannot be conducted.

Aside from the availability of leachate for use in treatability studies, several key questions must be answered as part of the leachate-treatment technology selection process. Among these are the following:

1. Does the leachate need to be considered in a hazardous waste?
2. What are the treated effluent discharge options and the corresponding performance or discharge limitations?
3. What pollutants are present in the leachate? What concentrations?
4. Are toxic or refractory compounds present?
5. What is the leachate flow rate? How will it vary with time (diurnal, seasonal, long term)?

LEACHATE TREATMENT SELECTION METHODOLOGY— LEACHATE UNAVAILABLE

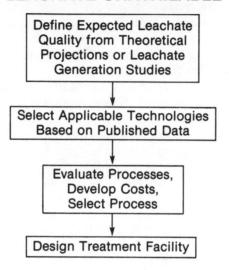

Define Expected Leachate Quality from Theoretical Projections or Leachate Generation Studies

↓

Select Applicable Technologies Based on Published Data

↓

Evaluate Processes, Develop Costs, Select Process

↓

Design Treatment Facility

LEACHATE TREATMENT SELECTION METHODOLOGY— LEACHATE EXISTS

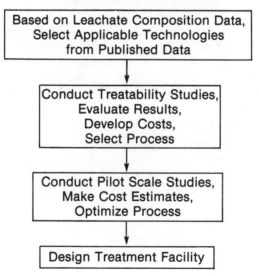

Based on Leachate Composition Data, Select Applicable Technologies from Published Data

↓

Conduct Treatability Studies, Evaluate Results, Develop Costs, Select Process

↓

Conduct Pilot Scale Studies, Make Cost Estimates, Optimize Process

↓

Design Treatment Facility

Figure 7-2. Pre-disposal management options: Effect of leachate.

6. Are there any other aqueous wastes generated at the site? Should they be combined with the leachate for treatment?
7. Will the leachate quality or quantity change (could be a function of disposal site operation)? Does the leachate treatment process need to be able to respond to such changes? If so, in what time frame?
8. How much land is available for the leachate treatment facility? Are there any special construction constraints?
9. Should leachate from different areas of the disposal site be combined or segregated for treatment? (Note that this will affect site and leachate collection system design.)
10. How will leachate treatment residues be managed?
11. What is required to support the leachate treatment operation (e.g., analytical testing, operations personnel)?
13. What skills and resources will be needed for post-closure operation?

Where a leachate exists, process selection methodology involves iden-tificating processes reported to be capable of treating the constituents pres-ent in the leachate; conducting laboratory-scale studies to identify viable pro-cess combinations; and conducting pilot-plant studies to develop design criteria. On the other hand, if a leachate does not exist or if its composition is expected to change greatly, the first major step is to determine what the leachate composition is expected to be. This could be accomplished by pre-dicting leachate characteristics from a knowledge of expected disposal site conditions or by similating leachate generation with small extraction columns or large-scale lysimeters.

Once leachate characteristics are projected, promising technologies would be subjected to detailed "desktop" analysis to evaluate their applicability to leachate treatment. In cases where treatment process design is based on the desktop approach, consideration should be given to contingency plans for leachate treatment and disposal in the event that the original design does not perform as required. The feasibility of adopting an interim measure such as leachate storage and/or recycle until the design can be confirmed by actual treatability studies also should be considered.

Process Train Formulation

Because hazardous-waste leachates are expected to vary widely in compo-sition and will often contain a variety of constitutents, in general, no single unit process will be capable of providing the necessary treatment. Process trains will be necessary to achieve high levels of treatment in a cost-effective manner. Presently, most process trains envisioned will be made up of the unit processes enumerated in Table 7-4. Detailed considerations for formulating process trains have been described elsewhere [1]. In general, these considerations include attention to

1. Pretreatment requirements
2. Post-treatment requirements
3. Process variations [e.g., aerobic or anaerobic biological treatment, granular or powdered activated carbon (PAC)]
4. Process interfaces
5. Process optimization
6. Residuals generated
7. Gaseous emissions
8. Reusable materials
9. Residue treatment and disposal

The next section describes sample treatment process trains for several selected hypothetical and actual leachate situations involving the most commonly expected problems.

SAMPLE TREATMENT PROCESS TRAINS

Leachate Containing Organic Contaminants

Sample process trains for treating high-strength organic-containing leachate are based upon two actual case studies, with a third train described or a hypothetical alternative method.

Love Canal Experience. McDougall and co-workers [3,4] have reported on temporary and permanent process trains used to treat leachate for Love Canal. Processes selected for the permanent facility are shown in the flowchart in Figure 7-3. Treated leachate is discharged to the city sewerage system and conveyed to a physical–chemical municipal sewage treatment plant. Treatment costs are reported to be $9.80/m³ (3.7¢/gal), which includes amortized carbon system capital costs, replacement carbon, and equipment maintenance.

Performance data from the temporary system that was similar to the permanent installation is given in Table 7-5.

The following considerations or actions taken during development of the Love Canal leachate treatment system illustrate factors to be taken into account when selecting a leachate treatment technology in any situation.

I. SITE CONTAINMENT - Capping and Leachate Collection

2. TREATMENT:

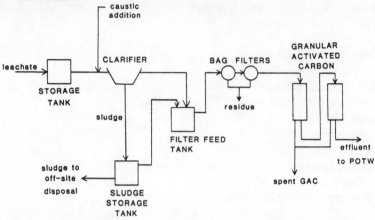

3. COST: > **$20 million**

Figure 7-3. Love canal permanent treatment system schematic flow diagram (3).

1. Love Canal was judged to be a public-health hazard, and immediate emergency actions were required. This *limited the time* that could be devoted to evaluating alternative approaches.
2. A *leachate existed* that could be used for treatability studies. These studies focused primarily on priority (organic) pollutant removals.
3. A physical–chemical POTW was in proximity. This provided not only a *discharge option* but also additional treatment and dilution, thus serving as a buffering device if the leachate treatment system ultimately selected fails to meet performance requirements.
4. Discharge to a POTW allowed for *performance requirements* likely to be less stringent than for direct discharge to surface waters.
5. Regular monitoring was practiced, and the system was constructed so that the system *modifications could be made* as needed at a future time.

Results of limited treatability and feasibility study efforts prior to treatment-system selection at Love Canal are summarized as follows:

1. A mobile treatment unit equipped for pH adjustment, clarification, sand filtration, and carbon adsorption was operated at the site and produced effluent that was found to be acceptable for discharge [3].
2. Because granular activated carbon (GAC) was believed to be the best available technology for priority polutant removal from the leachate, car-

Table 7-5. Performance Data on Temporary Treatment System at Love Canal

Pollutant	Raw Leachate ($\mu g/L$)	Carbon System Effluent ($\mu g/L$)
Anthracene and phenanthrene	29	ND
Benzene	28,000	<10
Carbon tetrachloride	61,000	<10
Chlorodibromomethane	29	ND
Chloroform	44,000	<10
2-chloronaphthalene	510	ND
Chloropenzene	50,000	12
1,2-dichlorobenzene	1,300	ND
1,3 and 1,4-dichlorobenzene	960	ND
1,1-dichloroethane	66	ND
1,2-dichloroethane	52	ND
1,1-dichloroethylene	16	ND
1,2-*trans*-dichloroethylene	3,200	12
2,4-dichlorophenol	5,100	ND
1,2-dichloropropane	130	ND
Ethylbenzene	590	10
Hexachlorobenzene	110	ND
Hexachlorobutadiene	1,500	ND
Methyl chloride	370	ND
Methylene chloride	140	46
Phenol	2,400	<10
1,1,2,2-tetrachloroethane	80,000	<10
Tetrachloroethylene	44,000	12
TOC		~30 mg/L
Toluene	25,000	<10
1,2,3-trichlorobenzene	870	ND
1,1,1-trichloroethane	23	ND
1,1,2-trichloroethane	780	<10
Trichloroethylene	5,000	ND
2,4,6-trichlorophenol	85	<10

Source: W. J. McDougall, S. D. Cifrulak, R. A. Fusco, and R. P. O'Brien, 1980, Treatment of chemical leachate at the Love Canal landfill site, in *Proceedings of the 12th Mid-Atlantic Industrial Waste Conference,* Bucknell University, Lewisburg, Pa., pp. 69–75.

ND = not detected.

bon isotherm, dynamic column, and carbon reactivation studies were undertaken [4]. Isotherms indicated that the treatment objective of 300 mg/L could be met with reasonable carbon usage. Dynamic colume studies indicated that the 300-mg/L TOC limit could be achieved, that only one organic compound—methanol—was found in the effluent in the mg/L range, that no traces of several priority pollutants were found in the effluent, and that pretreatment would be required to provide separation of oily, liquid, and sludge phases in the raw leachate. Carbon reactivation

studies indicated that high-temperature reactivation could restore most of the carbon adsorptive capacity, and that the reaction furnace and after-burner could be operated to provide total destruction of the organics.
3. Biological treatability studies were conducted in small-scale reactors [5]. Leachate was diluted with nutrient-supplemented tap water or sewage in these studies. Results indicated that biodegradation was possible at 1:5 dilution with either tap water or sewage, provided that nutrients were added and pH was controlled. Because oxygen demand was high and a possibility existed that off-gases might contain undesirable compounds, a closed pure-oxygen system with scrubbing or carbon adsorption of off-gases was thought promising [5]. These initial studies also concluded that carbon adsorption treatment of raw leachate was impractical because of the high carbon doses required and that pretreatment with activated sludge with cargob polishing might be reasonable. Leachate quality, however, was found to change substantially from that used in these early tests.

The Love Canal experience illustrates *a case where activated carbon treatment is an effective and relatively cost-effective method for removing organic contaminates* from a hazardous waste leachate. This approach may or may not have been the optimum choice but the emergency nature of the situation did not permit lengthy process optimization studies. Since the Love Canal leachate treatment system is an operating facility, additional experience should better define the effectiveness and cost associated with this approach.

Ott/Story Site Study.   Ongoing efforts to evaluate technology for treating groundwater contaminated by a variety of toxic and hazardous organic compounds have been reported in several references [2,6,7,8]. This experience is highlighted for several reasons even though the subject wastewater is groundwater rather than leachate:

1. Many of the compounds are the same as would be expected to occur in leachate.
2. Treatability studies have been conducted using groundwater obtained from the most concentrated part of the contamination plume. Therefore, contaminant concentrations may approach those of leachate.
3. Groundwater quality data indicate compounds likely to migrate.
4. The compounds present include toxic and hazardous pollutants as well as other organics. Thus, treatability results reflect the effects of the non-toxic, nonhazardous organics in the matrix.
5. Numerous technologies are being screened in the laboratory using actual wastewater.
6. The site is subject to ongoing remedial action work, so further information is likely to become available.

Table 7-6. Ott/Story Groundwater Characterization

Parameter	Composition Range
Chloride	3,800 mg/L
COD	5,400 mg/L
Conductivity	18,060 μmhos/cm
NH_3-N	64 mg/L
Organic N	110 mg/L
pH	10–12
TDS	12,000 mg/L
TOC	600–1,500 mg/L
Volatile organics	
Benzene[b]	6–7,800
Chlorobenzene[b]	< 5–140
Chloroform	1,400
1,1-dichloroethane[b]	< 5–14,280
1,2-dichloroethane[b]	0.350–111 mg/L
1,1-dichloroethylene[b]	60–19,850
Ethyl benzene[b]	< 5–470
Methylene chloride[b]	< 5–6,570
1,1,2,2-tecrachloroethane[b]	< 5–1,590
Tetrachloroethylene	110
1,1,2-trichloroethane[b]	< 5–790
Trichloroethylene	40
Trichlorofluoromethane[b]	< 5–18
Toluene[b]	< 5–5,850
Vinyl chloride[b]	140–32,500
Acid extractable organics	
m-acetonyhlanisol[c]	< 3–1,546
p-acetonylanisol[c]	< 3–86
Benzoic acid	< 3–12,311
p-2-oxo-*n*-butylphenol	< 3–1,357
o-sec-butylphenol[c]	< 3–83
p-sec-butylphenol[c]	< 3–48
o-chlorophenol[b]	< 3–20
Dimethylphenol[b]	< 3
1-ethylpropylphenol	< 3
p-isobutylanisol[c]	< 3–86
Isopropylphenol[c]	< 3–8
Methylethylphenol	20
Methylphenol	40
3,4-D-methylphenol	160
Methylprophylphenol	210
Phenol[b]	< 3–33
Base extractable organics	
Benzylamine or *o*-toluidine	< 10–471
Camphor	< 10–7,571
Chloroaniline	< 10–86
Dichlorobenzene[b]	< 10–172
Dimethylaniline	< 10–17,000
m-ethylaniline	< 10–7,640
Methylaniline	310
Methylnapthalene	< 10–290
Naphthalene[b]	< 10–66
Phenanthrene[b] or anthracene[b]	< 10–670
1,2,4-trichlorobenzene[b]	< 10–28

[a] All concentrations in μg/L except as noted.   [b] A priority pollutant.
[c] Structure not validated by actual compound.

457

Table 7-6 presents a summary of raw groundwater composition data as represented by composite samples from two wells in the contaminant plume being used in the treatability studies. Groundwater samples from other wells in the problem area differ widely in composition from those presented here.

Because the contamination problem is solely organic in nature, the following processes individually and in combination were selected for screening biological treatment (activated sludge, trickling filter, anaerobic filter), chemical precipitation, GAC and PAC adsorption, resin adsorption, air and steam stripping, and ozonation.

Results of completed studies are summarized:

1. Chemical coagulation of raw groundwater does not achieve significant removal of organics as measured by TOC reduction.
2. An aerobic biomass could not be acclimated to treat raw groundwater.
3. Adding trace elements and nutrients did not aid acclimation to raw groundwater.
4. Adding PAC to the aeration chamber at concentrations of about 10,000 mg/L neither aided acclimation to raw groundwater nor improved TOC removal or mixed-liquor appearance.
5. Batch adsorption studies for four different carbons and three resins indicated that no sorbent was able to reduce residual TOC to less than 230 mg/L.
6. Granular activated carbon (GAC) employed in continuous-flow small columns was not capable of sustaining high levels of TOC removal.
7. GAC adsorption was capable of sustaining high levels of organic priority pollutant removals even when TOC removal had declined to 35% and effluent TOC levels were approximately 600 mg/L.
8. Continuous-flow, small-column, resin adsorption studies demonstrated TOC breakthrough characteristics similar to those of GAC adsorption. However, TOC breakthrough occurred more rapidly with resin than with carbon.
9. GAC pretreatment of raw groundwater enabled development of a culture of aerobic organisms capable of further treating GAC effluent.
10. Several organic priority pollutants were detected in off-gas from activated sludge reactors.
11. Anaerobic treatment [upflow anaerobic filter (UAF)] of GAC-pretreated groundwater was possible, but the GAC–UAF process train performed more poorly than the GAC–activated sludge process train.

Based upon these results, one can make several observations:

1. The removal of priority pollutants by the GAC and the air-stripping unit processes generally corresponds with other published information.

2. A considerable fraction of the TOC is made up of nonpriority organic compounds. This fraction is more difficult to remove than the priority pollutants are.
3. The need for removal of the TOC attributed to the nonpriority pollutants needs to be closely assessed. A few static bioassay tests with *Daphnia Magna* or carbon-treated groundwater suggest significant residual toxicity; whether this is attributable to the compounds present or to low dissolved-oxygen levels needs to be determined.

Results of the treatability studies to date suggest that a process train consisting of GAC followed by aerobic biological treatment is the most feasible approach to groundwater treatment.

This process train, which, in general, is *applicable to high TOC wastewaters in situations where waste-stream components may be toxic to biological cultures, is illustrated in Figure 7-4.* The rationale is to utilize the activated carbon to protect the biological system from toxicity problems. Therefore, the carbon could be allowed to "leak" relatively high concentrations of TOC (organics) rather than be operated to achieve maximum reduction of organic compounds. Allowable leakage would be based upon determination of the point at which the carbon-treated effluent becomes toxic to the subsequent

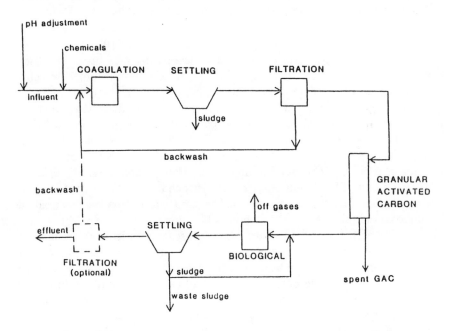

Figure 7-4. Schematic of carbon sorption for biological treatment system.

biological process. Thus, the selection of the allowable TOC or organics leakage (i.e., breakthrough) from the carbon contactors is crucial to the performance and cost effectiveness of this process train. Higher organic loads handled by the biological system result in greater service life of the granular carbon and, consequently, lower costs related to the carbon-treatment phase.

The flowchart in Figure 7-4 includes a chemical coagulation step (including settling and filtration). Although not necessary in the groundwater treatment situation discussed above, these processes could be used in situations where soluble inorganics removal or particulate removal to minimize head losses and frequent backwashing in carbon contact columns may be necessary.

Disadvantages that may be associated with the treatment system given in Figure 7-4, as illustrated by the above groundwater treatment case, are substantial carbon utilization rates to maintain effluent TOC levels below 100 mg/L, and stripping of the volatile compound in activated sludge off-gases. Other factors that must be considered when evaluating this approach include carbon regeneration feasibility and sludge disposal alternatives.

Other Possibilities.   Several other potentially effective process trains for treating leachates containing primarily organic contaminants have been postulated elsewhere [8]. One of these, believed to have high potential, is depicted in Figure 7-5, which illustrates a sequence of biological treatment followed by granular carbon sorption. This process train is *applicable to treatment of wastewaters high in TOC, low in toxic (to a biomass) organics, and containing refractory organics.* Chemical coagulation and pH adjustment are provided for heavy-metal removal and protecting the subsequent biological system. This may not be necessary if heavy-metal concentrations are below toxicity thresholds and if the moderate removal efficiencies typical of activated sludge are sufficient. Biological treatment such as activated sludge or anaerobic filters are included to reduce biochemical oxygen demand (BOD) as well as biodegradable toxic organics. This reduces the organic load to subsequent sorption processes. To prevent rapid head losses caused by accumulation of solids in the sorption columns, clarification and multimedia filtration are provided. THe intent is to reduce suspended solids to 25–50 mg/L. Granular carbon adsorption is included to remove refractory organic residuals and toxic organics. Activated carbon rather than polymeric or carbonaceous resins has been suggested because more full-scale experience exists and performance as well as design and operating criteria have been reported. This process train is expected to be highly effective and relatively economical when compared to other alternatives. Its success, however, depends on biological system performance. Moreover, the presence of high concentrations of volatile organic constituents may create a potential air contamination problem. Three by-product wastes are produced: chemical sludge, biological sludge, and spent carbon. Spent carbon can be regenerated, but the sludges must be disposed.

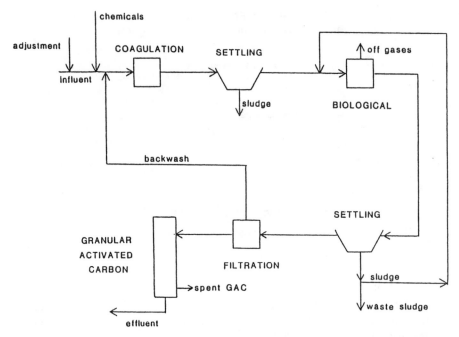

Figure 7-5. Schematic of system biological treatment system followed by carbon sorption.

Leachate Containing Inorganic Contaminants

Disposal sites or segregated portions of sites handling solely inorganic hazardous wastes, such as wastes from the metals plating and finishing industry, are likely to generate leachates of a predominantly inorganic nature. The most probable approach to treatment of this type of leachate would be chemical precipitation followed by sedimentation and, possibly, infiltration as well. However, it may be necessary to modify/supplement this approach if any of the following conditions pertain:

1. Hexavalent chromium present—addition of chemical reduction process
2. Cyanide present—addition of alkaline chlorination or ozonation
3. Total dissolved-solids (TDS) control required—addition of ion exchange if TDS level is less than about 5000 mg/L or reverse osmosis if TDS level is about 5000–50,000 mg/L
4. Ammonia present—addition of air stripping or ion exchange

Several examples of leachates containing only inorganic contaminants are discussed subsequently to illustrate process trains responsive to these four conditions. Cases discussed are heavy metals only (Fig. 7-6); heavy metals

GROUNDWATER TREATMENT AT SITE LIKE GRATIOT, MI.
(metals and low concentration of organics)

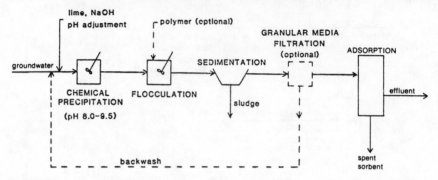

Figure 7-6. Process train for leachate containing metals and low concentrations of organics.

including hexavalent chromium (Fig. 7-7); heavy metals including hexavalent chromium and cyanide (Fig. 7-8); and heavy metals, ammonia, and TDS control (Fig. 7-9).

Figure 7-6 illustrates a process train for treating leachates containing several heavy metals. The treatment system includes chemical precipitation using lime or ferric chloride. Depending on the metals present, the pH should be adjusted to 8.0–9.5. Flocculation could be aided by polymer addition for more efficient precipitate removal in the subsequent sedimentation step. Polishing with granular media filtration also could be provided for better solids removal.

Figure 7-7 is a treatment-process schematic for leachate containing heavy metal, including hexavalent chromium. The first step in the process is chemical reduction at an acidic pH (pH reduced to 3.0 or less with sulfuric acid) to reduce hexavalent chromium to the trivalent state. Sulfur dioxide is used as the reducing agent, although sodium bisulfite or metabisulfite can be used. Following reduction, the pH is raised to 8.0–9.5 by using lime sodium hydroxide. This produces precipitation of trivalent chromium as well as other metals. The remainder of the process train is shown in Figure 7-6.

Figure 7-8 is a process schematic illustrating the treatment of a hazardous-waste leachate containing cyanide and heavy metals, including hexavalent chromium. Alkaline chlorination (with NaOCl or Cl_2 gas) at pH 9.0–10.5 for cyanide oxidation is provided first. Complete cyanide oxidation requires close pH control and an excess of chlorine. Reaction time and chlorine requirements depend greatly on operating pH. Ozone oxidation is a potential

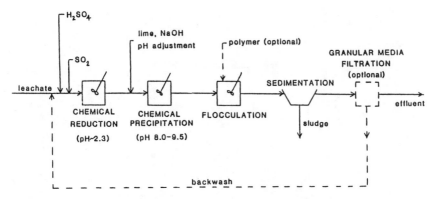

Figure 7-7. Process train for leachate containing metals including hexavalent chromium.

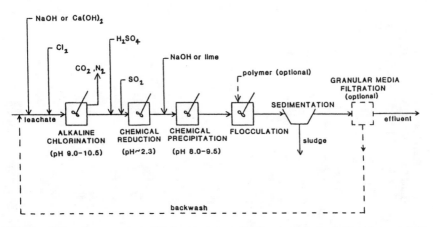

Figure 7-8. Process train for leachate containing metals including hexavalent chromium and cyanide.

alternative to alkaline chlorination, particularly for leachates containing organic compounds that might be converted to chlorinated forms.

Chemical reduction of hexavalent chromium to the trivalent state is accomplished next. For this step pH must be decreased to less than 3 by using sulfuric acid. Sulfur dioxide is added as the reducing agent. Care must be taken to assure complete cyanide removal prior to this process because acid conditions permit generation of toxic hydrogen cyanide gas. Following reduction, pH is raised to 8.0–9.5 for precipitation of trivalent chromium and other metals. The remainder of the process train is shown in Figure 7-6.

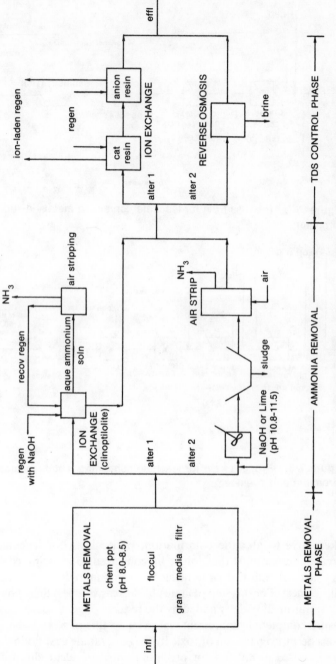

Figure 7-9. Process train for leachate containing metals and ammonia and requiring TDS control.

464

Several alternatives for treating leachates containing metals and ammonia and also requiring TDS control are illustrated in Figure 7-9. The first phase of the process train addresses removal of heavy metals by using chemical precipitation as depicted in Figure 7-6. Two alternatives for subsequent ammonia removal are then presented. Alternative 1 involves selective ion exchange using clinoptilolite (a natural zeolite). For removing ammonia concentrated in the regenerant stream, air stripping can be used, and the lime slurry regenerant can be reused. Alternative 2 uses an air-stripping tower operated under alkaline conditions; pH adjustment can be accomplished by using sodium hydroxide or lime. The latter, however, can generate large volumes of sludge.

The last phase of the process train in Figure 7-9 provides for TDS control by using either ion exchange or reverse osmosis. Ion-exchange resins include cationic and anionic species; the ions to be exchanged determine whether a strong acid–or a weak acid–base is used.

Each of the treatment systems discussed produces chemical sludges that may need to be handled as hazardous wastes. The primary disposal alternative is to landfill, preferably without dewatering or stabilization. However, site specifics and subsequent resolubilization concerns influence this decision.

The foregoing cases and example process trains do not encompass every conceivable leachate treatment situation involving hazardous-waste leachates containing only inorganic contaminants. However, the examples are applicable to a broad range of leachate concerns and illustrate the approach to formulating conceptual process flow charts.

Leachate Containing Organic and Inorganic Pollutants

Hazardous-waste leachate is expected frequently to be more complex than the previous cases, and many contain both inorganic and organic contaminants. Treatment of this leachate will involve some combinations of the treatment processes discussed previously. Because possible leachate composition variations are numerous, it is not feasible to illustrate the myriad potential treatment-process trains. Instead, an overview of important considerations is presented.

In general, when both inorganic and organic contaminants are present, the inorganics generally should be removed first to minimize effects on subsequent processes. Examples of such effects include metal toxicity to biological processes and corrosion, scaling, and inerts accumulation during carbon regeneration.

The two leading processes for treating organics are biological treatment and activated carbon adsorption. Leachate characteristics determine if these processes should be used separately or combined. If the organics consist solely of biodegradable compounds, then biological treatment alone would suffice, although a subsequent solids-removal polishing step could be necessary in some situations.

A leachate containing degradable organics only is not expected to occur frequently; consequently, the two processes will most frequently be used in series. They may be arranged such that the biological process precedes GAC to remove degradable organics and to reduce the organic load to the GAC process, which then is used to remove and polish refractory organics. To avoid GAC column plugging, locate a sedimentation or filtration step between the biological process and GAC. This treatment sequence could be applied when organics content is high and refractory but not when toxic organics are present.

A second arrangement is GAC preceding biological treatment. This sequence is used when toxic organics would interfere with the biological process. The GAC could be operated to leak the maximum concentration of organics that the biological system could tolerate and still meet performance requirements. This results in a longer sorption cycle for the carbon.

One additional process train that merits consideration is shown schematically in Figure 7-10. This biophysical treatment approach combines simultaneous biological (activated sludge) and PAC treatments in the biological process reactor. This approach is simpler than the previously described sequential carbon-biological treatments and has the potential of achieving comparable effluent quality. Potential advantages include the use of less costly carbon (powdered vs. granular) and minimization of physical facilities required. Spent carbon-biological sludge can be regenerated or dewatered and disposed directly. However, if the latter approach is considered, it is necessary to include cost for disposal of toxics-laden carbon when making economic comparisons.

REFERENCES

1. A. J. Shuckrow, A. P. Pajak, and C. J. Touhill, 1980 *Management of Hazardous Waste Leachate,* SW-871, U.S. Environmental Protection Agency, Office of Solid Waste and Emergency Response, Washington, D.C.
2. A. J. Shuckrow, A. P. Pajak, and J. W. Osheka, 1981 *Concentration Technologies for Hazardous Aqueous Waste Treatment,* EPA-600/2-81-019, U.S. Environmental Protection Agency, Cincinnati, Ohio.
3. W. J. McDougall, S. D. Cifrulak, R. A. Fusco, and R. P. O'Brien, 1980, Treatment of chemical leachate at the Love Canal landfill site, *Proceedings of the 12th*

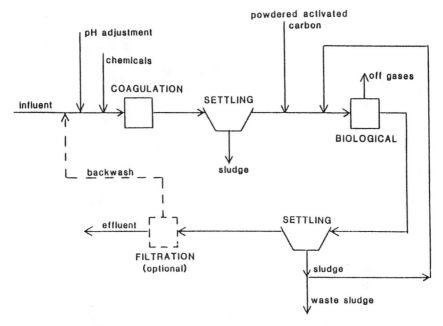

Figure 7-10. Schematic of biophysical process train.

Mid-Atlantic Industrial Waste Conference, Bucknell University, Lewisburg, Pa., pp 69–75.

4. W. J. McDougall, R. A. Fusco, and R. P. O'Brien, 1980, Containment and treatment of the Love Canal landfill leachate, *Water Pollution Control Federation Journal* **52**:2914–2924.

5. E. F. Barth, and J. M. Cohen, 1978 (Nov. 30), Evaluation of Treatability of Industrial Landfill Leachate, unpublished report, U.S. Environmental Protection Agency, Cincinnati, Ohio.

6. A. P. Pajak, A. J. Shuckrow, J. W. Osheka, and S. C. James, 1980, Concentration of hazardous constituents of contaminated groundwater, in *Proceedings of the 12th Mid-Atlantic Industrial Waste Conference,* Buckness University, Lewisburg, Pa. pp 82–87.

7. A. P. Pajak, A. J. Shuckrow, J. W. Osheka, and S. C. James, 1980 Assessment of technologies for contaminated groundwater treatment, in *Proceedings of the Industrial Waste Symposia, 53rd Annual Water Pollution Control Federation Conference,* Las Vegas, Nev.

8. A. J. Shuckrow, A. P. Pajak, J. W. Osheka, and S. C. James, 1980, Bench scale assessment of technologies for contaminated groundwater treatment, in *Proceedings of National Conference on Management of Uncontrolled Hazardous Waste Site,* Washington, D.C., pp 184–191.

8

Hazardous-Waste-Management Facility Siting

Michael Barclay
U. S. Environmental Protection Agency
Washington, D.C.

It is not hard to imagine what the typical developer might say after suffering through an unsuccessful initiative to site a hazardous-waste-management facility.

COMMENT: As soon as we announced plans to build our facility, the public was up in arms. We expected opposition, but we had no idea it could form so quickly and that it could be so strong. Organized groups sprang up overnight and within days issued press releases and circulated newsletters describing the disasters that would result if we opened our facility. By the time we prepared information to refute one of their claims, these groups had two others to replace it. We never had a chance. Before we knew it, the entire community opposed our facility, and misconceptions about it were too widely held to overcome.

QUESTION: How do we prevent this from happening to us again? How do we stop the spread of misconceptions about our future projects?

COMMENT: Our company was strongly committed to answering all of the public's questions about the facility we wanted to build. We organized a town meeting so that we could explain our facility plans to the townspeople and answer any of their questions. Our engineers didn't even get through their presentations, however, before the meeting began to fall apart on us. First, one or two people in the audience began to interrupt our presentations with shouts and jeers. Then others joined in. Pretty soon the whole mood of the audience changed. We had an angry, hostile crowd on our hands. We had lost control, and we didn't know what to do. Obviously, the meeting was a failure. Any townsperson who had an open mind about our facility before the meeting certainly didn't leave with one.

QUESTION: How do we get the public to listen to us? Does every meeting like this necessarily have to be adversarial in nature?

COMMENT: One of the biggest problems we faced in siting our facility was the local newspaper. In our view the reporters from the paper just didn't treat us fairly. They wrote articles almost daily that focused on every major hazardous-waste horror story of the past. These stories convinced the public that we were about to open up a "Love Canal" right on their doorsteps. The public was, of course, understandably upset when they read these stories in their newspaper.

QUESTION: How do we get the media to tell the whole story about hazardous-waste-management?

COMMENT: After we announced plans to build our facility, we ran an extensive public information program. Our hope was that if we answered everyone's questions about our facility, then the community would be less likely to oppose us during hearings on our permit application. We spent a lot of time and money working with the community. It was hard work trying to address all the concerns raised about our facility. We thought we had made a lot of headway, but at the public hearing for our permit application the people with whom we had spent so much time surfaced the same old arguments. It was very discouraging.

QUESTION: Is working with the public a productive use of our time and money? Will it make a difference? Will it help us get our site approved?

THE SITING DILEMMA

In 1976 Congress enacted the Resource Conservation and Recovery Act (RCRA). With the passage of RCRA Congress envisioned the development of a regulatory program that would control hazardous wastes from the "cradle"—the point where the wastes are first generated—to the "grave"— the point of their ultimate disposal. Congress directed the Environmental Protection Agency (EPA) to develop a manifest system under RCRA to track shipments from waste generators to waste treatment, storage, and disposal facilities (TSDFs). Congress also directed EPA to write specific regulations for owners/operators of TSDFs.

The RCRA regulatory control system developed by EPA is now in place. EPA has issued RCRA regulations for generators, transportors, and owners/operators of TSDFs. The states are quickly becoming equal partners with EPA in implementing the RCRA program. The character of the management industry will change dramatically over the next decade as the marketplace responds to this new regulatory system. EPA projects that the RCRA program will increase the need for new, commercial TSDFs so much so that unless at least 50 to 60 new TSDFs are built in the next few years, the country will face an acute TSDF capacity shortfall. EPA estimates that the marketplace has the capacity to absorb up to 75 new commercial treatment facilities and up to 50 commercial landfills and incinerators. The greatest need for new facilities, according to EPA, is in the Northeast, Southeast and Midwest.

Ideally, the supply of TSDFs will simply expand to meet the demand. The waste-management industry has found, however, that it is unable to respond to the need for new TSDFs due to *public opposition.* Communities are almost universally unwilling to accept TSDFs and will go to great lengths to block a developer's attempt to site one near them. The intensity and ferocity of this opposition has frustrated all but a handful of developers from siting major commercial TSDFs in recent years. Although many communities are willing to acknowledge the need for TSDFs, from a societal perspective most are unwilling to be the host for one. A 1979 EPA study concluded that the public fears major and long-term risks to the health and welfare of the surrounding community. The study also said that "opposition reflects a loss of faith by local residents in the ability of government and private industry to solve environmental problems and at the same time to consider and protect local interests" [1]. The study concluded that opposition is almost inevitable, and that if developers are unable to deal with it, they will not site many facilities.

The implications of the siting problem are grave. Inability to site new facilities will inhibit the growth of a mature waste-management industry. New facilities will not be built near the industrial markets they serve, and generators will have to continue to rely upon landfills as their predominant means of waste disposal. New, sophisticated treatment and incineration facilities needed by industry will not be built. It will also be harder for EPA and the states to close poorly designed or poorly operated facilities unless new, safer facilities are available to take their place. In those regions of the country with limited disposal capacity, capacity shortages may become worse and disposal costs will increase. High disposal costs could provide powerful incentives for firms to use other, illegal, means of waste disposal. Clearly, community opposition to the siting of TSDFs is a serious problem that must be solved if this nation is to manage its wastes efficiently and effectively.

THE DEVELOPER'S OBJECTIVES

Because the site developer must interact with the public and regulatory agencies, he should adopt the following objectives.

Obtain operating permits. The whole siting initiative is pointless unless the developer obtains the necessary permits to construct and operate the facility. Obtaining operating permits is the developer's chief objective.

Obtain favorable permits. Under RCRA program, regulatory officials have a great deal of latitude in writing permits. These officials will use their "best engineering judgment" to set permit conditions. The developer should seek reasonable but explicit permit conditions. He can and should question decisions made by the permit writer if he feels they are unreasonable or not based on legitimate regulatory principles. The more explicit the permit the less the chance that state or federal inspectors or the public will misinterpret its provisions. This does, of course, require the developer to have a clear idea of the market he plans to serve and his development plans.

Minimize permits costs. The developer should avoid delays contributing to opportunity costs. A protracted siting process—or even worse, litigation— is extremely expensive and diverts capital and management resources away from productive uses.

Select a site(s) requiring minimum engineering. A site needing little field modification to meet regulatory requirements lessens costs to the developer and probably has a smaller chance of future problems, such as groundwater contamination.

Select a site near the market. Selecting a site close to markets minimizes haul distances, thus lowering costs to customers and reducing the probability for transportation-related incidents.

Select a minimal-impact site. In addition to those impacts requiring consideration by regulatory agencies, the developer should consider the facility's effect on property values or the community's self-image, trucking routes and the probability of accidents, community fears of the short- and long-term risks posed by the facility and any other matter of concern to a potential host community.

Cultivate a favorable image. Most firms take great pains to cultivate a favorable public image. A few months of high-profile controversy can destroy years of good work by a firm's public relations office.

Manage opposition. The developer should strive to lessen, not eliminate, opposition. An unfocused attempt by the developer at managing opposition will please no one involved in the site development.

Avoid a pyrrhic victory. The developer should not try to "win at all costs." A community that feels it has been wronged by a developer or by regulatory agencies can become a formidable opponent. It can possibly succeed in closing the site, and the regulatory agency may reap the political consequences.

Maintain corporate credibility. Corporate credibility must be maintained at any cost. This is a corollary to the preceding objective. If a community distrusts a developer, it will work against him during the entire operation. A firm with no credibility will be blamed for all the community's problems.

WHY DO MOST DEVELOPERS FAIL?

Developers fail much more often than they succeed. In fact, developers fail so frequently that most feel siting a new major waste-management facility is practically impossible and have simply given up trying. Many waste-management firms are not intent on expanding or upgrading their existing facilities or purchasing existing waste-management facilities from others. Although these developers are correct in their assertion that siting new facilities is very difficult, the fact remains that some developers *are* successful in their siting initiatives. Why are they successful? Are they lucky? Do they have some secret formula for success? What separates them from those who fail? The biggest reason a developer is successful is that he communicates with the public.

Figure 8-1 is a simple model of how the developer and the public communicate during the siting initiative. The model's components are developer, information, techniques, public, timing, feedback, other information sources.

The failure of many past siting initiatives may be related to a breakdown at one point or another in the communications process. The case study shown in sidebars (all sidebars shown in Appendix 8-1) is an example of one such breakdown. It is possible to relate problems in past siting initiatives to each of the components of this simple model.

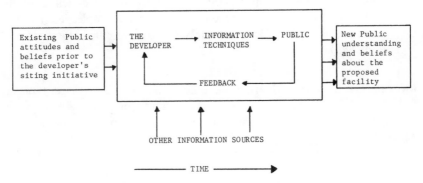

Figure 8-1. The communication process.

The Developer

Developers often fail because they are unable to establish their credibility with the public. A developer's credibility is a function of the public's perception of his *technical expertise,* his *trustworthiness,* and his *personal qualities.*

In the highly technical and competitive business of hazardous-waste-management, the developer must understand the business environment he competes in and he must plan to gain an acceptable share of his market. To apply for a permit, he has to know many technical details regarding the design and operation of the facility—for example, how the incineration or landfill groundwater monitoring system will work, what environmental safeguards are needed, and so on. Most developers are comfortable with the technical aspects of their jobs. By their personal credentials, by preparing a thorough permit application, and/or by retaining consultants with proper technical credentials, most developers can demonstrate to the public satisfaction that they are technically capable of running a safe operation.

It is very hard, however, to gain the public's trust, especially in a contentious situation such as a siting initiative. Because many developers are "outsiders" to the community or represent large faceless corporations only exacerbates the problem. When a developer faces a community that is by nature suspicious and resentful of his presence, the chances are very high that the community will attack his corporate and, perhaps, his personal honor. Few developers have been able to prevent this from happening. It usually starts a spiral of conflict that dooms the developer's siting initiative.

Information

In terms of the sample communications model, the information is the message the developer wishes to transmit to the public. The effectiveness of his information is a function of its *content* (facts, ideas, examples, etc.) and its *presentation* (the sequence in which messages are presented, the degree of repetition, etc.).

Information Content.   During the siting initiative, the developer typically follows one of these actions when deciding what information to give the public about the facility. While disseminating information about the positive and negative aspects of the facility, he highlights positive and downplays the negative; he stresses the negative consequences to the society if the community does not accept the facility.

The public will only believe the developer if what he says is accurate, even-handed, and comprehensive. In many past siting attempts—at least from the public's point of view—the developer was unwilling or unable to tell the public enough about the facility's operation and risks the public would face from the facility. The developer often was unable or unwilling to identify limitations in the state-of-the-art of disposal technologies, which prevented him from answering some questions or admitting that some questions could have conflicting answers. In almost all of these cases the public subsequently exposed inaccuracies or perceived inaccuracies in the developer's proposal and took advantage of the opportunity to discredit him. The developer could have avoided these problems if he had accepted the responsibility of indicating the limitations of the information he disseminated about the facility and had assisted the public with understanding why there were limitations in his information and what were the consequences of those limitations.

In past siting initiatives the developer often tried to avoid giving information that he felt could be damaging to his proposal. For example, few developers were willing to discuss problems of waste-management such as Love Canal and the Valley of the Drums. These developers feared that discussion of these examples would only convince the public that their facilities would present the same problem. When this happened, opposition groups inevitably advertised these problem facilities. The unwillingness of the developer to acknowledge that problems with other hazardous-waste-management facilities exist, or to respond to the issue when it arose, prevented him from gaining the public's trust. This was particularly true if the developer refused to acknowledge past problems and to offer confirmable assurance that they would not be repeated at his proposed facility.

Many developers try to ignore what they consider to be nontechnical issues that arise during the siting initiative, such as the impact of the facility on property values and the contribution of the facility to negative community image ("the dumping ground of the state"). These issues are usually of great public concern, but many developers refuse to focus on them because they feel they are issues irrelevant to regulatory decisions to be made regarding the facility. Worse yet, some developers feel they can ignore public concerns expressed in an emotional manner, regardless of the merit of those concerns. When developers refuse to discuss nontechnical issues or disseminate information to answer questions of this nature, the public enevitably focuses its full attention on these problems.

Developers have often tried to disseminate fear-invoking information during a siting initiative. They assumed that frightening the public with the dangers of illegal dumping or improperly designed and monitored facilities in their area was an effective means of motivating them to accept his facility. The problem with this approach is that most people have no personal knowledge of cases of illegal dumping or improper waste disposal (except when the site developer, attempting to establish a facility, was the one charged with improper disposal). Consequently, the threat implied by the alternative to the developer's facility is not directly relevant to the local community. In most cases the developer's facility will import wastes into a nongenerating area. In this situation the public sees the developer's proposed facility as a direct and relevant threat to it, but it sees illegal dumping as only a remote possibility. The inability of the site developer to establish illegal dumping rather than his proposed hazardous-waste-management facility as the direct and relevant threat explains why fear appeals generally do not work. The public, often in a state of emotional arousal at the prospect of hosting the developer's facility, generally seizes upon the course of action that is most effective in minimizing the relevant threat. That is, they oppose the siting of the facility in their community.

Most developers have provided information to the local community about the benefits of their proposed facilities. This information seldom substantially contributed to the successful siting of a facility. A community will likely view the benefits of a facility as not enough to overcome the risks it poses. The offering of compensation and incentives by a developer may contribute to a realignment of the public's perception of this risk-benefit equation. However, another factor undermining the effectiveness of stressing such benefits has been the timing of the introduction of a discussion of benefits in the public-consultation process. Often the facility's benefits were only stressed after concentrating on the potential risk and risk-avoidance technologies incorporated into the facility's design and management. Similarly, compensatory changes were often made in proposals only after significant public pressure was exerted on the developer. Positive information is probably much less effective in influencing the opinions of the public if it follows a detailed discussion of the facility's risks, or if it comes after opposing sides have taken positions on the issue.

When discussing negative information, developers often discuss the benefits of their facilities from the societal perspective (i.e., the need for a facility to protect the environment and/or the risks of illegal dumping). This societally focused information has little value in contributing to attitude change among the public. From the local community's perspective, the site developer is in control of the risks and benefits from societal perspective. The issue for the public is a local one, and information relevant to that perspective should be provided.

Presenting Information.   The way a developer presents information will help the developer and enhance his credibility; contribute to the retention, comprehension, and acceptance of information by the public; and contribute resolving issues by helping to clarify negotiable points and by facilitating conflict-resolution strategies.

When organizing information, it is important for the developer to present correct information in an understandable and forthright manner and to segregate information according to the type of issue to be addressed.

In most past siting initiatives developers seldom planned a strategy for presenting information. They let the regulatory process dictate their actions; therefore, in public hearings they simply make a formal presentation describing the positive aspects of the proposed facility and then respond to public criticism. The presentation of the site developer's information can often be characterized as one-sided (i.e., they discussed the positive aspects of the proposal and responded to the negative aspects only when raised by the public) and sequential (i.e., information contrary to the public opinions was presented first).

The presentation of information can influence the public's attitudes and opinions. The developer seldom presents both sides of an argument even when the public disagrees with him and the public is familiar with the issue and/or likely to hear opposing arguments from another information source. When the public and the developer disagree, the developer will be more persuasive if he allows opinions held by the public to be expressed first. Such two-sided argument will help the developer to ensure that greater attention is paid to his arguments by acknowledging the opposition's position at the outset, and to significantly contribute to credibility and trustworthiness by demonstrating a sensitivity to the public's point of view.

A well-organized argument is generally more convincing. Information lacking clarity, detail, and coherence produces little attitude change and information retention and reflects poorly on the developer's credibility. In many past siting intiatives information was apparently exchanged verbally in response to random questions. In these situations the argument was probably poorly organized.

Information organization has influenced issue resolution. This occurs only when the developer segregated information according to the type of issue and the nature of the expected public response. There are many similarities between the various classifications of public concerns related to hazardous-waste-management facility sitings. The essential point is that not all issues produce the same reaction from the public. Some issues are open to discussion and negotiations (e.g., site monitoring or closure provisions, access routes). Many times negotiations between the developer and the public have led to acceptance by the local community.

Issues related to the policy of hazardous-waste-management facility siting generally invite debate rather than negotiation. These issues are not negotiable in a local siting attempt because they are determined by federal or state legislatures. Usually issues dealing with the risk or safety of the facility provoke confrontation. The public views its position on these as nonnegotiable and reacts strongly whenever the issues come up in discussion. In past siting initiatives, few developers attempted to segregate information according to whether it was likely to provoke discussion, debate or confrontation. Segregating information in the public-consultation program can help to

Ensure that the program does not get held up dealing with policy issues that cannot be resolved locally

Identify issues for negotiation and thereby contribute to compensation and/or mediation in the siting process

Allow the developer to select information dissemination techniques to accommodate the anticipated public response

Techniques

The success of a developer's information program ultimately depends on the effectiveness of the techniques selected to transmit information to the public. For facility sitings the techniques must be appropriate for communicating technical and often complex information, and working in contentious situations where the public is predisposed to reject the information (i.e., the techniques must have the maximum possible persuasive potential). Developers have seldom selected communication techniques to maximize interaction opportunities with the public; maximize opportunities for public involvement in the siting process; and effectively use every communication mode (e.g., visual, verbal, written).

Maximizing interaction opportunities demands selecting techniques that allow the public to ask the developer questions and to probe the developer's responses for clarification. Ideally, each member of the public should meet with the site developer to discuss the facility until all of the individual's concerns and questions are exhausted. Although this ideal is seldom possible, the site developer's information will be more persuasive, suffer less distortion, be more easily understood by the public, if techniques are adopted that approach this ideal.

Greater public involvement in the consultation program entails a more active public role in information analysis reaching conclusions, and making recommendations. Members of the public actively involved in the siting process are more likely to accept recommendations they helped formulate. They will usually retain more information for a longer time.

Selecting the right communication modes can greatly enhance the public's comprehension of information and their perception of the developer's credibility. Visual displays can be particularly effective to clarify concepts and increase the public's understanding. Complex material is usually more readily understood and more persuasive if presented in a written format. Consequently, a combination of modes is usually most effective for communicating information to the public.

Developers seldom select the best communication techniques. With few exceptions the public meeting or hearing has been the mainstay of every siting initiative. Large-meeting formats do not work well because they minimize the interaction between the site developer and the public, minimize the public's involvement in the program, and generally rely on verbal communication only. The many effective alternatives such as information centers, briefings, workshops, and small-group meetings have not been capitalized on.

The Public

The developer has usually treated the public as a single homogenous entity and supplied the same information to all groups. There was no variety or modification of information or techniques. But different publics* with different information needs do exist. The following publics can be identified in many of the past siting attempts:

The unorganized public—individuals interested in the initiative but not affiliated with any organization

The organized public—existing community organizations and associations and those formed specifically in response to the developer's proposal

Political decision-makers

Environmental groups, generally operating at a regional or national scale

Local experts—individuals with specific expertise in the technology associated with the hazardous-waste-management facility siting

Economically affected parties—waste generators, local industry, real estate developers, and so on

Regulatory officials

The information needs of these publics are different. In addition, each public has a unique organizational structure that should affect the techniques the developer will select to disseminate information. Neither of these points have generally been recognized by developers.

---

* "Publics" will be used from now on in lieu of "the public" to make it clear that the public consists of many diverse groups and individuals.

The organized and unorganized publics probably do not differ significantly in their concerns and information requirements. However, the organized public has a formal organizational process that can be capitalized on to disseminate information and resolve issues. If this is possible, the complexity and detail of information the developer can disseminate to the organization will increase. In addition, leaders of such organizations can frequently speak for their membership, and this provides the site developer with an appropriate focus for negotiation efforts.

Political decision-makers share many of the publics' concerns but their mandate entrusts them with a larger public interest that may expand their information requirements. In addition, the politicians are a small enough group to be dealt with using a variety of small-group discussion techniques.

The information requirements of regional and national environmental groups are much more likely to focus on policy than are the requirements of the local community. Local experts will probably require much more detailed information of a technical nature than will the other groups involved in the public-consultation program. Similarly, the economically affected parties will require specific information regarding the proposal's impacts on their interests.

Timing

The developer should ask himself the following questions about the information to be disseminated during the siting process. At what point in the siting process should information be disseminated? Should information be provided to different publics at different times during the program?

The developer has usually provided information only at the public hearing. The timing of the hearing was, of course, beyond the control of the developer. The hearing date often fell longer after the time the public first learned of the developer's proposal. However, it is evident that the developer's credibility was particularly damaged if the public first heard of the proposed facility from the media.

The publics have a wide range of concerns and information needs. The difference between the public's and their requirements over the amount and complexity of information suggests the need for different levels and duration of involvement with the developer. Consequently, the program must operate over time if it is to maximize opportunities for information exchange.

The sequence of providing information to the publics is a special timing problem. Opinion setters (e.g., local politicians, respected and influential community leaders) will be looked to by the general public for an early reaction to the facility. An informed response by opinion setters can contribute greatly to the site developer's credibility and the success of the siting initiative. However, this demands that opinion setters receive information sufficiently in advance of other publics to be informed but not so far in advance as to lead to charges of secrecy and collusion between the site developer and the local decision-makers. This principle of the staggered dissemination of information to different publics has seldom been respected in past siting initiatives. Frequently, in past cases, the first that community leaders heard of the proposed facility was when local media asked them for their reaction. Not surprisingly, despite having no specific information on the proposal, the opinion setters opposed a facility in their community. Having publicly committed themselves to this position, it was virtually impossible to get them to reverse it.

Feedback

The feedback loop in the communications model describes the monitoring process available to the developer. If the site developer establishes an information monitoring process he can determine if the public has received and comprehended the information, and if additional information needs to be provided; also if information is required to deal with misconceptions or incorrect information emerging from other sources.

Very few developers have established mechanisms for monitoring public response. When opportunities for feedback existed, they were generally limited to a public meeting, for example, or to one group (e.g., a task force or repeated meetings with local politicians).

The problems involved in not providing for public feedback were most apparent in cases where significant misconceptions arose that the site developer could not identify or had no means to effectively respond to.

Other Information

During a siting initiative, the public will receive information, from sources other than the developer. Site-specific information relates to the developer's proposed facility. General information relates to everything else available to the public on hazardous-waste-management facilities or other relevant issues. In most siting initiatives, the media has been the major source of site-specific information. However, the developer usually was unable to build a working relationship with the media. Most developers chose to avoid the media. This is, however, neither possible nor desirable. A facility siting initiative is a newsworthy event, and if the media does not have information from the developer on which to base its reports it will seek other sources. These other sources are frequently opposition groups or media archives on problem sites such as Love Canal. Lack of working relationship has been a major factor in escalating conflict in past siting initiatives.

PUBLIC CONSULTATION

As we have already said, the developer must generate and disseminate information, monitor the effect of the information on the publics, counter misconceptions and misstatements, diffuse or minimize hostility, and eventually bring the various publics together and try to resolve differences with them. The *public consultation* can achieve these things.

Public consultation is a structured process that consists in listening to public complaints and allowing the public and its leaders to help devise solutions. This establishes the facility developer as a responsible and concerned citizen rather than a cold and faceless outsider bent solely on profit making. Public consultation promotes direct and comprehensive two-way communication between the public and the developer. Its approach is comprehensive, but its focus is site specific. A public relations program alone will not work because it is too one-sided and cannot address specific public fears. Nor can a public education program alone succeed for it usually addresses concepts so broad that everyone can agree with it. No one opposes protecting the environment, but when the issue comes down to how and where to site a new landfill, the disagreements become sharp and fears, often unfounded, become focused.

A public-consultation program as used in a TSDF siting initiative may be characterized as

Voluntary. Designed and operated by a developer, is not required by regulatory authorities nor does it replace opportunities for public input in the siting process required by regulatory authorities.

Scheduled before final permit review. It begins when site(s) are selected for a facility and ends when regulatory authorities begin deliberations on the permit application.

Site specific. It is not a vehicle for educating society at large on hazardous-waste-management. The proposed host community is the program's audience.

Useful in a contentious environment. Its value increases directly with the amount and degree of opposition the developer will face in a community.

Useful for complex issues. It is most useful when the issues are complex and technical.

A successful public-consultation program can

Help a community overcome its misconceptions about a developer's proposal and contribute to a less emotional examination of the facility and its impacts on community

Help the developer establish open and honest dialogue with the public. This face-to-face, constructive dialogue will identify legitimate public concerns that the developer can address with modifications to the design or operation of the facility.

Support negotiation or mediation efforts by identifying the central areas of disagreement between the developer and the public

Contribute to a positive corporate image and to the developer's credibility as a concerned citizen. A positive corporate image may hasten the process of receiving regulatory approvals

Help the developer avoid delays—litigation, extended regulatory scrutiny, and so on

A public-consultation program's contribution to the developer's siting effort does have limitations, however. The program cannot

Overcome all public opposition to a proposed facility

Overcome differences of opinion held by affected parties on the impacts of a proposed facility

Replace existing mechanisms for dealing with regulatory officials. A decision on the technical acceptability of the proposed facility is the responsibility of a regulatory agency

HOW TO DESIGN AND OPERATE A PUBLIC-CONSULTATION PROGRAM

Figure 8-2 illustrates six discrete stages in a public-consultation program.

1. *Planning* (the planning of the program prior to implementation)
2. *Initial announcement* (the activities surrounding the first announcement of a proposed facility)
3. *Information dissemination* (ongoing information exchange with the public)
4. *Monitoring* (recording public response and reaction)
5. *Documentation* (documenting public response and reaction)
6. *Evaluation* (reporting on the results of the program)

Planning

No two communities are exactly the same; consequently, the developer cannot rely on standard approaches. The public-consultation program must be

Figure 8-2. Stages in the public consultation program.

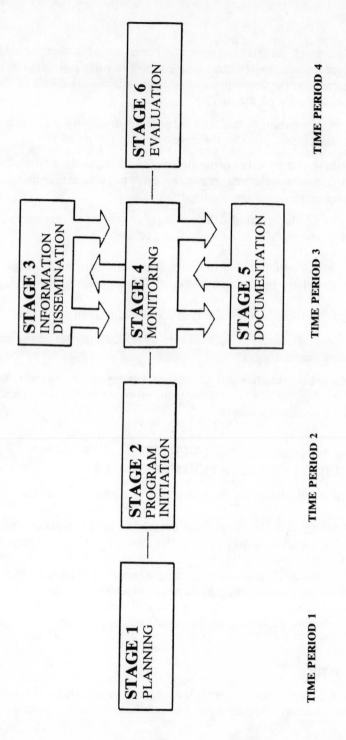

STAGE 1	STAGE 2	STAGE 3	STAGE 6
PLANNING	PROGRAM INITIATION	INFORMATION DISSEMINATION	EVALUATION
		STAGE 4 MONITORING	
		STAGE 5 DOCUMENTATION	

TIME PERIOD 1   TIME PERIOD 2   TIME PERIOD 3   TIME PERIOD 4 tailored to fit each community. The developer must collect data in the planning stage to determine program information content, the range of publics to be dealt with, and the techniques that should be used to disseminate information.

A community should learn of the developer's facility from the public-consultation program. Therefore, the developer must design the program before siting plans become public knowledge. Doing so allows allows the consultation program to be in place and operating when facility plans are announced. Although this timing places some limitations on the information the site developer can gather in the planning stage of the consultation program, much of the program can be completed before announcing the proposed siting.

Objectives.   When planning the program, the site developer collects information to prepare a community profile, prepares public distribution materials for the initial announcement of the proposed facility, and develops a preliminary public-consultation-program design for implementation immediately following the intitial announcement.

Implementation Approach—The Community Profile.   The community profile is a comprehensive description of the community compiled from data that is available in secondary sources such as census data, newspapers, reports, and so on. It gives the site developer a chance to get to know the community before directly interacting with its population through his consultation program. The community profile tells the developer who will be involved in the public-consultation program; what information should be included in the program; and what techniques should be used to get the information to the publics.

Who Will Be Involved?   The community profile describes the people living in the vicinity of the developer's facility. The developer's public consultation program is intended this population. To ensure that his consultation program serves all of the people, the site developer should collect the following information:

The size and distribution of the population. The size of the community's population indicates the size of the potential public body and consequently the magnitude of the program. The distribution of the population (i.e., concentrated or dispersed) indicates the appropriateness of various information dissemination techniques. For example, information available at a single location (e.g., a meeting or a site office) is more appropriate for concentrated populations. (source of information: census data)

The family and the household structure of the community. The family and household structure of the community may indicate public concerns. The level of concern and the types of concerns should vary somewhat between a community of young couples with children owning their own homes and a childless and/or transient community (e.g., student tenants, retirement community). (Source of information: census data.)

The existence of organized interest groups in the community. Organized groups, particularly if they are concerned with environmental issues, may constitute a major portion of the developer's public. The concerns of these groups may differ from those of the general public, giving rise to unique information requirements. The organizational structure of the groups may recommend the use of certain types of information dissemination techniques. (Sources of information: community directories, national organizational directories.)

Municipal politicians and other identifiable community leaders or opinion setters. Local politicians. community leaders, and opinion setters are an important audience for the public-consultation program. If identified before implementing the program, the developer can provide them with information early in the process so that their opinions will be informed ones. (Sources of information: local newspapers, municipal directories.)

Basic economic characteristics. The developer may find some differences in community response if jobs depend on industrial activity producing many waste by-products, or if the community has high unemployment. (Sources of information: census data, industrial directories.)

What Information Should Be Included?

The information content of most consultation programs will be similar. However, the community profile may identify unique characteristics of the proposed host community that the developer can address in his consultation program. The site developer should investigate the following community characteristics:

The economic base and viability of the community. The community's reaction to the developer's facility depends to some degree on the compatibility of the facility with the cummunity's economic activity. For example, if the facility does not serve local industry but imports wastes into the area, experience suggests this will be a major concern to be addressed in the consultation program.

Public reaction also depends on the contribution the facility would make to an improved tax base, stable employment, and other factors. This information would also be relevant to compensation/incentive strategies the developer may adopt. (sources of information: industrial directories, municipal budgets, planning documents)

The community's previous experience with contentious facility sitings or waste-management issues. The community's reaction to other attempts to site contentious facilities or to deal with waste-management issues may provide insights on the community's concerns and the needs to address these in the consultation program. (Source of information: local newspapers.)

The community's development or planning goals. If the community has documented long-range development goals (e.g., an industrial development strategy), the proposed facility's effect on those goals should be a part of the program's information content. (Sources of information: local newspapers, planning or policy documents of local political administrations.)

The community's self-image of predominant character. The community will react to the introduction of any new facilities largely in terms of its self-image. A community that views itself as a tourist or cultural center or prides itself on its scenic beauty will react to a facility differently than an industrial center. The developer must use different information to respond to these differences in reaction. (Source of information: local newspapers, promotional material, tourist material.)

In addition to data collected on the local community curing the planning stage, the developer should investigate the following issues so that information can be provided on them in this consultation program:

The regulatory process. The developer's consultation program should describe the nature of the approvals process and the opportunities for public input to regulatory bodies. An appreciation of the timing of the regulatory process will provide an indication of the consultation program's timetable.

The timetable for the completion of site specific technical studies. The developer should provide information on the results of local site studies. An indication of when these might be available may influence the timing of the public-consultations program

The history and reputation of the site developer. An objective analysis of the developer's activities in the area may identify questions related to his credibility that he will have to answer in his program.

What Techniques Should Be Used?   The community profile will allow the developer to select appropriate public-consultation-program techniques for his program by identifying the publics he needs to reach through his consultation program, and the media available to him in the local community.

As discussed earlier, the developer needs to use different public-consultation techniques to reach concentrated and dispersed populations, organized and unorganized segments of the community, and so on.

The local community will rely heavily on the media for information on the developer's facility, particularly in the second stage of the public-consultation program (the initial announcement fo the proposed facility). The developer must understand the opportunities available to him to reach the public through the media. Consequently, he should collect information on the types of local media available (newspaper, radio, TV) and the characteristics of each medium (weekly or daily newspaper, circulation, readership patterns, etc.).

Preliminary Design. When the community profile is complete, the developer can design a preliminary consultation program using the following steps.

Step l: identify information content. The developer must identify the issues and concerns to be addressed in the consultation program from a review of issues commonly raised by the public in TSDF sitings and any unique local issues identified in the community profile (see Appendix 8-1).

Step 2: identify publics. The developer must identify the publics to be reached with the consultation program from the community profile. If possible, he should identify the particular concerns of each of these publics.

Step 3: select probable public-consultation techniques. Based on data from the community profile, the developer should select appropriate public-consultation techniques for the program. The distribution of the population, the number and types of organized interest groups, and the availability of local media will suggest the most appropriate techniques to the developer.

Step 4: establish a method for developing material for public distribution. The developer's staff responsible for operating the public-consultation program may be unfamiliar with the technical aspects of siting TSDFs. Consequently, it is important for the developer to establish a procedure for his consultation program staff to collect information from his technical experts, translate it into program material, and verify its accuracy.

Step 5: develop material required for initial announcement. The developer must plan the initial announcement stage of the public-consultation program, select specific techniques to provide the public with information in the initial stages of the program, and prepare the required materials (press releases, information sheets, etc.).

Termination of Stage One. At the conclusion of the planning stage, the developer has produced a complete community profile, a preliminary design of the entire public-consultation program, and materials for public distribution in stage two of the program.

Initial Announcement

The developer begins active exchange of information with the public starting with the initial announcement of the proposed facility. The initial announcement is a unique stage in the consultation program. At this stage the site developer's relationship with the public is significantly different from subsequent stages. Public interest and concern will likely be at its height just after the site is announced. The community's political representatives and the media will probably be most actively involved at this time. The community will expect its leaders to take a position on the developer's facility. A limited number of key questions will monopolize the public's attention in the initial stages of the program. The public will begin to ask more detailed questions later in the siting process. (Appendix 8-2 lists questions the public has asked owners/operators in past siting initiatives.) Intensive activity characterizes the initial announcement. However, this stage occupies a relatively short period of time. It is followed by a lengthier exchange of detailed information in the main body of the public-consultation program.

Objectives. In this stage the site developer must provide information that will address the public's dominant concerns and questions in the period of time immediately after they first hear of the proposed facility, and convince the public that the consultation program is a credible and available source of further information.

Implementation Approach. The initial announcement will be effective if the developer provides the public with the answers to the questions they have at this stage in the siting process, schedules the announcement carefully, establishes the credibility of the consultation program, and uses appropriate techniques to disseminate information.

Providing Information to the Public.   This information must be carefully chosen. The public's ability to comprehend a great deal of specific technical information in a short period of time is limited, and their attention will be focused on issues of pressing concern. Consequently, the developer must provide basic information in an easily understood manner, and address the questions of most immediate concern to the public.

The issues most likely to command the public's attention at this stage are (a) rationale for the proposed location, (b) opportunities for public input to the decision-making process, (c) basic information on the TSDFs operation, and (d) major impact(s) of the facility on the community.

The developer must be able to justify the proposed location of the facility with information available at this stage of the consultation program. The public will want to know how he selected the site (physical factors, accessibility and transportation factors, social and economic factors, etc.).

Opposition to the TSDF is likely to materialize immediately after the initial announcement. Those opposed to the facility as well as certain segments of the public with specific interests pertaining to the facility will quickly formulate questions that go beyond the information that can be provided in the initial announcement. They will also be anxious to have their views and opinions heard. Consequently, it is very important that the developer describe the entire public-consultation program and the opportunities for public input during the regulatory process.

At this stage the developer should describe the facility *in general terms*. The description should include a discussion of the types of waste to be handled at the facility, such as toxic waste by-products of industry, radioactive wastes, and so forth; the potential hazards of these wastes if not properly disposed; the origin of the wastes (by industry and geographic location); the quantity of wastes to be treated; the life of the facility; and a general description of the disposal or treatment process (this includes how wastes are transported to the site, disposal by incineration, landfill, etc.).

The major impacts of the proposed facility will probably not be the dominant concern of most segments of the public immediately after the initial announcement. These impacts are likely to be most closely scrutinized during later stages of the program. At this point the public may simply assume the impacts are negative and focus their attention on why the developer selected their community as the facility's location and on the opportunities for opposing the selection during the regulatory process. However, the developer's initial information should address the major impacts of the proposed facility—for example, the potential for environmental damage to the locality hosting the proposed facility, the disruption and nuisance caused by the facility's operation, the possible direct and indirect economic impacts of the facility, and other factors.

The credibility of the developer's consultation program depends on the public's perception of the quality and accuracy of the program's information. (The guidelines for preparing information for the public distribution are presented in Appendix 8-1.) The site developer should do two things: (1) avoid misleading oversimplifications resulting from summarizing much material because of limitations to the amount of information that can be provided during this program stage. The developer must state technical information in laymen's terms. The rewriting of technical reports for public distribution must not result in information that oversimplifies impacts in a way that the public may consider misleading. (2) maintain an appropriate balance in the information he provides to the public by not ignoring the potential negative impacts of his proposed facility.

Scheduling the Announcement. The developer must carefully consider the sequence in which he provides information to various segments of the public at this stage of the consultation program. He must provide comprehensive information to local decision-makers and community leaders early in the process so that they will be prepared to respond in a knowlegeable fashion when questioned by their constituents or by the local media. He must also quickly put the rest of the consultation program to accommodate the reaction to the initial announcement.

Local community leaders will quickly respond to the initial announcement of the proposed facility. If they possess relevant information concerning the facility, they can respond from a position of knowlege. Without prior information their rejection of the proposal is almost certain. Providing community leaders with advance information does not guarantee their support for the facility, but they may adopt a noncommittal stance and support the principle of providing the public with information prior to the final decision-making. In addition, a knowlegeable community leadership may help contain the spread of misconceptions concerning the facility.

Briefing local decision-makers and community leaders before the initial announcement is a difficult task. In-camera meetings, even if they can be arranged, may undermine the credibility of the site developer in the public's view. Furthermore, there is a strong possibility that information would be leaked that would damage the credibility of subsequent stages of the public-consultation program and would destroy the rationale and objectives of the initial announcement. The best strategy for the developer· is to brief community leaders and then immediately announce the facility to the media and begin the public-consultation program.

Establishing Credibility. The site developer's actions at this stage are crucial to establishing his consultation program's credibility with the public.

The developer should incorporate the following safeguards into his program to protect and enhance credibility:

The information content of the program must not be vulnerable to criticism because it lacks objectivity or balance.

The opportunities for the public to obtain information and to make an input to the decision-making process must be stressed. Information should be available on (a) the comprehensiveness of the public-consultation program that is to follow the initial announcement; (b) the various techniques to insure that every segment of the public has ample opportunity to have all their questions answered; (c) the opportunities provided by law in the regulatory process for public input to the final decision.

The separation of the public-consultation program and the regulatory process must be stressed. The following points should be made: (a) The regulatory agencies will act as *independent* evaluators of his facility. (b) The public has definite opportunities during the siting process to express their views to the regulatory agencies. The opportunities are independent of the developer's consultation program; however, the developer will also relay public concerns and opinions expressed through the program to the regulatory agencies.

Disseminating Information.   The developer's information dissemination effort should be directed at the media and local decision-makers. The public will learn of the facility from the local media. Consequently, the developer must involve the media in the consultation program at this stage. The site developer should make information available to the media through press releases, news conferences, and briefings. The necessity of relying on community media to make the initial announcement removes much of the site developer's control over information available to the public at this program stage. Consequently, it is very important that carefully prepared information be available for the media if the message is to be accurately conveyed to the community.

Termination of Stage Two.   The second stage of the program concludes after initial contact with the media and local decision-makers. From this point the proposed facility is public knowledge, and the focus of the program is redirected to responding to questions and concerns. At the end of the second stage the developer should have announced the facility without jeopardizing his credibility or that of the public-consultation program. If successful this stage results in a high level of awareness for the program, an acceptance of the program and its role as a source of objective information, an established relationship with the media, and a desire on the part of the public for more information on the proposed facility.

Information Dissemination

The developer establishes a dialogue with the community in this stage. He answers questions as the public scrutinizes the proposed facility. This stage is usually the longest stage. The site developer uses a variety of techniques to identify and address public concerns. The stage is complete when, for all practical purposes, the developer has answered all of the public's questions concerning his facility.

Objectives. The developer should ensure that all individuals and organizations in the community have the opportunity to raise questions; provide comprehensive and relevant information on all of the questions and concerns raised by the public; and deal with the community in a manner that protects and, if possible, contributes to his credibility as a facility operator.

Implementation Approach. This stage of the developer's consultation program is effective if it (1) provides relevant information to all segments of the public, answering the public's questions in a clear, understandable manner; and (2) counteracts incorrect or misleading information from other sources to contribute to the resolution of issues over which the developer and the community disagree.

Providing Relevant Information. The developer cannot treat the public as a single group. The public consists of organized and unorganized segments of the community, all of whom the developer must serve through his consultation program. At a minimum the following publics are potential participants in the consultation program:

The local, organized public (existing community associations or clubs that may take an interest in the proposed facility, local groups formed specifically in reaction to the facility, etc.)

The general public (local residents unaffiliated with organized groups, but seeking information on the facility)

Decision-makers (appointed and elected officials responsible for making decisions that could affect the proposed facility)

National or regional public-interest groups (environmental or industry groups operating in the state or nationally with an interest in the developer's facility)

The media (local or national media requiring information on the facility)

Economically affected parties (e.g., waste generators)

Local experts (parties with a specific interest and detailed understanding of waste-management technology).

These publics may not all be present in every siting process, and individuals may be members of more than one public group. However, the site developer must identify all of the public group. However, the site developer should be prepared for other publics to emerge following the initial announcement.

What information is required by each public? All publics require information on the facility that explains in general terms its purpose, operations, and impacts. The developer should provide this information in the initial announcement and continue to make it available for latecomers to the consultation prc gram. However, during this stage of his consultation program, the developer's emphasis should be on information that describes the implications of the proposed facility for the specific interests of each public. The site developer provides specific pieces of information to the publics expected to be most interested in that information. For example, environmental groups with a national constituency may be interested in policy issues for the process by which the site was selected; local resident associations in the vicinity of the site and local real estate boards may be interested in the facility's impact on property values; the chamber of commerce and local politicians may be interested in the facility's implications for industrial development in the area; professors from the local university's school of engineering may be interested in a detailed discussion of the facility technology, and so on. Although the site developer should target information for each public in this stage of the program, he should not restrict the public's access to any information.

The developer will increase his credibility by providing relevant information to each public, but there is no guarantee that this information will win support for his facility. In past siting attempts the publics that developers expected to support their facilities (waste generators, chambers of commerce, politicians concerned about industrial development) often refused to take an active part in the siting process or retracted initial support for the facility in the face of mounting community opposition.

What is the best way of getting information to the publics? The site developer should consider the characteristics of each public before selecting techniques for disseminating information to them. Items to consider include: the type of information to communicate, the publics can be easily identified and/or expected to come forward for information, and how each public is organized.

Detailed and complex information is best communicated with techniques involving high levels of personal interaction and public involvement. Publics requesting technical data or an opportunity to examine issues in detail can best be dealt with personally in small-group meetings. The developer should rely on techniques stressing face-to-face interaction to serve these publics. The developer can reach publics requiring only general information about the facility through mass distribution of written materials or other techniques involving less interaction with the public.

The developer will be able to identify most of the organized publics easily after he announces his plans, and most of these publics will not hesitate to seek information. However, some members of the general public may hesitate to seek information from the site developer. If the developer ignores these unorganized individuals in his public-consultation program, they may seek answers to their questions elsewhere, and their views on the developer's facility may be based on inaccurate information. The developer must use techniques appropriate to communicating with the general public. The appropriate techniques make information accessible, provide effective explanations to complex questions, and offer anonymity to those desiring it, particularly when the public is asked to express an opinion on the facility.

A final consideration that will influence the developer's selection techniques is the organizational structure of the public. Many publics have an organizational structure that the site developer can use to maximize the dissemination of information. The site developer can wisely invest public-consultation resources by dealing in depth with an organization's representatives and relying on the organization's own methods for keeping in touch with its membership for the further dissemination of information. The site developer can increase the likelihood that the information reported to the membership is accurate and complete by preparing information packages appropriate for distribution through the organization's reporting mechanism (e.g., fact sheets for reproduction in the organization's newletter).

Counteracting Incorrect Information. The site developer's consultation program is not the public's sole source of information on his proposed facility. The public may also receive information from the media, consultation programs run by groups opposing his facility, or by word-of-mouth and informal discussion with acquaintances. The site developer must monitor each of these information sources to identify common misconceptions or inaccurate information. The method by which the site developer responds to inaccurate information should depend on its source. The sources of misconceptions spread by word-of-mouth cannot be traced. However, the developer can identify these misconceptions with monitoring techniques and counter them with information in his consultation program. If the developer establishes an effective monitoring system, he should identify misconceptions before they are widely accepted by the public and may try to set the record straight with an immediate response in his consultation program.

The developer's most effective strategy for minimizing inaccurate news reports is to establish early in the consulatation a working relationship with media representatives. However, inaccurate or misleading reports may still appear in the media, and the developer should respond. If the developer has a good relationship with the media, he can usually obtain a correction or follow-up story providing another point of view. This is usually the developer's best course of action because the response is not directly attributed to the developer and appears in the same medium as the inaccurate or misleading item. However, if this is not possible, the developer's only resort is to respond through his consultation program.

Groups opposed to the developer's facility may establish their own public information program and challenge the developer's information. The opposition's information program is likely to stress the negative aspects of the proposed facility and highlight areas of uncertainty in the technical studies supporting the TSDF. This emphasis on negtive impacts and uncertainty may not necessarily be based on inaccurate information. The opposition may be accurately reporting on negative facility impacts. If the information in the site developer's initial announcement and in this stage of the program is objective and well balanced (i.e., if negative impacts and areas of uncertainty have already been openly acknowledged), the developer may minimize the damage of an opposing group's information program. If the opposition's program is the first and/or the only source of negative information on the facility, the credibility of the developer will be jeopardized, and the consultation program may be ineffective.

The developer should respond differently to an opposition public information program that is misleading or simply inaccurate. The site developer has little to gain from publicly denouncing the opposition's information. In most cases this can only lead to a prolonged series of accusations and counteraccusations that lends credibility and draws attention to the opponent's information. At best, the public is confused by this exchange. At worst, a public that is not generally sympathetic to the site developer will reject the developer's consultation program in favor of the opposition's program. Although it is unwise for the site developer to become entangled in a war of words waged with conflicting information, he must respond to the inaccuracies. He can best accomplish this by expanding those parts of his consultation program that contradict or clarify the misleading information in the opposition's program. The developer can thus respond to inaccurate information without acknowledging its source. It is, of course, not always possible to avoid commenting on the opposition's information. It is likely that other audiences or the media will ask the developer to respond to the opposition's charges. The developer should avoid denouncing the opposition's program and simply refer the questioner to information in his program that contradicts the opposition's charges.

Termination of Stage Three. Finally, the developer should continue to make information available to the opposing group. It is not wise for the developer to restrict access to his consultation program or refuse to deal with the opposition publics. At a later time the site developer may benefit from demonstrating that accurate information was available to all publics at all times.

Monitoring

The developer monitors public response to his facility to decide how to modify his consultation program and to identify public concerns he may be able to address through changes in his proposal. *It is only through monitoring that the site developer can establish the dialogue with the public that separates consultation programs from more narrowly defined information and public-relations programs.* Monitoring is the site developer's steering mechanism for the consultation program and the public's guarantee that their concerns are being considered by the developer.

Objectives. The developer should incorporate monitoring mechanisms into the public consultation program to (a) record public response to the public consultation program, (b) record public response to his facility, (c) identify public questions or concerns not being addressed in the consultation program so that information can be provided in these areas, (d) determine if the public is receiving and understanding the information in the consultation program, (e) identify issues of concerns to the public that can be reduced or eliminated with alterations to the proposed facility.

Implementation Approach. Monitoring mechanisms should be established for information available in the developer's public-consultation program and information available to the local public from other information sources (e.g., the media or consultation programs operated by groups opposing the facility).

Monitoring provides the developer with information on both the public-consultation program and the proposed facility. Specifically, data is collected to answer the following questions?

Is the program reaching all interested parties?

Are all of the public's questions being answered?

Is the public understanding the information in the program?

Is the public-consultation program well balanced?

Should the emphasis of the program's information be altered?

What is the extent and basis of the public's opposition to the proposed facility?

Are there public concerns that can be reduced or overcome by the actions of the site developer?

The cost of monitoring and the availability of the results of the monitoring process are two issues the developer should consider before selecting a monitoring mechanism. Monitoring is only useful if the developer analyzes public response. The developer should be aware of this and the potential analysis costs when selecting a monitoring mechanism. Considerable staff time can be invested in analyzing questionnaires or the content of newspapers. Furthermore, the results of some monitoring mechanisms will generally have to be made available to the public. When the public is asked to submit briefs or questionnaires, the developer is committed to analyzing the results and making them available to the community. To not do so would seriously damage his credibility.

The developer will have to determine the extent of monitoring that is feasible and appropriate for the consultation program to be undertaken. However, some type of monitoring is essential if the developer hopes to effectively respond to public concerns and/or to resolve disputes through negotiations with the public prior to the approvals process. Finally, monitoring allows the developer to record the public's response to the proposed facility.

Termination of Stage Four.   Monitoring concludes at the termination of the public-consultation program. This will generally be just before initiating the formal approvals process. The monitoring undertaken throughout the program is the basis for the documentation and evaluation report discussed in the following section.

Documentation and Evaluation

In the fifth and sixth stages of the public-consultation program, the site developer compiles data and reports on the program. The data is documented throughout the period fo the consultation program. A report describing the public's involvement in the siting process is prepared for the regulatory agencies.

Objectives. The developer should document and report on the public-consultation program to

Indicate the efforts made to inform the public of the impacts of his facility

Establish that mandatory and recommended requirements for public involvement have been met and/or exceeded in the program

Demonstrate an understanding of the public's concerns and objections to his facility

Establish that every attempt has been made to mitigate these concerns in the proposed design and management of his facility

Public-consultation programs of the type described in this report are unlikely to be required of the developer. If they are undertaken, it will be done voluntarily by developers anxious to constructively deal with community opposition in a manner that does not jeopardize their siting attempt. The consultation program may contribute to this end by allowing the developer to enter into an informed and reasoned discussion of issues with the public. In addition, the consultation program should allow the developer to identify those public concerns that are within his power to resolve with changes to the design and operation of the facility. Finally, the developer who operates a public-consultation program cannot be faulted for ignoring the public's concerns or failing to inform them of the facility's impacts.

The developer should not expect public consultation to overcome opposition to his facility. However, even in the event of continued opposition, the developer should be received more favorably by a regulatory agency if he has done everything within his power to inform the public, to understand their concerns, and to respond to these concerns in the proposed design and operation of the facility. The developer must document and evaluate the public-consultation program if he hopes to prove this to the regulatory agency.

Implementation Approach. During the documentation stage the developer collects data to describe the public-consultation program and its results. This data is collected as a standard part of the program monitoring. During the evaluation stage the developer prepares a report to indicate

The availability of information to the public

The comprehensiveness of the available information

A thorough understanding of public concerns

The steps taken to respond to public concerns

Alternatives proposed by the public

Termination of Stages Five and Six.   Documentation concludes with the termination of the public-consultation program. The evaluation report is written between the termination of the program and the initiation of the regulatory process.

RELATION OF THE PUBLIC-CONSULTATION PROGRAM TO THE SITING PROCESS

The developer must integrate his public-consultation program into the overall siting process. In addition to public consultation, there are two other major sreams of activity in the siting process: *Technical studies* and *regulatory review*. The developer's public-consultation program will be most effective if he coordinates its activities with the technical studies he will perform to support his permit application and with the various steps of the regulatory process. The most important thing the developer should keep in mind is that his formal program should start before he actually enters into the regulatory process.

As discussed earlier, there are six stages in the public-consultation program. Stages may also be identified in the other activity streams also.

Technical Studies

1. *Site selection*   Identifying candidate areas on a broad scale concluding with the selection of a specific site(s)
2. *Impact statements and detailed facility design*   Site-specific studies documenting impacts and facility designs including the identification of alternatives
3. *Feasibility of alternatives*   Examination of design and/or operational alternatives
4. *Report preparation*   Preparation of impact reports and permit application

Regulatory Review

1. *Notice of intent*   Initial contact with the regulatory agency to indicate that a facility is or may be proposed
2. *Permit application*   Application to state regulatory agencies and any other necessary authorities for approvals
3. *Establishment of approvals boards*   If required, appointments to any approvals bodies responsible for hearing the case

4. *Notice of hearing* Notifications of a public hearing on the permit application
5. *Public input* If required, provision of information to the public and response to questions in a setting other than the formal hearing process (e.g., a public meeting)
6. *Public hearing* Interested parties present their case concerning the permit application
7. *Approvals* A decision on the permit application
8. *Appeals* Provisions for appealing the decision of the approvals body

The developer must ensure that the various stages of the three streams of activity are properly integrated. The relative timing of these stages is crucial to the success of the developer's consultation program. Proper scheduling of the consultation program will maximize the developer's opportunities to provide information to the public and to respond to public concerns in a timely fashion, and facilitate the timely exchange of information between the public and other actors in the siting process. Table 8-1 illustrates the optimal scheduling of the stages of the public-consultation program relative to the other stages in the siting process.

Table 8-1. Timing of the Three Streams of Activity

Timing: Sequence of Events *Streams of Activity*	*Period 1*	*Period 2*	*Period 3*	*Period 4*
Public consultation	Planning	Initial Announcement	Information dissemination Monitoring Documentation Evaluation	
Technical studies	Site Selection		Impact statement and detailed design Feasibility of alternatives Report	
Regulatory process		Notice of intent	Permit application Establish approvals body Notice of hearing Public Input	Public hearing Approvals Appeals

Relationship to Technical Studies

Prior to the site developer's public announcement of his plan to develop a facility, he generally identifies suitable sites with technical studies undertaken on a regional scale. He begins the initial planning of his public-consultation program (period 1) when he identifies these sites. He compiles readily available information describing the local community to design a preliminary public-consultation program. Some of the information on the local community he needs for the preliminary design of the consultation program will probably be available from his site selection studies. However, the developer may have to generate additional information before the initial announcement.

Following the identification of a potential site but before the preparation of specific impact statements and local site studies, the developer brings his facility plans to the attention of the community (period 2). This is the initial announcement stage of the consultation program.

As the developer completes local studies, he makes information on the site's suitability for his facility available to the public through the consultation program (period 3). In addition, he relays the community's concerns and questions to his technical experts through the public-consultation program. He identifies topics for further investigation in local impact studies. If requests by certain segments of the public for detailed technical information emerge, the developer must accommodate them. During this period the developer promotes an ongoing exchange of information between the public and his technical experts through the consultation program.

Relationship to Regulatory Process

The public should learn of the developing facility from the public-consultation program's initial announcement. Consequently, this stage of the program must occur before or simultaneous with the initiation of the regulatory process (period 2). As soon as the developer files a notice of intent and/or a permit application, the public consultation program should begin.

Period 3 encompasses the main body of the program. In this period the developer explains the nature of the facility and its impacts to the community in an ongoing program of information dissemination. This phase of the public-consultation program must immediately follow the initial announcement of the facility. Public interest and concern is at its height just after the community learns of the facility, and a program must be in place to answer questions. If the developer delays the initiation of the program, the risk of irreversible misconceptions developing among the public is greatly increased. In addition, the developer's credibility largely depends on the activities undertaken to address the community's concerns in the early stages of the siting process.

The public-consultation program runs parallel with other activities in the regulatory process. A considerable amount of time may elapse between the initial announcement of the facility and public meetings or hearings in the regulatory process. The developer's program should operate throughout this period. Public involvement in the regulatory process (usually in the form of the public hearings) generally will not meet the informational needs of the public because hearings occur too late in the siting process. In addition, the hearing is a decision-making forum rather than a vehicle for information dissemination. The developer's public-consultation program cannot and should not duplicate opportunities for public input in the regulatory process. The developer's program complements the regulatory process by providing information to the public prior to hearings.

The time between the initial announcement and the public hearing constitutes the bulk of the developer's consultation program. The developer's level of activity may not be constant throughout this time and if necessary, he may have to shut down the consultation program. (However, if resources permit, the public should always have access to an information source.) As the developer completes his technical studies, he should make the results available through his public-consultation program. He should always make his technical experts available to participate in the consultation program (at briefings and open houses, verifying information for public distribution, etc.).

The developer's public-consultation program continues until the public hearing but will probably be terminated at this point (period 4). It is not advisable for the site developer to make information available to the community when the issue is being considered at a public hearing. This might be interpreted as an attempt by the developer to prejudice the formal decision-making process.

Relationship to Negotiation and Mediation Efforts

Public consultation, as discussed here, is *not* a process for resolving disputes between the developer and the public. However, a variety of dispute-resolution strategies may be appropriate in the siting process and may work well in conjunction with a public-consultation program. Dispute resolution generally involves negotiation between the developer and the public. This negotiation may or may not be undertaken with the assistance of an impartial and independent third party unconnected to the siting attempt (a mediator). Before negotiation can occur, the following prerequisites must be fulfilled.

All parties must recognize the need for each of the others to participate.

It is most desirable if all parties have some influence on the outcome, more preferable if they each can exercise some power in the negotiation.

The parties conducting the negotiations must, in fact, be able to commit the support of their constituencies to any settlement.

There must be a sense of urgency. If one party can defeat the negotiation process by delay tactics, meaningful discussion will not occur.

It is not possible to predict when in the negotiation process these prerequisites will be met. Usually meaningful negotiation occurs late in the siting process and the outcomes are not acceptable to any one party. It would be most desirable if the issues are defined, constituencies mobilized, and parties convinced of the necessity for compromise soon after the application.

Without being able to accurately predict when negotiation will begin, it is impossible to locate these activities relative to consultation, technical studies, and the regulatory process in Table 8-1. Consequently, the consideration of the consultation-negotiation link in the siting process will focus on the contribution the consultation program can make to the process of negotiation during the siting process, and the implications that ongoing negotiations may have for the simultaneous operation of consultation programs.

The consultation program may contribute to the negotiation process in a siting process by identifying when the prerequisites for negotiation have occurred, identifying issues in dispute, and publicizing negotiated settlements.

The developer's public-consultation staff is likely to be in continual contact with all parties interested in the proposed facility and may be in a position to evalute whether the negotiation prerequisites have occurred. Consequently, they may be able to advise the developer if and when he should approach the parties involved in the dispute for the purposes of negotiation or seeking the assistance of a mediator.

By monitoring public reaction to the TSDF through the consultation program, the developer should succeed in identifying major issues of dispute to be dealt with in the negotiation process. The monitoring may also identify the priority various publics attach to these issues and the acceptability of various strategies for their resolution.

A negotiated settlement must be acceptable to all of the major affected parties in the siting process. However, the representatives of these affected parties will negotiate the settlement, and the consultation program may have a role in publicizing and monitoring the reaction of the wider public to the terms of the compromise.

If the developer is negotiating with the public during the period when his consultation program is operating, steps may have to be taken to ensure that the consultation program does not jeopardize the negotiation or the mediator's efforts. The negotiator should be relied upon for direction on steps to be taken to avoid conflict between consultation and negotiation. This could involve temporary embargoes on information at key points in the negotiations. This possibility, along with the likelihood that all groups in the community will not be directly involved in the consultation program, may contradict the principle of open access to information which is the basis of the consultation program. Consequently, the developer contemplating negotiation should be aware of these possible contradictions. Guarantees of complete accessibility to information should only be made if they do not interfere with potential negotiation strategies. In addition, the conflicting principles suggest that the consultation program should be separate from any negotiation activities. The staff responsible for the consultation program should probably not be involved in the negotiation activities.

APPENDIX 8-1
CASE STUDY [1]

Case

In late February 1979, Industrial Environmental Services (IES) withdrew a permit application for a secure landfill. The withdrawal was made after a particularly long and heated public meeting held in Kirksville, Missouri. By that action IES scuttled work to develop the site that had been in progress for over 18 months.

Opposition arose from major groups within the area and was based on a broad range of issues. County officials, faculty and students at Northeastern Missouri State University (NMSU), state elected officials, and others were among those that expressed outright opposition or major concerns about the proposed facility. These concerns covered the site, its design, and facility operations as well as the credibility of Missouri's Department of Natural Resources (DNR), which regulates hazardous-waste management and the hazardous-waste management industry.

IES and, to a lesser extent, DNR attempted to generate support and to allay public concerns by providing more information to the community. Although these attempts did address some concerns, they did not produce any significant change in the public response. On the contrary, most of those interviewed, on both sides of the dispute, felt that these attempts only increased public opposition.

Background Information

The proposed IES site was about three miles north of Kirksville, Missouri, just off U.S. 63. IES planned to build a 192-acre secure landfill and an adjacent 33-acre municipal landfill. The secure landfill was to be bounded by a ½-mi-strip of undeveloped land owned by IES's president and by the municipal landfill. Surrounding land use was primarily agricultural. The soil directly underlying the site is Kansas till, which IES test borings indicated was over 150 ft thick. Tests showed the till's permeability to be in the range from 10^{-8} to 10^{-9} cm/s. The Missouri Geologic Survey indicated no usable groundwater within Adair County, although the possibility of small perched aquifers beneath the site existed. IES's consulting engineers concluded that the site was "well suited for its proposed use," and any anomalies at the site could be corrected by engineering.

IES planned to develop its secure landfill in three phases over 20 years. Total capacity would approach 2.5 million cubic yards of hazardous waste. IES, the facility sponsor, was incorporated for the specific purpose of developing the secure landfill. As a corporation IES had no other experience in hazardous-waste management. IES's president, however, owned and operated Missouri Dispose-All, a solid-waste collection and hauling company in Kirksville, and had formerly operated a sanitary landfill in the Kirksville area. IES planned to serve Northeastern Missouri, St. Louis, and Kansas City. The city of Kirksville had a 1979 population of 19,000; it serves as a trade center for the multicounty area in Northeastern Missouri. Kirksville is about 220 mi north of St. Louis and 165 mi northeast of Kansas City. The area is predominantly agricultural, although several industries in Kirksville employ almost 2,000 persons. Kirksville is the home of NMSU and the Kirksville College of Osteopathic Medicine. Because of the presence of these institutions, and because of the town's role as a regional service center, the city has a disproportionately high number of professional and technical workers.

History Facility Development and Public Response

At the time of IES's siting initiative the owner/operator of a TSDF had to obtain a special operating permit from the Missouri DNR. The permit is based on case-by-case negotiations between DNR and applicants. DNR's regulations and authority did not overrule zoning. Adair County, however, had no zoning. Thus, IES required only the DNR permit to operate.

IES decided to try to develop a TSDF for two reasons: (a) they felt a market existed for a TSDF in the Kirksville area; and (b) the development of a municipal landfill at the site would complement Missouri Dispose-All's collections and hauling business at the Kirksville site. Although these facilities would be two distinct operations, IES planned to subsidize the operation of the municipal landfill with revenues from the TSDF.

IES began planning the project in July 1977. The first 10 months were spent in selecting the site, acquiring options on the site, and seeking preliminary approval of DNR. The site was selected on the basis of accessibility to U.S. 63, proximity to Kirksville, and the geology of the site. IES considered no other sites. By May 1978, IES obtained options for a 682-acre site. At that time IES applied to DNR for preliminary approval of its secure and municipal landfill.

DNR's technical analyses indicated no geological conditions to prevent the site from being used as the location for either a solid- or hazardous-waste disposal facility. With preliminary approval given in August 1978, DNR indicated that IES must make a formal application complete with a detailed engineering plan.

In June 1978, in anticipation of DNR's preliminary approval, IES retained a consulting engineering firm to prepare the detailed engineering plan. The firm had previous design experience in both solid- and hazardous-waste engineering. Field work began that same month. Work continued through the fall, and by December the permit application was completed. IES submitted its permit application to DNR on December 12, 1978.

In December 1978, IES approached the city of Kirksville to discuss the proposed facility. IES and the city focused on the proposed municipal landfill.

Relatively little attention was given to the hazardous-waste municipal facility (HWMF), although IES told the city that revenues from the HWMF were needed to make the municipal landfill. The city manager explained that although the city-owned landfill could be expanded at its current site, the city would prefer to get out of the landfill business if possible. The city landfill, which served a multicounty area, was a costly operation to the city and a continuing source of management problems, including problems with meeting DNR regulations. IES asked city officials to hold these discussions in confidence until DNR issued a public notice of IES's permit application. The city agreed to this request; however, some local leaders had already learned of the proposed municipal landfill through informal channels.

On January 3, 1979, DNR publicly announced that the permit application had been received and invited citizens to comment on the proposed facility. DNR planned to accept comments until January 26. The notification listed wastes that would be accepted at the "industrial waste disposal site" as "wastewater treatment plant sludges, industrial sludges, industrial liquids and other potentially hazardous wastes." The announcement briefly described existing state regulations and the fact that DNR approval was required for each type of industrial waste accepted by the HWMF. It also stated that if requested or considered necessary, DNR would hold a public meeting in Kirksville before making a final decision on the permit.

DNR's announcement was the first public knowledge of the proposed facility, and it generated an immediate response from area residents. In January, DNR received eight letters of concern and opposition, many from land-owners near the proposed site. On February 5, 1979, DNR announced that a public meeting would be held February 22 at NMSU. DNR's public announcements touched off a general, increasingly heated discussion of the proposal in the Kirksville area. The first major public forum was a local radio station's morning talk show, which asked listeners to phone in with comments and concerns. The proposed site sparked heated discussions. People spoke of their concerns and fears and solicited the support of others to fight the site. During one show a resident suggested that a petition against the facility be circulated and volunteered to be responsible for the petition. By the time of the public meeting (about one month later), more than 3,000 signatures had been collected.

In early February the pace of events quickened and opposition began to solidify. Within a week of the announcement of the public meeting, the Adair County Court (i.e., the elected officials of the county) passed a resolution formally opposing the HWMF, although the court did indicate its support for the municipal landfill. Faculty and students of NMSU began to review the permit application and other materials pertaining to hazardous-waste disposal. (Because of public pressure, DNR made a copy of the permit application available to the public at its Macon office, about 35 mi south of Kirksville. Some local residents charged that the application could not be photocopied and that only handwritten notes could be made of its contents.) Two county-level special-purpose government agencies in the area—a water district and a soil conservation service—went on record against the facility primarily because of fears of water-supply and soil contamination from leachate. The city of Kirksville never took a formal position, although like the county court, it supported the proposed municipal landfill.

On February 13, the Community Betterment Council held a meeting at city hall to present issues surrounding the proposed facility. The council is one of several such councils set up across Missouri. Its city-appointed members review a broad range of community concerns, seek to inform the public of these issues, and advise elected officials. The council invited IES's president and a representative of the opposition to speak at the meeting. At the meeting IES's president acknowledged the public concerns that had been raised and stated that residents should be concerned. He also stated that he felt community suport was necessary for the project and that without such support he would not pursue his plans. He qualified his statement by saying that he did not want to give up the project because of emotional objections that were not based on substantive issues. At the meeting he explained the project and its safeguards, the state regulations that would apply to his facility, and his intention to serve the northeastern Missouri market. The project opponents expressed concerns about the facility, including the fear that buried drums would leak and threaten water supplies.

After this meeting a local radio station invited IES's president and two NMSU professors opposed to the facility to speak on a local radio program. All three accepted the invitation and discussed the project and responded to listeners' questions phoned in during the show. The council meeting and the radio show were the two major opportunities for IES and opponents to discuss and debate issues prior to the public meeting. Both events resulted in increased opposition to the IES proposal. As more information became available to the public, the public asked more questions and requested that more information be made available, but it was considered vague, incomplete, and unreassuring. On February 18, the *Kirksville Daily Express,* in continuing its coverage of the proposed site, ran an article entitled "Proposed Waste Disposal Site Becomes Heated Controversy." The article served to describe the sense of the community prior to the public meeting. On February 22, DNR held the public meeting at the NMSU student union. DNR officials from the solid-waste program, a state geologist, IES's president, and representatives of IES's consulting engineers attended, as did an EPA representative. (EPA had no official involvement in the permit or the meeting and was present as an observer at the invitation of the area's state representative.) Also in attendance was a standing-room-only crowd of several hundred local residents and officials.

The public meeting provided visible evidence of public opposition. DNR officials explained both the permit application procedures and the technical qualifications of the site. The state geologist explained that any new geological data presented at the meeting would be considered before a final decision was made. The public response involved four hours of prepared statements followed by several more hours of questions and answers.

Many issues of concern to local residents were raised, these issues covered almost every conceivable topic. They ranged from questioning the legitimacy of the site's geology to attacking the integrity of DNR and IES. Opponents pointed to experiences at an area mine and a quarry that suggested the existence of substantial underground water supplies. The permeability of soils was questioned, based on the area's history of leaky ponds.

NMSU faculty members criticized the permit application for being incomplete, vague, and open-ended, particularly in terms of wastes (i.e., the application listed in effect all organic and inorganic materials as being acceptable). Many opponents felt that DNR had endorsed the IES application even though no formal DNR decision had been made. These opponents saw the vagueness of the application as a clear sign of DNR's inability to understand hazardous-waste-management and the implications of this particular proposal. Local officials charged that wastes would leach into public water supplies and demanded 100% assurance that this would not happen. They felt that the monitoring provisions proposed for the facility were inadequate. Added to these and other technical concerns were fears of property devaluation and the notoriety of being known as the "dumping ground" of the Midwest. Opponents' reviews of the permit application led to the belief that nothing would prevent out-of-state wastes from being brought in. They felt the site's large capacity would guarantee a multistate market area. Comments were not confined to technical, political, or economic issues; opponents also attacked the integrities of president and the state officials.

The public meeting produced fierce and emotional opposition to the project. The public response was heightened by a sense of powerlessness on the part of local officials and residents. Some felt that DNR would make a decision regardless of local response and "ram it down the throats" of local residents. During the meeting DNR announced that it would accept public comment on the permit application until March 6.

The day following the public meetings, IES withdrew its application. The opposition at the public meeting convinced IES's president that there was no public support for his proposal. He also announced that he would sell the land for the proposed site. He felt that even if DNR had approved his application, public opposition would have probably continued and increased. He predicted legal action to close the facility and ongoing harassment by opponents. In fact, several local officials interviewed after the withdrawal of IES's permit application stated that county judges were likely to enact county-wide zoning as a means of increasing local control over development in the county.

Evaluation

The attempts to generate support for IES's permit application failed completely. Indeed, the argument can be made that most of these attempts only added to the opposition. The more IES or DNR tried to anticipate concerns or to respond to questions, the more they became mired in the increasingly widespread and vehement opposition. The only major exception to this is the favorable response city and county officials gave to the proposed municipal landfill. The issues surrounding the HWMF, however, overwhelmed any advantages that the municipal landfill may have lent the overall project.

The failure of these efforts is undoubtedly the result of numerous interrelated causes and conditions. A number of these stand out. Opposition began and quickly developed before any major attempts by IES or DNR to discuss or explain in detail the proposal and state regulations. This opposition was based on limited information and appears to have mushroomed because of unanswered fears. Some accusations (e.g., that nuclear waste would be accepted by IES) were completely unfounded, but nevertheless had the apparent effect of solidifying a deep-rooted opposition. IES and DNR were then in a position of defending (as opposed to explaining) the proposal at the time they responded publicly. Although IES's president went to some lengths to explain the restrictions that DNR would place on the site's design and operation, opponents were not satisfied. They either raised questions unanswered by those regulations or placed different interpretations on them than those made by IES or DNR. The fact that DNR's regulations were pending and that the same was true of U.S. EPA regulations intensified public unease. From the local perspective, information that was made available was belated, unwillingly shared, and, most important, incomplete. In spite of this information, too many "what ifs" remained unanswered. The public response and attempts to address issues were made in an atmosphere lacking in dialogue and compromise. A reconciliation of divergent viewpoints appears never to have been seriously considered by all interested parties. Coupled with this was the absence of a trusted and knowledgeable party with no direct interest in the result of the conflict. For the most part every party was seen as having its own ulterior motive. A great many issues were raised by opponents. These emotional objections did not specifically pertain to the siting and operation of the proposed facility. They did, however, reflect the general hostility that (regardless of justification) people felt toward the facility sponsor and regulatory agencies. The hostility became a concrete factor in the controversy that characterized the siting process. The major issues raised by opponents were the following:

Suitability of the Site.   Opponents questioned the soil's ability to contain wastes and the claims that there was no significant underground water supply beneath the site. They feared contamination of public and private water supplies and of the soil itself.

Operational Risks.   Opponents objected to the range of wastes that IES wanted to handle and to the importation of wastes from outside the northeastern Missouri region. They feared that containerized wastes would leak and pollute soil and water. They felt that monitoring of the facility was insufficient.

Impact on Local Image and Land Values.   Opponents did not want Kirksville to be known as the Midwest's hazardous-waste dump, and they felt that the proposed size of the facility guaranteed this result. They believed land values, particularly those of land adjacent to the site, would fall.

Credibility of the Hazardous-Waste-Management Industry.   The notoriety of Love Canal and the publicity associated with Wilsonville, Illinois, as well as other information, contributed to a local image of the hazardous-waste-management industry as irresponsible. Local residents also saw public information as vague and euphemistic, and thus they saw the industry as being secretive and evasive.

Credibility of Regulators.   For some residents DNR was seen as a previously ineffectual agency that would not do its job of regulating IES properly. Although not all felt that DNR had a bad track record, many saw DNR's performance with respect to the IES application as inept and indicative of a staff without sufficient qualifications to judge the applications or the resources to do a thorough job regardless of qualification. DNR's information was seen as being just as vague and unreassuring as that provided by IES.

Status of Regulations and Research.   Coupled with the previous issue was the fact that neither DNR nor EPA hazardous-waste regulations were promulgated. This gave rise to a sense that DNR had no basis for permitting or regulating IES and, for some, a feeling that there would be no controls over the facility. Opponents also questioned the state-of-the-art of hazardous-waste disposal technology, which they saw as environmentally unsound. Present knowlege about the nature of hazardous waste (e.g., the degree of hazard and length of time specific wastes would be hazardous) was considred uncertain and therefore unreassuring.

Local Powerlessness.   The lack of any local regulatory power angered some local officials and residents. Comments from those interviewed concerning the siting process were primarily concerned with information made publicity available. Local officials and leaders invariably felt that more information should have been available earlier in the process. This would have helped to reduce concerns that the public was not being fully informed and that sufficient time was not available to study the proposal carefully. Had this been available, some felt that there might have been greater opportunity to discuss the IES proposal in a reasonable nonadversarial manner. The public meeting was the most dramatic event in the siting process. It was praised by opponents because it allowed all opponents to express their concerns and because it was conducted reasonably well, given the highly emotional atmosphere. On the other hand, the city and IES felt that the meeting was allowed to get out of control.

Kirksville's city manager thought that the meeting's purpose was never made clear and that the issues to be discussed should have been clarified before public comments were made. The timing of the meeting was also criticized.

Opponents believed that not enough time elapsed between the announcement of the meeting and the meeting itself. IES's president thought the meeting should have been held at the time of DNR's preliminary approval so that any information on anomalies of site hydrogeology would have been available before developing detailed engineering plans.

APPENDIX 8-2
QUESTIONS PEOPLE ASK

A Compilation of Actual Questions Asked by Interested People Relative to the Siting of Hazardous-Waste-Management Facilities in Their Communities

Although TSDF developers are generally aware that the public will ask questions relative to the siting of their proposed facility, they are usually unprepared for either the diversity or the depth of the questions actually asked. More importantly, developers are almost always unprepared for the degree of seriousness that the public attaches to the developer's answers.

To illustrate this important aspect of the public-consultation process a representative sample of the questions asked at hearings, in interviews, in conversations, and through telephone answering services has been compiled. For clarity and convenience the questions have been grouped into four categories: (1) waste, (2) site planning concerns, (3) health and environmental concerns, and (4) policy concerns.

Waste

1. What type of waste will be handled at the site? How dangerous are these materials?
2. How were the wastes produced (i.e., by-products of which industries or manufacturing processes)?
3. Where were the wastes generated (i.e., distance from the site or geographical area)?
4. What proportion of waste will be imported? What proportion will be generated locally?
5. How much waste is likely to be handled at the facility each year or over the life of the site?
6. How will waste be transported to the site? Will special hazardous-waste routes be designated?
7. How will the waste be disposed of or treated at the site?
8. Have new types of wastes likely to be handled in the future been considered in the facility's design?

Site Planning Concerns

Facility operation

1. Why are hazardous-waste-management facilities required? Why is this facility being built? Can the wastes be handled elsewhere or some other way? Why don't you put it somewhere else?
2. Who owns the facility? Who will manage the facility? Who will this organization be responsible to?
3. Who will act as a contact person to the community to ensure public access to information?
4. Will there be a continuous stream of trucks to and from the facility? How many trucks? What time of day? Will this flow of trucks cause noise, traffic congestion, and accidents?
5. Who will ensure that transport vehicles and equipment are properly maintained?
6. Will the facility operate 24 hours a day? Will there be noise 24 hours a day? Will any of the buildings be insulated against sound?
7. Will odors result from the operation of the facility?
8. How much waste will be in storage at the site at any one time? Can this waste catch fire or blow up?
9. Will the facility's operation have a detrimental impact on the natural environment (e.g., water tables, pollution, wildlife) and on the community's economic base (farming, fishing, tourist industry, etc.)?
10. How will the site be monitoried? For what? By whom? Where will funds for monitoring come from?
11. What happens if there is a malfunction of the processing equipment?

Facility appearance

1. Why is it necessary to develop a new facility in this general region of the country? Why in this area?
2. What general criteria did the developer consider in selecting the site? (environmental concerns, transportation links, geological criteria, etc.)
3. Will the location of the facility affect the property values of adjacent landowners?
4. What land-use controls will be implemented to prevent incompatible development (e.g., residential) near the site?
5. Will the facility hurt my community's image?
6. Will the facility have a detrimental impact on the character of my community? The quality of life in my community?

Facility security

1. Will the facility be properly secured? Will there be high fences and security guard(s)?
2. What is the nature and extent of emergency measures in case of accidental spill or fire? What are the specifics of a contingency plan?
3. Can the local hospitals handle any problems that may arise in the event of a mishap at the site or a mishap from an in-transit waste spill?
4. How will you keep children and pets away from the site?

Health and Environmental Concerns

1. How can you guarantee this facility will be safe when others haven't been?
2. What guarantees does my community have that you won't make the same mistakes and run this facility as poorly as the one Company X runs elsewhere?
3. Do these materials need to be treated specially? If improperly disposed of, are they hazardous to the environment and human health?
4. What is the precise impact of the facility and the waste on human health? What are the risks involved for human health and safety?
5. What level of waste could escape from the facility during normal operations and be carried into the community? What level is safe? How do you know what level is safe? What is the effect of long-term low-level exposures? How will we know if we've been exposed to toxic chemicals? Will it be too late by the time we know?
6. What is the possibility of leaks or ruptures in storage tanks, pipelines, and other structures in the facility? What level of waste would escape to the community? What level is safe?
7. What is the possibility of spills from trucks or trains transporting wastes to the facility? Where are these spills most likely to occur? What is the impact of spills at these locations?
8. What is the possibility of seepage or escape from geological formations used to store wastes?
9. What is the possibility of earthquakes resulting from the storage of wastes under pressure in geological formations?
10. Can a reaction between chemical wastes occur causing a fire or explosion?
11. What is the possibility of human error leading to a spill?
12. Are procedures adequate to ensure treated wastes are no longer toxic prior to their discharge? What safeguards are there to ensure this?
13. What is the possibility that treated wastes or emissions will still constitute pollutants (toxic or nontoxic) in the environment?

14. What is the possibility of groundwater pollution? If there was a spill and groundwater was contaminated, who would pay? Who would be responsible for ensuring water is supplied?
15. What assurances will the developer provide that his facility is safe? Who will be responsible for monitoring and maintaining the facility when it is closed down?
16. What effect would an accidental spill have on living organisms? (fish in lakes, livestock, etc.)
17. Should accidental spill or seepage occur, would there be compensation for damaged farmland? For effects on innocent bystanders? How would it be determined? Who takes the immediate responsibility for cleanup?
18. What employee safety and orientation measures will be instituted?

Policy Concerns

1. Have options to the storage and treatment of hazardous wastes been adequately explored? Particularly the potential for recovery? Given these options and their development in the future, can you prove this facility is really required?
2. What will be the process for public input to the regulatory decision-making process? Who will reimburse the public for the costs of their participation? Will anyone listen?
3. Who is legally liable in case of damage to the environment or human health and what are the limits of that liability?
4. What guarantees will the government make to ensure that the safety and livelihood of those in the area will not be adversely affected by the facility?
5. Do regulatory agencies really understand how these facilities work? Can they control them? Will they control them? What guarantees do we have? What recourse do we have?
6. Will the facility generate jobs for local residents?
7. What government guidelines or regulations govern generators, haulers, and disposers of waste? Can't industry recycle or reuse their wastes? Why does industry make things which generate so much wastes?
8. What about the larger societal issues of generation of wastes? What steps are being taken to restrict the reproduction of hazardous wastes?
9. What should be the role of the local municipality regarding costs, fiscal impact, and so on?

REFERENCE

1. Industrial Environmental Services, 1981 (ca.), *Public Opposition to the Siting of Hazardous Waste Management Facilities,* Kirksville, MO.

Index